APPLIED PARAMETER ESTIMATION FOR CHEMICAL ENGINEERS

CHEMICAL INDUSTRIES

A Series of Reference Books and Textbooks

Consulting Editor

HEINZ HEINEMANN

1. *Fluid Catalytic Cracking with Zeolite Catalysts,* Paul B. Venuto and E. Thomas Habib, Jr.
2. *Ethylene: Keystone to the Petrochemical Industry,* Ludwig Kniel, Olaf Winter, and Karl Stork
3. *The Chemistry and Technology of Petroleum,* James G. Speight
4. *The Desulfurization of Heavy Oils and Residua,* James G. Speight
5. *Catalysis of Organic Reactions,* edited by William R. Moser
6. *Acetylene-Based Chemicals from Coal and Other Natural Resources,* Robert J. Tedeschi
7. *Chemically Resistant Masonry,* Walter Lee Sheppard, Jr.
8. *Compressors and Expanders: Selection and Application for the Process Industry,* Heinz P. Bloch, Joseph A. Cameron, Frank M. Danowski, Jr., Ralph James, Jr., Judson S. Swearingen, and Marilyn E. Weightman
9. *Metering Pumps: Selection and Application,* James P. Poynton
10. *Hydrocarbons from Methanol,* Clarence D. Chang
11. *Form Flotation: Theory and Applications,* Ann N. Clarke and David J. Wilson
12. *The Chemistry and Technology of Coal,* James G. Speight
13. *Pneumatic and Hydraulic Conveying of Solids,* O. A. Williams
14. *Catalyst Manufacture: Laboratory and Commercial Preparations,* Alvin B. Stiles
15. *Characterization of Heterogeneous Catalysts,* edited by Francis Delannay
16. *BASIC Programs for Chemical Engineering Design,* James H. Weber
17. *Catalyst Poisoning,* L. Louis Hegedus and Robert W. McCabe
18. *Catalysis of Organic Reactions,* edited by John R. Kosak
19. *Adsorption Technology: A Step-by-Step Approach to Process Evaluation and Application,* edited by Frank L. Slejko
20. *Deactivation and Poisoning of Catalysts,* edited by Jacques Oudar and Henry Wise
21. *Catalysis and Surface Science: Developments in Chemicals from Methanol, Hydrotreating of Hydrocarbons, Catalyst Preparation, Monomers and Polymers, Photocatalysis and Photovoltaics,* edited by Heinz Heinemann and Gabor A. Somorjai
22. *Catalysis of Organic Reactions,* edited by Robert L. Augustine

23. *Modern Control Techniques for the Processing Industries,* T. H. Tsai, J. W. Lane, and C. S. Lin
24. *Temperature-Programmed Reduction for Solid Materials Characterization,* Alan Jones and Brian McNichol
25. *Catalytic Cracking: Catalysts, Chemistry, and Kinetics,* Bohdan W. Wojciechowski and Avelino Corma
26. *Chemical Reaction and Reactor Engineering,* edited by J. J. Carberry and A. Varma
27. *Filtration: Principles and Practices, Second Edition,* edited by Michael J. Matteson and Clyde Orr
28. *Corrosion Mechanisms,* edited by Florian Mansfeld
29. *Catalysis and Surface Properties of Liquid Metals and Alloys,* Yoshisada Ogino
30. *Catalyst Deactivation,* edited by Eugene E. Petersen and Alexis T. Bell
31. *Hydrogen Effects in Catalysis: Fundamentals and Practical Applications,* edited by Zoltán Paál and P. G. Menon
32. *Flow Management for Engineers and Scientists,* Nicholas P. Cheremisinoff and Paul N. Cheremisinoff
33. *Catalysis of Organic Reactions,* edited by Paul N. Rylander, Harold Greenfield, and Robert L. Augustine
34. *Powder and Bulk Solids Handling Processes: Instrumentation and Control,* Koichi Iinoya, Hiroaki Masuda, and Kinnosuke Watanabe
35. *Reverse Osmosis Technology: Applications for High-Purity-Water Production,* edited by Bipin S. Parekh
36. *Shape Selective Catalysis in Industrial Applications,* N. Y. Chen, William E. Garwood, and Frank G. Dwyer
37. *Alpha Olefins Applications Handbook,* edited by George R. Lappin and Joseph L. Sauer
38. *Process Modeling and Control in Chemical Industries,* edited by Kaddour Najim
39. *Clathrate Hydrates of Natural Gases,* E. Dendy Sloan, Jr.
40. *Catalysis of Organic Reactions,* edited by Dale W. Blackburn
41. *Fuel Science and Technology Handbook,* edited by James G. Speight
42. *Octane-Enhancing Zeolitic FCC Catalysts,* Julius Scherzer
43. *Oxygen in Catalysis,* Adam Bielanski and Jerzy Haber
44. *The Chemistry and Technology of Petroleum: Second Edition, Revised and Expanded,* James G. Speight
45. *Industrial Drying Equipment: Selection and Application,* C. M. van't Land
46. *Novel Production Methods for Ethylene, Light Hydrocarbons, and Aromatics*, edited by Lyle F. Albright, Billy L. Crynes, and Siegfried Nowak
47. *Catalysis of Organic Reactions*, edited by William E. Pascoe
48. *Synthetic Lubricants and High-Performance Functional Fluids*, edited by Ronald L. Shubkin
49. *Acetic Acid and Its Derivatives*, edited by Victor H. Agreda and Joseph R. Zoeller
50. *Properties and Applications of Perovskite-Type Oxides*, edited by L. G. Tejuca and J. L. G. Fierro

51. *Computer-Aided Design of Catalysts*, edited by E. Robert Becker and Carmo J. Pereira
52. *Models for Thermodynamic and Phase Equilibria Calculations*, edited by Stanley I. Sandler
53. *Catalysis of Organic Reactions*, edited by John R. Kosak and Thomas A. Johnson
54. *Composition and Analysis of Heavy Petroleum Fractions*, Klaus H. Altgelt and Mieczyslaw M. Boduszynski
55. *NMR Techniques in Catalysis,* edited by Alexis T. Bell and Alexander Pines
56. *Upgrading Petroleum Residues and Heavy Oils*, Murray R. Gray
57. *Methanol Production and Use*, edited by Wu-Hsun Cheng and Harold H. Kung
58. *Catalytic Hydroprocessing of Petroleum and Distillates*, edited by Michael C. Oballah and Stuart S. Shih
59. *The Chemistry and Technology of Coal: Second Edition, Revised and Expanded,* James G. Speight
60. *Lubricant Base Oil and Wax Processing*, Avilino Sequeira, Jr.
61. *Catalytic Naphtha Reforming: Science and Technology*, edited by George J. Antos, Abdullah M. Aitani, and José M. Parera
62. *Catalysis of Organic Reactions*, edited by Mike G. Scaros and Michael L. Prunier
63. *Catalyst Manufacture,* Alvin B. Stiles and Theodore A. Koch
64. *Handbook of Grignard Reagents,* edited by Gary S. Silverman and Philip E. Rakita
65. *Shape Selective Catalysis in Industrial Applications: Second Edition, Revised and Expanded*, N. Y. Chen, William E. Garwood, and Francis G. Dwyer
66. *Hydrocracking Science and Technology*, Julius Scherzer and A. J. Gruia
67. *Hydrotreating Technology for Pollution Control: Catalysts, Catalysis, and Processes*, edited by Mario L. Occelli and Russell Chianelli
68. *Catalysis of Organic Reactions*, edited by Russell E. Malz, Jr.
69. *Synthesis of Porous Materials: Zeolites, Clays, and Nanostructures,* edited by Mario L. Occelli and Henri Kessler
70. *Methane and Its Derivatives*, Sunggyu Lee
71. *Structured Catalysts and Reactors*, edited by Andrzei Cybulski and Jacob Moulijn
72. *Industrial Gases in Petrochemical Processing*, Harold Gunardson
73. *Clathrate Hydrates of Natural Gases: Second Edition, Revised and Expanded*, E. Dendy Sloan, Jr.
74. *Fluid Cracking Catalysts*, edited by Mario L. Occelli and Paul O'Connor
75. *Catalysis of Organic Reactions*, edited by Frank E. Herkes
76. *The Chemistry and Technology of Petroleum, Third Edition*, *Revised and Expanded*, James G. Speight
77. *Synthetic Lubricants and High-Performance Functional Fluids, Second Edition: Revised and Expanded,* Leslie R. Rudnick and Ronald L. Shubkin

78. *The Desulfurization of Heavy Oils and Residua, Second Edition, Revised and Expanded,* James G. Speight
79. *Reaction Kinetics and Reactor Design: Second Edition, Revised and Expanded,* John B. Butt
80. *Regulatory Chemicals Handbook*, Jennifer M. Spero, Bella Devito, and Louis Theodore
81. *Applied Parameter Estimation for Chemical Engineers*, Peter Englezos and Nicolas Kalogerakis
82. *Catalysis of Organic Reactions,* edited by Michael E. Ford

ADDITIONAL VOLUMES IN PREPARATION

The Chemical Process Industries Infrastructure: Function and Economics, James R. Couper, O. Thomas Beasley, and W. Roy Penney

Elements of Transport Phenomena, Joel Plawsky

APPLIED PARAMETER ESTIMATION FOR CHEMICAL ENGINEERS

Peter Englezos
University of British Columbia
Vancouver, Canada

Nicolas Kalogerakis
Technical University of Crete
Chania, Greece

MARCEL DEKKER, INC. NEW YORK • BASEL

ISBN: 0-8247-9561-X

This book is printed on acid-free paper.

Headquarters
Marcel Dekker, Inc.
270 Madison Avenue, New York, NY 10016
tel: 212-696-9000; fax: 212-685-4540

Eastern Hemisphere Distribution
Marcel Dekker AG
Hutgasse 4, Postfach 812, CH-4001 Basel, Switzerland
tel: 41-61-261-8482; fax: 41-61-261-8896

World Wide Web
http://www.dekker.com

The publisher offers discounts on this book when ordered in bulk quantities. For more information, write to Special Sales/Professional Marketing at the headquarters address above.

Current printing (last digit):
10 9 8 7 6 5 4 3 2 1

PRINTED IN THE UNITED STATES OF AMERICA

Dedicated to
Vangie, Chris, Kelly, Gina & Manos

Preface

Engineering sciences state relations among measurable properties so that a technological system or process can be analyzed mathematically (Ferguson, 1992). The term *model* is adopted here to refer to the ensemble of equations that describes and interrelates the variables and parameters of a system or process (Basmadjan, 1999). In chemical, biochemical, environmental and petroleum engineering these models are based on the principles of chemistry, physics, thermodynamics, kinetics and transport phenomena. As most engineering calculations cannot be based on quantum mechanics as of yet, the models contain a number of quantities the value of which is not known a priori. It is customary to call these quantities *adjustable parameters*. The determination of suitable values for these adjustable parameters is the objective of parameter estimation, also known as *data regression*. A classic example of parameter estimation is the determination of kinetic parameters from a set of data.

Parameter estimation is essentially an optimization problem whereby the unknown parameters are obtained by minimizing a suitable objective function. The structure of this objective function has led to the development of particularly efficient and robust methods. The aim of this book is to provide students and practicing engineers with straightforward tools that can be used directly for the solution of parameter estimation problems. The emphasis is on applications rather than on formal development of the theories. Students who study chemical, biochemical, environmental or petroleum engineering and practicing engineers in these fields will find the book useful. The following table summarizes how the book can be used:

Subject	*Chapters from this book*
Regression Analysis & Applications	All chapters
Chemical Kinetics & Reactor Design	1, 2, 3, 4, 6, 8, 10, 11, 12, 16
Biochemical Engineering	1, 2, 3, 4, 6, 7, 8, 11, 12, 17
Petroleum Reservoir Engineering	1, 2, 3, 6, 8, 10, 11, 18
Computational Thermodynamics	1, 2, 4, 8, 9, 11, 12, 14, 15
Optimization Methods	1, 2, 3, 4, 5, 6, 7, 8, 9, 10, 11, 12

With this book the reader can expect to learn how to formulate and solve parameter estimation problems, compute the statistical properties of the parameters, perform model adequacy tests, and design experiments for parameter estimation or model discrimination.

A number of books address parameter estimation (Bard, 1974; Bates and Watts, 1988; Beck and Arnold, 1977; Draper and Smith, 1981; Gans, 1992; Koch, 1987; Lawson and Hanson, 1974; Seber and Wild, 1989; Seinfeld and Lapidus, 1974; Sorenson, 1980). However, the majority of these books emphasize statistics and mathematics or system identification and signal processing. Furthermore, most of the existing books pay considerable attention to linear and nonlinear regression for models described by *algebraic equations*. This book was conceived with the idea of focusing primarily on chemical engineering applications and on systems described by nonlinear algebraic and *ordinary differential equations* with a particular emphasis on the latter.

In Chapter 1, the main areas where chemical engineers encounter parameter estimation problems are introduced. Examples from chemical kinetics, biochemical engineering, petroleum engineering, and thermodynamics are briefly described. In Chapter 2, the parameter estimation problem is formulated mathematically with emphasis on the choice of a suitable objective function. The subject of linear regression is described in a succinct manner in Chapter 3. Methodologies for solving linear regression problems with readily available software such as Microsoft Excel^{TM} and SigmaPlot^{TM} for Windows^{TM} are presented with examples.

In Chapter 4 the Gauss-Newton method for systems described by algebraic equations is developed. The method is illustrated by examples with actual data from the literature. Other methods (indirect, such as Newton, Quasi-Newton, etc., and direct, such as the Luus-Jaakola optimization procedure) are presented in Chapter 5.

In Chapter 6, the Gauss-Newton method for systems described by ordinary differential equations (ODE) is developed and is illustrated with three examples formulated with data from the literature. Simpler methods for estimating parameters in systems described by ordinary differential equations known as *shortcut methods* are presented in Chapter 7. Such methods are particularly suitable for systems in the field of biochemical engineering.

Chapter 8 provides practical guidelines for the implementation of the Gauss-Newton method. Issues such as generating initial guesses and tackling the issues of overstepping and matrix ill-conditioning are presented. In addition, guidelines are provided on how to utilize "prior" information and selecting a suitable weighting matrix. The models described by ODE require special attention to deal with stiffness and enlargement of the region of convergence.

Chapter 9 deals with estimation of parameters subject to equality and inequality constraints whereas Chapter 10 examines systems described by partial differential equations (PDE). Examples are provided in Chapters 14 and 18.

Procedures on how to make inferences on the parameters and the response variables are introduced in Chapter 11. The design of experiments has a direct impact on the quality of the estimated parameters and is presented in Chapter 12. The emphasis is on sequential experimental design for parameter estimation and for model discrimination. Recursive least squares estimation, used for on-line data analysis, is briefly covered in Chapter 13.

Chapters 14 to 18 are entirely devoted to applications. Examples and problems for solution by the reader are also included. In Chapter 14 several applications of the Gauss-Newton method are presented for the estimation of adjustable parameters in cubic equations of state. Parameter estimation in activity coefficient models is presented in Chapter 15. Chemical kinetics has traditionally been the main domain for parameter estimation studies. Examples formulated with models described by algebraic equations or ODE are presented in Chapter 16. The increasing involvement of chemical engineers in biotechnology motivated us to devote a chapter to such applications. Thus Chapter 17 includes examples from enzyme kinetics and mass transfer coefficient determination in bioreactors. The last chapter (Chapter 18) is devoted to applications in petroleum engineering. Thus the modeling of drilling data is a linear regression problem whereas oil reservoir simulation presents an opportunity to demonstrate the application of the Gauss-Newton method for systems described by partial differential equations.

It is a pleasure to acknowledge those individuals who helped us indirectly in preparing this book: our colleagues Professors L.A. Behie, P.R. Bishnoi, R.A. Heidemann and R.G. Moore and our graduate students who over the years as part of their M.Sc. and Ph.D. thesis have gathered and analyzed data.

We sincerely thank Professor Hoffman of the Institute of Technical Chemistry, Friedrich-Alexander University, Germany for providing us with the raw data for the hydrogenation of 3-hydroxypropanol.

Professor Englezos acknowledges the support of the University of British Columbia for a sabbatical leave during which a major part of this book was completed. Professor Englezos also acknowledges the support from the Technical University of Crete and Keio University where he spent parts of his leave.

Professor Kalogerakis acknowledges the support of the Technical University of Crete in completing this book; Professor Luus for his encouragement and help with direct search procedures; and all his colleagues at the University of Calgary for the many discussions and help he received over the years.

Finally, both of us would like to sincerely thank our wives Kalliroy Kalogerakis and Evangeline Englezos for their patience and understanding while we devoted many hours to completing this book.

Peter Englezos
Vancouver, Canada

Nicolas Kalogerakis
Chania, Crete

Contents

1

Introduction

During an experiment, measurement of certain variables e.g. concentrations, pressures, temperatures, etc. is conducted. Let $\hat{\mathbf{y}} = [\hat{y}_1, \hat{y}_2, ..., \hat{y}_m]^T$ be the *m-dimensional* vector of measured variables during an experiment. In addition, during each experiment, certain conditions are set or fixed by the experimentalist e.g. substrate concentration in an enzyme kinetics experiment, time and temperature in a kinetics experiment, etc. Let $\mathbf{x} = [x_1, x_2, ..., x_n]^T$ be the *n-dimensional* vector of these input variables that can be assumed to be known precisely.

The experimentalist often formulates a mathematical model in order to describe the observed behavior. In general, the model consists of a set of equations based on the principles of chemistry, physics, thermodynamics, kinetics and transport phenomena and attempts to predict the variables, **y**, that are being measured. In general, the measured variables **y** are a function of **x**. Thus, the model has the following form

$$\hat{\mathbf{y}} = \textit{Function of}(\mathbf{x}, \mathbf{k}) + \textit{Random Error} \tag{1.1}$$

The random error arises from the measurement of y the true value of which is not known. The measurements are assumed to be free of systematic errors. The modeling equations contain adjustable parameters to account for the fact that the models are phenomenological. For example, kinetic rate expressions contain rate constants (parameters) the value of which is unknown and not possible to be obtained from fundamental principles.

Parameter estimation is one of the steps involved in the formulation and validation of a mathematical model that describes a process of interest. Parameter estimation refers to the process of obtaining values of the parameters from the matching of the model-based calculated values to the set of measurements (data). This is the classic *parameter estimation* or *model fitting* problem and it should be distinguished from the *identification* problem. The latter involves the development of a model from input/output data only. This case arises when there is no a priori information about the form of the model i.e. it is a black box.

When the model equations are linear functions of the parameters the problem is called *linear estimation*. Nonlinear estimation refers to the more general and most frequently encountered situation where the model equations are nonlinear functions of the parameters.

Parameter estimation and identification are an essential step in the development of mathematical models that describe the behavior of physical processes (Seinfeld and Lapidus, 1974; Aris, 1994). The reader is strongly advised to consult the above references for discussions on what is a model, types of models, model formulation and evaluation. The paper by Plackett that presents the history on the discovery of the least squares method is also recommended (Plackett, 1972).

A smooth function f(x) is used quite often to describe a set of data, (x_1,y_1), (x_2,y_2), ...,(x_N,y_N). Fitting of a smooth curve, f(x), through these points is usually done for interpolation or visual purposes (Sutton and MacGregor, 1977). This is called *curve fitting*. The parameters defining the curve are calculated by minimizing a measure of the *closeness of fit* such as the function $S(\mathbf{k}) = \sum [y_i - f(x_i)]^2$. The parameters have no physical significance.

The scope of this book deals primarily with the parameter estimation problem. Our focus will be on the estimation of adjustable parameters in nonlinear models described by algebraic or ordinary differential equations. The models describe processes and thus explain the behavior of the observed data. It is assumed that the structure of the model is known. The best parameters are estimated in order to be used in the model for predictive purposes at other conditions where the model is called to describe process behavior.

The unknown model parameters will be obtained by minimizing a suitable objective function. The objective function is a measure of the discrepancy or the departure of the data from the model i.e., the *lack of fit* (Bard, 1974; Seinfeld and Lapidus, 1974). Thus, our problem can also be viewed as an optimization problem and one can in principle employ a variety of solution methods available for such problems (Edgar and Himmelblau, 1988; Gill et al. 1981; Reklaitis, 1983; Scales, 1985). Finally it should be noted that engineers use the term *parameter estimation* whereas statisticians use such terms as *nonlinear* or *linear regression analysis* to describe the subject presented in this book.

In parameter estimation, the general problem we have to solve is:

> *Given the structure of the model (i.e. the governing model equations) and a set of measured data points, the problem is to find the unknown model parameters so that the values calculated by the model match the data in some optimal manner (e.g., by minimizing the sum of squares of errors)*

The specific issues that we have tried to address in this book are:

(i) *Structure of the Model* ("What kind of models can be used? Linear or nonlinear? Algebraic or differential equation models?")

(ii) Selection of the *Objective Function* ("What do we minimize to estimate the parameters?")

(iii) *Solution Techniques* ("How do we minimize the objective function?")

(iv) *Statistical Properties of Parameter Estimates* ("How accurate are the estimated parameters?")

(v) *Statistical Properties of Model-Based Calculated Values* ("Given the uncertainty in the model parameters, what is the uncertainty in the calculated values?")

(vi) *Tests for Model Adequacy* ("Is the model good enough?")

(vii) *Tests for Model Discrimination* ("Among several rival models, which is the best one?")

(viii) *Factorial Experimental Design* ("What is the first set of experiments I should run?")

(ix) *Sequential Experimental Design* ("What should my next experiment be so that I gather maximum information?") for *model discrimination* (to select the best model among several ones that fit the data,) or for *precise parameter estimation* (to minimize further the parameter uncertainty in a particular model).

Issues such as the ones above are of paramount importance and interest to practicing engineers, researchers and graduate students. In the next paragraphs we mention several examples that cover many important areas of chemical, biochemical and petroleum engineering.

Chemical kinetics is an area that received perhaps most of the attention of chemical engineers from a parameter estimation point of view. Chemical engineers need mathematical expressions for the intrinsic rate of chemical reactions in order to design reactors that produce chemicals in an industrial scale. For example, let us consider the following consecutive first order reactions that take place (Smith, 1981; Froment and Bischoff, 1990)

$$A \xrightarrow{k_1} B \xrightarrow{k_2} D \tag{1.2}$$

During an experiment, the concentrations of A and B are usually measured. If it is assumed that only component A is present in the reactor initially ($C_{A0}\neq 0$, $C_{B0}=0$, $C_{D0}=0$), we do not need to measure the concentration of chemical D because $C_A+C_B+C_D=C_{A0}$. The rate equations for A and B are given by

$$-\frac{dC_A}{dt} = k_1 C_A \tag{1.3a}$$

$$\frac{dC_B}{dt} = k_1 C_A - k_2 C_B \tag{1.3b}$$

where C_A and C_B are the concentrations of A and B, t is reaction time, and k_1, k_2 are the rate constants.

The above rate equations can be easily integrated to yield the following algebraic equations

$$C_A = C_{A0} e^{-k_1 t} \tag{1.4a}$$

$$C_B = C_{A0} \frac{k_1}{k_2 - k_1}\left(e^{-k_1 t} - e^{-k_2 t}\right) \tag{1.4b}$$

We can then obtain C_D from the concentration invariant $C_D=C_{A0}-C_A-C_B$.

$$C_D = C_{A0}\left(1 + \frac{k_1 e^{-k_2 t} - k_2 e^{-k_1 t}}{k_2 - k_1}\right) \tag{1.4c}$$

The values of the rate constants are estimated by fitting equations 1.4a and 1.4b to the concentration versus time data. It should be noted that there are kinetic models that are more complex and integration of the rate equations can only be done numerically. We shall see such models in Chapter 6. An example is given next. Consider the gas phase reaction of NO with O_2 (Bellman et al. 1967):

$$2NO + O_2 \leftrightarrow 2NO_2 \tag{1.5}$$

The model is given by the following kinetic rate equation

$$\frac{dC_{NO_2}}{dt} = k_1(\alpha - C_{NO_2})(\beta - C_{NO_2})^2 - k_2 C_{NO_2}^2 \; ; \quad C_{NO_2}(0) = 0 \tag{1.6}$$

where α=126.2, β=91.9, and k_1, k_2 are the rate constants to be estimated. The data consists of a set of measurements of the concentration of NO_2 versus time. In this case, fitting of the data requires the integration of the governing differential equation.

In biochemical engineering one often employs parametric models to describe enzyme-catalyzed reactions, microbial growth, nutrient transport and metabolic rates in biological systems. The parameters in these models can be readily obtained though the methods described in this book. In addition, we present *shortcut estimation methods* for the estimation of the average specific uptake or production rates during cell cultivation in batch, fed-batch, chemostat or perfusion cultures. For example, such an analysis is particularly important to scientists screening cell lines to identify high producers.

Petroleum and chemical engineers perform *oil reservoir simulation* to optimize the production of oil and gas. Black-oil, compositional or thermal oil reservoir models are described by sets of differential equations. The measurements consist of the pressure at the wells, water-oil ratios, gas-oil ratios etc. The objective is to estimate through *history matching of the reservoir* unknown reservoir properties such as porosity and permeability.

Volumetric *equations of state* (EoS) are employed for the calculation *of fluid phase equilibrium and thermo-physical properties* required in the design of processes involving non-ideal fluid mixtures in the oil, gas and chemical industries. Mathematically, a volumetric EoS expresses the relationship among pressure, volume, temperature, and composition for a fluid mixture. The next equation gives the Peng-Robinson equation of state, which is perhaps the most widely used EoS in industrial practice (Peng and Robinson, 1976).

$$P = \frac{RT}{v-b} - \frac{a(T)}{v(v+b)+b(v-b)} \tag{1.7}$$

where P is the pressure, T the temperature and v the molar volume. The mixture parameters a and b are defined as follows

$$a = \sum_i \sum_j x_i x_j \left(1 - k_{ij}\right)\sqrt{a_i a_j} \tag{1.8a}$$

$$b = \sum_i x_i b_i \tag{1.8b}$$

where a_i, b_i and a_j are parameters specific for the individual components i and j, and k_{ij} is an empirical interaction parameter characterizing the binary formed by component i and component j. It is well known that the introduction of empirical parameters, k_{ij}, in the equations of state enhances their ability as tools for process design. The interaction parameters are usually estimated from the regression of binary vapor-liquid equilibrium (VLE) data. Such data consist of sets of temperature, pressure and liquid and vapor phase mole fractions for one of the components.

Typical examples such as the ones mentioned above, are used throughout this book and they cover most of the applications chemical engineers are faced with. In addition to the problem definition, the mathematical development and the numerical results, the implementation of each algorithm is presented in detail and computer listings of selected problems are given in the attached CD.

Finally, it is noted that even though we are not concerned with model development in this book, we should always keep in mind that:

(i) The development of models is not necessarily in the direction of greater complexity or an increasing number of adjustable parameters.

(ii) A higher degree of confidence is gained if the adjustable parameters are found from independent sources.

2

Formulation of the Parameter Estimation Problem

The formulation of the parameter estimation problem is equally important to the actual solution of the problem (i.e., the determination of the unknown parameters). In the formulation of the parameter estimation problem we must answer two questions: (a) what type of mathematical model do we have? and (b) what type of objective function should we minimize? In this chapter we address both these questions. Although the primary focus of this book is the treatment of mathematical models that are nonlinear with respect to the parameters (*nonlinear regression*) consideration to linear models (*linear regression*) will also be given.

2.1 STRUCTURE OF THE MATHEMATICAL MODEL

The primary classification that is employed throughout this book is *algebraic* versus *differential* equation models. Namely, the mathematical model is comprised of a set of algebraic equations or by a set of ordinary (ODE) or partial differential equations (PDE). The majority of mathematical models for physical or engineered systems can be classified in one of these two categories.

2.1.1 Algebraic Equation Models

In mathematical terms these models are of the form

$$\mathbf{y} = \mathbf{f}(\mathbf{x}, \mathbf{k}) \tag{2.1}$$

where

$\mathbf{k} = [k_1, k_2,\ldots,k_p]^T$ is a *p-dimensional* vector of parameters whose numerical values are unknown;

$\mathbf{x} = [x_1, x_2,\ldots,x_n]^T$ is an *n-dimensional* vector of *independent* variables (also called *regressor* or *input* variables) which are either fixed for each experiment by the experimentalist or which are measured. In the statistical treatment of the problem, it is often assumed that these variables are known precisely (i.e., there is no uncertainty in their value even if they are measured experimentally);

$\mathbf{y} = [y_1, y_2,\ldots,y_m]^T$ is an *m-dimensional* vector of *dependent* variables (also often described as *response* variables or the *output* vector); these are the model variables which are actually measured in the experiments; note that **y** does not necessarily represent the entire set of measured variables in each experiment, rather it represents the set of dependent variables;

$\mathbf{f} = [f_1, f_2,\ldots,f_m]^T$ is an *m-dimensional* vector function of known form (these are the actual model equations).

Equation 2.1, the mathematical model of the process, is very general and it covers many cases, namely,

(i) The single response with a single independent variable model (i.e., m=1, n=1)

$$y = f(x; k_1, k_2,\ldots,k_p) \tag{2.2a}$$

(ii) The single response with several independent variables model (i.e., m=1, n>1)

$$y = f(x_1, x_2,\ldots,x_n; k_1, k_2,\ldots,k_p) \tag{2.2b}$$

(iii) The multi-response with several independent variables model (i.e., m>1, n>1)

$$\begin{bmatrix} y_1 \\ y_2 \\ \cdot \\ \cdot \\ \cdot \\ y_m \end{bmatrix} = \begin{bmatrix} f_1(x_1,x_2,...,x_n;k_1,k_2,...,k_p) \\ f_2(x_1,x_2,...,x_n;k_1,k_2,...,k_p) \\ \cdot \\ \cdot \\ \cdot \\ f_m(x_1,x_2,...,x_n;k_1,k_2,...,k_p) \end{bmatrix} \tag{2.2c}$$

A single experiment consists of the measurement of each of the *m* response variables for a given set of values of the *n* independent variables. For each experiment, the measured output vector which can be viewed as a random variable is comprised of the *deterministic* part calculated by the model (Equation 2.1) and the *stochastic* part represented by the error term, i.e.,

$$\hat{\mathbf{y}}_i = \mathbf{f}(\mathbf{x}_i,\mathbf{k}) + \boldsymbol{\varepsilon}_i \quad ; \quad i=1,2,\ldots,N \tag{2.3}$$

If the mathematical model represents adequately the physical system, the error term in Equation 2.3 represents only measurement errors. As such, it can often be assumed to be normally distributed with zero mean (assuming there is no bias present in the measurement). In real life the vector $\boldsymbol{\varepsilon}_i$ incorporates not only the experimental error but also any inaccuracy of the mathematical model.

A special case of Equation 2.3 corresponds to the celebrated *linear systems.* Linearity is assumed with respect to the unknown parameters rather than the independent variables. Hence, Equation 2.3 becomes

$$\hat{\mathbf{y}}_i = \mathbf{F}(\mathbf{x}_i)\mathbf{k} + \boldsymbol{\varepsilon}_i \quad ; \quad i=1,2,\ldots,N \tag{2.4}$$

where $\mathbf{F}(\mathbf{x}_i)$ is an *m×p dimensional* matrix which depends only on $\mathbf{x}_i$ and it is independent of the parameters. Quite often the elements of matrix **F** are simply the independent variables, and hence, the model can be further reduced to the well known *linear regression* model,

$$\hat{\mathbf{y}}_i = \mathbf{X}_i\,\mathbf{k} + \boldsymbol{\varepsilon}_i \quad ; \quad i=1,2,\ldots,N \tag{2.5}$$

A brief review of *linear regression analysis* is presented in Chapter 3.

In algebraic equation models we also have the special situation of *conditionally linear systems* which arise quite often in engineering (e.g., chemical kinetic models, biological systems, etc.). In these models some of the parameters enter in a linear fashion, namely, the model is of the form,

$$\hat{\mathbf{y}}_i = \mathbf{F}(\mathbf{x}_i, \mathbf{k}_2)\,\mathbf{k}_1 + \boldsymbol{\varepsilon}_i \quad ; \qquad i=1,2,\ldots,N \tag{2.6}$$

where the parameter vector has been partitioned into two groups,

$$\mathbf{k} = \begin{bmatrix} \mathbf{k}_1 \\ \mathbf{k}_2 \end{bmatrix} \tag{2.7}$$

The structure of such models can be exploited in reducing the dimensionality of the nonlinear parameter estimation problem since, the conditionally linear parameters, $\mathbf{k}_1$, can be obtained by linear least squares in one step and without the need for initial estimates. Further details are provided in Chapter 8 where we exploit the structure of the model either to reduce the dimensionality of the nonlinear regression problem or to arrive at consistent initial guesses for any iterative parameter search algorithm.

In certain circumstances, the model equations may not have an *explicit* expression for the measured variables. Namely, the model can only be represented *implicitly*. In such cases, the distinction between dependent and independent variables becomes rather fuzzy, particularly when all the variables are subject to experimental error. As a result, it is preferable to consider an augmented vector of measured variables, **y**, that contains both regressor and response variables (Box, 1970; Britt and Luecke, 1973). The model can then be written as

$$\boldsymbol{\varphi}(\mathbf{y}; \mathbf{k}) = \mathbf{0} \tag{2.8}$$

where, $\mathbf{y} = [y_1, y_2, \ldots, y_R]^T$ is an *R-dimensional* vector of *measured* variables and $\boldsymbol{\varphi} = [\varphi_1, \varphi_2, \ldots, \varphi_m]^T$ is an *m-dimensional* vector function of known form.

If we substitute the actual measurement for each experiment, the model equation becomes,

$$\boldsymbol{\varphi}(\hat{\mathbf{y}}_i ; \mathbf{k}) = \boldsymbol{\varepsilon}_i \quad ; \qquad i=1,2,\ldots,N \tag{2.9}$$

As we shall discuss later, in many instances *implicit estimation* provides the easiest and computationally the most efficient solution to this class of problems.

The category of algebraic equation models is quite general and it encompasses many types of engineering models. For example, any *discrete dynamic model* described by a set of *difference equations* falls in this category for parameter estimation purposes. These models could be either *deterministic* or *stochastic* in nature or even combinations of the two. Although on-line techniques are available for the estimation of parameters in sampled data systems, off-line techniques such as the ones emphasized in this book are very useful and they can be employed when all the data are used together after the experiment is over.

Finally, we should refer to situations where both independent and response variables are subject to experimental error regardless of the structure of the model. In this case, the experimental data are described by the set $\{(\hat{\mathbf{y}}_i, \hat{\mathbf{x}}_i), i=1,2,...N\}$ as opposed to $\{(\hat{\mathbf{y}}_i, \mathbf{x}_i), i=1,2,...,N\}$. The deterministic part of the model is the same as before; however, we now have to consider besides Equation 2.3, the error in $\mathbf{x}_i$, i.e., $\hat{\mathbf{x}}_i = \mathbf{x}_i + \boldsymbol{\varepsilon}_{xi}$. These situations in nonlinear regression can be handled very efficiently using an implicit formulation of the problem as shown later in Section 2.2.2

2.1.2 Differential Equation Models

Let us first concentrate on dynamic systems described by a set of ordinary differential equations (ODEs). In certain occasions the governing ordinary differential equations can be solved analytically and as far as parameter estimation is concerned, the problem is described by a set of algebraic equations. If however, the ODEs cannot be solved analytically, the mathematical model is more complex. In general, the model equations can be written in the form

$$\frac{d\mathbf{x}(t)}{dt} = \mathbf{f}(\mathbf{x}(t), \mathbf{u}, \mathbf{k}) \quad ; \quad \mathbf{x}(t_0) = \mathbf{x}_0 \tag{2.10}$$

$$\mathbf{y}(t) = \mathbf{C}\mathbf{x}(t) \tag{2.11}$$

or

$$\mathbf{y}(t) = \mathbf{h}(\mathbf{x}(t),\mathbf{k}) \tag{2.12}$$

where

$\mathbf{k}=[k_1,k_2,...,k_p]^T$ is a *p-dimensional* vector of parameters whose numerical values are unknown;

$\mathbf{x}=[x_1,x_2,...,x_n]^T$ is an *n-dimensional* vector of state variables;

$\mathbf{x}_0=[x_{10},x_{20},...,x_{n0}]^T$ is an *n-dimensional* vector of initial conditions for the state variables and they are assumed to be known precisely (in Section 6.4 we consider the situation where several elements of the initial state vector could be unknown);

u=$[u_1,u_2,\ldots,u_r]^T$ is an *r-dimensional* vector of manipulated variables which are either set by the experimentalist and their numerical values are precisely known or they have been measured;

f=$[f_1,f_2,\ldots,f_n]^T$ is a *n-dimensional* vector function of known form (the differential equations);

y=$[y_1,y_2,\ldots,y_m]^T$ is the *m-dimensional* output vector i.e., the set of variables that are measured experimentally; and

C is the $m \times n$ observation matrix which indicates the state variables (or linear combinations of state variables) that are measured experimentally.

h=$[h_1,h_2,\ldots,h_m]^T$ is a *m-dimensional* vector function of known form that relates in a nonlinear fashion the state vector to the output vector.

The state variables are the minimal set of dependent variables that are needed in order to describe fully the state of the system. The output vector represents normally a subset of the state variables or combinations of them that are measured. For example, if we consider the dynamics of a distillation column, in order to describe the condition of the column at any point in time we need to know the prevailing temperature and concentrations at each tray (the state variables). On the other hand, typically very few variables are measured, e.g., the concentration at the top and bottom of the column, the temperature in a few trays and in some occasions the concentrations at a particular tray where a side stream is taken. In other words, for this case the observation matrix **C** will have zeros everywhere except in very few locations where there will be 1's indicating which state variables are being measured.

In the distillation column example, the manipulated variables correspond to all the process parameters that affect its dynamic behavior and they are normally set by the operator, for example, reflux ratio, column pressure, feed rate, etc. These variables could be constant or time varying. In both cases however, it is assumed that their values are known precisely.

As another example let us consider a complex batch reactor. We may have to consider the concentration of many intermediates and final products in order to describe the system over time. However, it is quite plausible that only very few species are measured. In several cases, the measurements could even be pools of several species present in the reactor. For example, this would reflect an output variable which is the summation of two state variables, i.e., $y_1=x_1+x_2$.

The measurements of the output vector are taken at distinct points in time, t_i with i=1,...,N. The initial condition $\mathbf{x}_0$, is also chosen by the experimentalist and it is assumed to be precisely known. It represents a very important variable from an experimental design point of view.

Again, the measured output vector at time t_i, denoted as $\hat{\mathbf{y}}_i$, is related to the value calculated by the mathematical model (using the true parameter values) through the error term,

$$\hat{\mathbf{y}}_i = \mathbf{y}(t_i) + \boldsymbol{\varepsilon}_i \quad ; \quad i=1,2,\ldots,N \tag{2.13}$$

As in algebraic models, the error term accounts for the measurement error as well as for all model inadequacies. In dynamic systems we have the additional complexity that the error terms may be autocorrelated and in such cases several modifications to the objective function should be performed. Details are provided in Chapter 8.

In dynamic systems we may have the situation where a series of *runs* have been conducted and we wish to estimate the parameters using all the data simultaneously. For example in a study of isothermal decomposition kinetics, measurements are often taken over time for each run which is carried out at a fixed temperature.

Estimation of parameters present in partial differential equations is a very complex issue. Quite often by proper discretization of the spatial derivatives we transform the governing PDEs into a large number of ODEs. Hence, the problem can be transformed into one described by ODEs and be tackled with similar techniques. However, the fact that in such cases we have a system of high dimensionality requires particular attention. Parameter estimation for systems described by PDEs is examined in Chapter 11.

2.2 THE OBJECTIVE FUNCTION

What type of objective function should we minimize? This is the question that we are always faced with before we can even start the search for the parameter values. In general, the unknown parameter vector **k** is found by minimizing a scalar function often referred to as the *objective function.* We shall denote this function as S(**k**) to indicate the dependence on the chosen parameters.

The objective function is a suitable measure of the *overall* departure of the model calculated values from the measurements. For an individual measurement the departure from the model calculated value is represented by the *residual* $\mathbf{e}_i$. For example, the ith residual of an explicit algebraic model is

$$\mathbf{e}_i = [\hat{\mathbf{y}}_i - \mathbf{f}(\mathbf{x}_i, \mathbf{k})] \tag{2.14}$$

where the model based value $\mathbf{f}(\mathbf{x}_i,\mathbf{k})$ is calculated using the estimated parameter values. It should be noted that the residual ($\mathbf{e}_i$) is not the same as the error term ($\boldsymbol{\varepsilon}_i$) in Equation 2.13. The error term ($\boldsymbol{\varepsilon}_i$) corresponds to the true parameter values that are never known exactly, whereas the residual ($\mathbf{e}_i$) corresponds to the estimated parameter values.

The choice of the objective function is very important, as it dictates not only the values of the parameters but also their statistical properties. We may encounter two broad estimation cases. *Explicit estimation* refers to situations where the output vector is expressed as an explicit function of the input vector and the parameters. *Implicit estimation* refers to algebraic models in which output and input vector are related through an implicit function.

2.2.1 Explicit Estimation

Given N measurements of the output vector, the parameters can be obtained by minimizing the *Least Squares* (LS) objective function which is given below as the *weighted sum of squares of the residuals*, namely,

$$S_{LS}(\mathbf{k}) = \sum_{i=1}^{N} \mathbf{e}_i^T \mathbf{Q}_i \mathbf{e}_i \tag{2.15a}$$

where $\mathbf{e}_i$ is the vector of residuals from the ith experiment. The LS objective function for algebraic system takes the form,

$$S_{LS}(\mathbf{k}) = \sum_{i=1}^{N} [\hat{\mathbf{y}}_i - \mathbf{f}(\mathbf{x}_i, \mathbf{k})]^T \mathbf{Q}_i [\hat{\mathbf{y}}_i - \mathbf{f}(\mathbf{x}_i, \mathbf{k})] \tag{2.15b}$$

where in both cases $\mathbf{Q}_i$ is an $m \times m$ user-supplied weighting matrix.

For systems described by ODEs the LS objective function becomes,

$$S_{LS}(\mathbf{k}) = \sum_{i=1}^{N} [\hat{\mathbf{y}}_i - \mathbf{y}(t_i, \mathbf{k})]^T \mathbf{Q}_i [\hat{\mathbf{y}}_i - \mathbf{y}(t_i, \mathbf{k})] \tag{2.15c}$$

As we mentioned earlier a further complication of systems described by ODEs is that instead of a single run, a series of runs may have been conducted. If we wish to estimate the parameters using all the data simultaneously we must consider the following objective function

$$S_{LS}(\mathbf{k}) = \sum_{j=1}^{N_R} \left[\sum_{i=1}^{N} [\hat{\mathbf{y}}_i - \mathbf{y}(t_i, \mathbf{k})]^T \mathbf{Q}_i [\hat{\mathbf{y}}_i - \mathbf{y}(t_i, \mathbf{k})] \right]_j \tag{2.15d}$$

where N_R is the number of runs. Such a case is considered in Chapter 6 where three isobaric runs are presented for the catalytic hydrogenation of 3-hydroxypropanol to 1,3-propanediol at 318 and 353 *K*. In order to maintain simplicity and clarity in the presentation of the material we shall only consider the case of a single run in the following sections, i.e., the LS objective function given by equation 2.15c. Depending on our choice of the weighting matrix $\mathbf{Q}_i$ in the objective function we have the following cases:

2.2.1.1 Simple or Unweighted Least Squares (LS) Estimation

In this case we minimize the sum of squares of errors (SSE) without any weighting factor, i.e., we use $\mathbf{Q}_i=\mathbf{I}$ and Equation 2.14 reduces to

$$S_{LS}(\mathbf{k}) = \sum_{i=1}^{N} \mathbf{e}_i^T \mathbf{e}_i \tag{2.16}$$

2.2.1.2 Weighted Least Squares (WLS) Estimation

In this case we minimize a weighted SSE with constant weights, i.e., the user-supplied weighting matrix is kept the same for all experiments, $\mathbf{Q}_i=\mathbf{Q}$ for all $i=1,\ldots,N$.

2.2.1.3 Generalized Least Squares (GLS) Estimation

In this case we minimize a weighted SSE with non-constant weights. The user-supplied weighting matrices differ from experiment to experiment.

Of course, it is not at all clear how one should select the weighting matrices $\mathbf{Q}_i$, $i=1,\ldots,N$, even for cases where a constant weighting matrix $\mathbf{Q}$ is used. Practical guidelines for the selection of $\mathbf{Q}$ can be derived from *Maximum Likelihood (ML)* considerations.

2.2.1.4 Maximum Likelihood (ML) Estimation

If the mathematical model of the process under consideration is adequate, it is very reasonable to assume that the measured responses from the ith experiment are normally distributed. In particular the joint probability density function conditional on the value of the parameters ($\mathbf{k}$ and $\boldsymbol{\Sigma}_i$) is of the form,

$$p(\hat{\mathbf{y}}_i \mid \mathbf{k},\boldsymbol{\Sigma}_i) \sim det(\boldsymbol{\Sigma}_i)^{-1/2} \, exp\left\{-\frac{\mathbf{e}_i^T \boldsymbol{\Sigma}_i^{-1} \mathbf{e}_i}{2}\right\} \tag{2.17}$$

where $\mathbf{\Sigma}_i$ is the covariance matrix of the response variables $\mathbf{y}$ at the ith experiment and hence, of the residuals $\mathbf{e}_i$ too.

If we now further assume that measurements from *different experiments are independent*, the joint probability density function for the all the measured responses is simply the product,

$$p(\hat{\mathbf{y}}_1, \hat{\mathbf{y}}_2, ..., \hat{\mathbf{y}}_N \mid \mathbf{k}, \mathbf{\Sigma}_1, \mathbf{\Sigma}_2, ..., \mathbf{\Sigma}_N) \sim det(\mathbf{\Sigma}_1)^{-1/2}\, det(\mathbf{\Sigma}_2)^{-1/2} \ldots det(\mathbf{\Sigma}_N)^{-1/2}$$

$$\times\, exp\left\{-\frac{\mathbf{e}_1^T \mathbf{\Sigma}_1^{-1} \mathbf{e}_1}{2}\right\} exp\left\{-\frac{\mathbf{e}_2^T \mathbf{\Sigma}_2^{-1} \mathbf{e}_2}{2}\right\} \ldots exp\left\{-\frac{\mathbf{e}_N^T \mathbf{\Sigma}_N^{-1} \mathbf{e}_N}{2}\right\} \qquad (2.18)$$

grouping together similar terms we have,

$$p(\hat{\mathbf{y}}_1, \hat{\mathbf{y}}_2, ..., \hat{\mathbf{y}}_N \mid \mathbf{k}, \mathbf{\Sigma}_1, \mathbf{\Sigma}_2, ..., \mathbf{\Sigma}_N) \sim \left(\prod_{i=1}^{N} det(\mathbf{\Sigma}_i)\right)^{-\frac{1}{2}} exp\left\{-\frac{1}{2}\sum_{i=1}^{N} \mathbf{e}_i^T \mathbf{\Sigma}_i^{-1} \mathbf{e}_i\right\} \qquad (2.19)$$

The *Loglikelihood* function is the *log* of the joint probability density function and is regarded as a function of the parameters conditional on the observed responses. Hence, we have

$$L(\mathbf{k}, \mathbf{\Sigma}_1, \mathbf{\Sigma}_2, ..., \mathbf{\Sigma}_N \mid \hat{\mathbf{y}}_1, \hat{\mathbf{y}}_2, ..., \hat{\mathbf{y}}_N) = A - \frac{1}{2}\sum_{i=1}^{N} log\left(det(\mathbf{\Sigma}_i)\right) - \frac{1}{2}\sum_{i=1}^{N} \mathbf{e}_i^T \mathbf{\Sigma}_i^{-1} \mathbf{e}_i \qquad (2.20)$$

where A is a constant quantity. The maximum likelihood estimates of the unknown parameters are obtained by maximizing the *Loglikelihood* function.

At this point let us assume that the covariance matrices ($\mathbf{\Sigma}_i$) of the measured responses (and hence of the error terms) during each experiment are *known* precisely. Obviously, in such a case the ML parameter estimates are obtained by minimizing the following objective function

$$S_{ML}(\mathbf{k}) = \sum_{i=1}^{N} \mathbf{e}_i^T \mathbf{\Sigma}_i^{-1} \mathbf{e}_i \qquad (2.21)$$

Therefore, on statistical grounds, if the error terms (ε_i) are normally distributed with zero mean and with a *known covariance matrix*, then $\mathbf{Q}_i$ should be the inverse of this covariance matrix, i.e.,

$$\mathbf{Q}_i = [COV(\varepsilon_i)]^{-1} = \mathbf{\Sigma}_i^{-1} \quad ; \; i=1,2,\ldots,N \qquad (2.22)$$

However, the requirement of *exact* knowledge of all covariance matrices ($\boldsymbol{\Sigma}_i$, i=1,2,...,N) is rather unrealistic. Fortunately, in many situations of practical importance, we can make certain quite reasonable assumptions about the structure of $\boldsymbol{\Sigma}_i$ that allow us to obtain the ML estimates using Equation 2.21. This approach can actually aid us in establishing guidelines for the selection of the weighting matrices $\mathbf{Q}_i$ in least squares estimation.

Case I: Let us consider the stringent assumption that the error terms in each response variable and for each experiment (ε_{ij}, i=1,...N; j=1,...,*m*) are all identically and independently distributed (i.i.d) normally with zero mean and variance, σ_e^2. Namely,

$$\boldsymbol{\Sigma}_i = \sigma_e^2\,\mathbf{I} \quad ; \quad i=1,2,\ldots,N \tag{2.23}$$

where **I** is the *m*×*m* identity matrix. Substitution of $\boldsymbol{\Sigma}_i$ into Equation 2.21 yields

$$S_{ML}(\mathbf{k}) = \frac{1}{\sigma_e^2}\sum_{i=1}^{N}\mathbf{e}_i^T\mathbf{e}_i \tag{2.24}$$

Obviously minimization of $S_{ML}(\mathbf{k})$ in the above equation does not require the prior knowledge of the common factor σ_e^2. Therefore, under these conditions the ML estimation is equivalent to simple LS estimation ($\mathbf{Q}_i=\mathbf{I}$).

Case II: Next let us consider the more realistic assumption that the variance of a particular response variable is constant from experiment to experiment; however, different response variables have different variances, i.e.,

$$\boldsymbol{\Sigma}_i = \begin{bmatrix} \sigma_{e_1}^2 & 0 & \cdots & 0 \\ 0 & \sigma_{e_2}^2 & \cdots & 0 \\ \vdots & \vdots & \ddots & \vdots \\ 0 & 0 & \cdots & \sigma_{em}^2 \end{bmatrix} \quad ; \quad i=1,2,\ldots,N \tag{2.25}$$

Although we may not know the elements of the diagonal matrix $\boldsymbol{\Sigma}_i$ given by Equation 2.25, *we assume that we do know their relative value.* Namely we assume that we know the following ratios (v_i, i=1,2,...,*m*),

$$v_1 = \frac{\sigma_{e1}^2}{\sigma^2} \tag{2.26a}$$

$$v_2 = \frac{\sigma_{e_2}^2}{\sigma^2} \tag{2.26b}$$

$$v_m = \frac{\sigma_{e_m}^2}{\sigma^2} \tag{2.26c}$$

where σ^2 is an unknown scaling factor. Therefore $\mathbf{\Sigma}_i$ can be written as

$$\mathbf{\Sigma}_i = \sigma^2 \, diag(v_1, v_2, \ldots, v_m) \quad ; \quad i=1,2,\ldots,N \tag{2.27}$$

Upon substitution of $\mathbf{\Sigma}_i$ into Equation 2.21 it becomes apparent that the ML parameter estimates are the same as the weighted LS estimates when the following weighting matrices are used,

$$\mathbf{Q}_i = \mathbf{Q} = diag(v_1^{-1}, v_2^{-1}, \ldots, v_m^{-1}) \tag{2.28}$$

If the variances in Equation 2.25 are totally unknown, the ML parameter estimates can only be obtained by the *determinant criterion* presented later in this chapter.

<u>*Case III:*</u> Generalized LS estimation will yield ML estimates whenever the errors are distributed with variances that change from experiment to experiment. Therefore, in this case our choice for $\mathbf{Q}_i$ should be $[COV(\varepsilon_i)]^{-1}$ for i=1,...,N.

An interesting situation that arises often in engineering practice is when the errors have a *constant, yet unknown, percent error for each variable*, i.e.

$$\sigma_{\varepsilon ij}^2 = \sigma^2 y_{ij}^2 \quad ; \quad i=1,2,\ldots,N \text{ and } j=1,2,\ldots,m \tag{2.29}$$

where σ^2 is an unknown scaling factor. Again, upon substitution of $\mathbf{\Sigma}_i$ into Equation 2.21 it is readily seen that the ML parameter estimates are the same as the generalized LS estimates when the following weighting matrices are used,

$$\mathbf{Q}_i = diag(y_{i1}^{-2}, y_{i2}^{-2}, \ldots, y_{im}^{-2}) \quad ; \quad i=1,2,\ldots,N \tag{2.30}$$

The above choice has the computational disadvantage that the weights are a function of the unknown parameters. If the magnitude of the errors is not excessive, we could use the measurements in the weighting matrix with equally good results, namely

$$\mathbf{Q}_i = diag(\hat{y}_{i1}^{-2}, \hat{y}_{i2}^{-2}, \ldots, \hat{y}_{im}^{-2}) \quad ; \quad i=1,2,\ldots,N \tag{2.31}$$

2.2.1.5 The Determinant Criterion

If the covariance matrices of the response variables are unknown, the maximum likelihood parameter estimates are obtained by maximizing the *Loglikelihood* function (Equation 2.20) over **k** and the unknown variances. Following the distributional assumptions of Box and Draper (1965), i.e., assuming that $\boldsymbol{\Sigma}_1=\boldsymbol{\Sigma}_2=\ldots=\boldsymbol{\Sigma}_N=\boldsymbol{\Sigma}$, it can be shown that the ML parameter estimates can be obtained by minimizing the determinant (Bard, 1974)

$$S_{det}(\mathbf{k}) = det\left(\sum_{i=1}^{N} \mathbf{e}_i \mathbf{e}_i^T\right) \tag{2.32}$$

In addition, the corresponding estimate of the unknown covariance is

$$\hat{\boldsymbol{\Sigma}} = \frac{1}{N}\sum_{i=1}^{N} \mathbf{e}_i \mathbf{e}_i^T \tag{2.33}$$

It is worthwhile noting that Box and Draper (1965) arrived at the same determinant criterion following a Bayesian argument and assuming that $\boldsymbol{\Sigma}$ is unknown and that the prior distribution of the parameters is noninformative.

The determinant criterion is very powerful and it should be used to refine the parameter estimates obtained with least squares estimation if our assumptions about the covariance matrix are suspect.

2.2.1.6 Incorporation of Prior Information About The Parameters

Any prior information that is available about the parameter values can facilitate the estimation of the parameter values. The methodology to incorporate such information in the estimation procedure and the resulting benefits are discussed in Chapter 8.

2.2.2 Implicit Estimation

Now we turn our attention to algebraic models that can only be represented *implicitly* though an equation of the form,

$$\boldsymbol{\varphi}(\mathbf{y}; \mathbf{k}) = \mathbf{0} \tag{2.34}$$

In implicit estimation rather than minimizing a weighted sum of squares of the residuals in the response variables, we minimize a suitable implicit function of the measured variables dictated by the model equations. Namely, if we substitute the actual measured variables in Equation 2.8, an error term arises always even if the mathematical model is exact.

$$\varphi(\hat{\mathbf{y}}_i;\mathbf{k}) = \varepsilon_i \quad ; \quad i=1,2,\ldots,N \tag{2.35}$$

The residual is not equal to zero because of the experimental error in the measured variables, i.e.,

$$\hat{\mathbf{y}}_i = \mathbf{y}_i + \boldsymbol{\varepsilon}_{y,i} \quad ; \quad i=1,2,\ldots,N \tag{2.36}$$

Even if we make the stringent assumption that errors in the measurement of each variable ($\varepsilon_{y,ij}$, i=1,2,...,N, j=1,2,...,R) are independently and identically distributed (i.i.d.) normally with zero mean and constant variance, it is rather difficult to establish the exact distribution of the error term ε_i in Equation 2.35. This is particularly true when the expression is highly nonlinear. For example, this situation arises in the estimation of parameters for nonlinear thermodynamic models and in the treatment of potentiometric titration data (Sutton and MacGregor, 1977; Sachs, 1976; Englezos et al., 1990a, 1990b).

If we assume that the residuals in Equation 2.35 (ε_i) are normally distributed, their covariance matrix ($\boldsymbol{\Sigma}_i$) can be related to the covariance matrix of the measured variables ($COV(\boldsymbol{\varepsilon}_{y,i})=\boldsymbol{\Sigma}_{y,i}$) through the error propagation law. Hence, if for example we consider the case of independent measurements with a constant variance, i.e,

$$\boldsymbol{\Sigma}_{y,i} = \boldsymbol{\Sigma}_y = \begin{bmatrix} \sigma_{y1}^2 & 0 & \cdots & 0 \\ 0 & \sigma_{y2}^2 & \cdots & 0 \\ \vdots & \vdots & \ddots & \vdots \\ 0 & 0 & \cdots & \sigma_{yR}^2 \end{bmatrix} \tag{2.37}$$

The diagonal elements of the covariance matrix ($\boldsymbol{\Sigma}_i$) can be obtained from,

$$\sigma_{ij}^2 = \sum_{k=1}^{R} \left(\frac{\partial \varphi_j}{\partial y_k}\right)_i^2 \sigma_{yk}^2 \quad ; \quad i=1,2,\ldots,N\ ,\ j=1,2,\ldots,R \tag{2.38}$$

where the partial derivatives are evaluated at the conditions of the ith experiment. As a result, the elements of the covariance matrix ($\boldsymbol{\Sigma}_i$) change from experiment to experiment even if we assume that the variance of the measurement error is constant.

Similarly, we compute the off-diagonal terms of $\boldsymbol{\Sigma}_i$ from,

$$\sigma_{ijh} = \sum_{k=1}^{R} \left(\frac{\partial \varphi_j}{\partial y_k}\right)_i \left(\frac{\partial \varphi_h}{\partial y_k}\right)_i \sigma_{yk}^2$$

$$i=1,2,\ldots,N,\ h=1,2,\ldots,R,\ j=1,2,\ldots,R \qquad (2.39)$$

Having an estimate (through the error propagation law) of the covariance matrix $\boldsymbol{\Sigma}_i$ we can obtain the ML parameter estimates by minimizing the objective function,

$$S_{ML}(\mathbf{k}) = \sum_{i=1}^{N} \boldsymbol{\varphi}(\hat{\mathbf{y}}_i, \mathbf{k})^T \boldsymbol{\Sigma}_i^{-1} \boldsymbol{\varphi}(\hat{\mathbf{y}}_i, \mathbf{k}) \qquad (2.40)$$

The above implicit formulation of maximum likelihood estimation is valid only under the assumption that the residuals are normally distributed and the model is adequate. From our own experience we have found that implicit estimation provides the easiest and computationally the most efficient solution to many parameter estimation problems.

Furthermore, as a first approximation one can use implicit least squares estimation to obtain very good estimates of the parameters (Englezos et al., 1990). Namely, the parameters are obtained by minimizing the following *Implicit Least Squares (ILS)* objective function,

$$S_{ILS}(\mathbf{k}) = \sum_{i=1}^{N} \boldsymbol{\varphi}(\hat{\mathbf{y}}_i, \mathbf{k})^T \boldsymbol{\varphi}(\hat{\mathbf{y}}_i, \mathbf{k}) \qquad (2.41)$$

If the assumption of normality is grossly violated, ML estimates of the parameters can only be obtained using the *"error-in-variables" method* where besides the parameters, we also estimate the true (error-free) value of the measured variables. In particular, assuming that $\boldsymbol{\Sigma}_{y,i}$ is known, the parameters are obtained by minimizing the following objective function

$$S_{ML}(\mathbf{k}) = \sum_{i=1}^{N} (\hat{\mathbf{y}}_i - \mathbf{y}_i)^T \boldsymbol{\Sigma}_{y,i}^{-1} (\hat{\mathbf{y}}_i - \mathbf{y}_i) \qquad (2.42)$$

subject to

$$\boldsymbol{\varphi}(\mathbf{y}_i; \mathbf{k}) = \mathbf{0} \quad ; \quad i=1,2,\ldots,N \qquad (2.43)$$

The method has been analyzed in detail by several researchers, for example, Schwetlick and Tiller (1985), Seber and Wild (1989), Reilly and Patino-Leal (1981), Patino-Leal and Reilly (1982) and Duever et al. (1987) among others.

2.3 PARAMETER ESTIMATION SUBJECT TO CONSTRAINTS

In parameter estimation we are occasionally faced with an additional complication. Besides the minimization of the objective function (a weighted sum of errors) the mathematical model of the physical process includes a set of constrains that must also be satisfied. In general these are either equality or inequality constraints. In order to avoid unnecessary complications in the presentation of the material, constrained parameter estimation is presented exclusively in Chapter 9.

3

Computation of Parameters in Linear Models - Linear Regression

Linear models with respect to the parameters represent the simplest case of parameter estimation from a computational point of view because there is no need for iterative computations. Unfortunately, the majority of process models encountered in chemical engineering practice are nonlinear. *Linear regression* has received considerable attention due to its significance as a tool in a variety of disciplines. Hence, there is a plethora of books on the subject (e.g., Draper and Smith, 1998; Freund and Minton, 1979; Hocking, 1996; Montgomery and Peck, 1992; Seber, 1977). The majority of these books has been written by statisticians.

The objectives in this chapter are two. The first one is to briefly review the essentials of linear regression and to present them in a form that is consistent with our notation and approach followed in subsequent chapters addressing nonlinear regression problems. The second objective is to show that a large number of linear regression problems can now be handled with readily available software such as Microsoft Excel™ and SigmaPlot™.

3.1 THE LINEAR REGRESSION MODEL

As we have already mentioned in Chapter 2, assuming linearity with respect to the unknown parameters, the general algebraic model can be reduced to the following form

$$\hat{\mathbf{y}}_i = \mathbf{F}(\mathbf{x}_i)\mathbf{k} + \boldsymbol{\varepsilon}_i \quad ; \quad i=1,2,\ldots,N \tag{3.1}$$

which can be readily handled by linear least squares estimation. In this model, **k** is the *p-dimensional* parameter vector, **x** is the *n-dimensional* vector of independent variables (*regressor* variables), **y** is the *m-dimensional* vector of dependent variables (*response* variables) and $\mathbf{F}(\mathbf{x}_i)$ is an *m×p dimensional* matrix which depends only on $\mathbf{x}_i$ and is independent of the parameters. This matrix is also the *sensitivity coefficients matrix* that plays a very important role in nonlinear regression. Quite often the elements of matrix **F** are simply the independent variables themselves, and hence, the model can be further reduced to the well known *linear regression* model,

$$\hat{\mathbf{y}}_i = \mathbf{X}_i \, \mathbf{k} + \varepsilon_i \quad ; \quad i=1,2,\ldots,N \tag{3.2}$$

where N is the number of measurements. The above model is very general and it covers the following cases:

The *simple linear regression model* which has a single response variable, a single independent variable and two unknown parameters,

$$\hat{y}_i = k_1 x_i + k_2 + \varepsilon_i \tag{3.3a}$$

or in matrix notation

$$\hat{y}_i = [x_i , 1] \begin{bmatrix} k_1 \\ k_2 \end{bmatrix} + \varepsilon_i \tag{3.3b}$$

(ii) The *multiple linear regression model* which has a single response variable, *n* independent variables and *p (=n+1)* unknown parameters,

$$\hat{y}_i = k_1 x_{1i} + k_2 x_{2i} + \ldots + k_{p-1} x_{p-1,i} + k_p + \varepsilon_i \tag{3.4a}$$

or in matrix notation

$$\hat{y}_i = [x_{1i}, x_{2i}, \ldots, x_{p-1,i} , 1] \begin{bmatrix} k_1 \\ k_2 \\ \vdots \\ k_p \end{bmatrix} + \varepsilon_i \tag{3.4b}$$

or more compactly as

$$\hat{y}_i = \mathbf{x}_i^T \mathbf{k} + \varepsilon_i \tag{3.4c}$$

where $\mathbf{x}_i = [x_{1i}, x_{2i}, \ldots, x_{p-1,i}, 1]^T$ is the augmented *p-dimensional* vector of independent variables ($p=n+1$).

(iii) The *multiresponse linear regression model* which has m response variables, $(m\times n)$ independent variables and p $(=n+1)$ unknown parameters,

$$\begin{aligned}
\hat{y}_{1i} &= k_1 x_{11i} + k_2 x_{12i} + \ldots + k_{p-1} x_{1,p-1,i} + k_p + \varepsilon_{1i} \\
\hat{y}_{2i} &= k_1 x_{21i} + k_2 x_{22i} + \ldots + k_{p-1} x_{2,p-1,i} + k_p + \varepsilon_{2i} \\
&\vdots \\
\hat{y}_{mi} &= k_1 x_{m1i} + k_2 x_{m2i} + \ldots + k_{p-1} x_{m,p-1,i} + k_p + \varepsilon_{mi}
\end{aligned} \tag{3.5a}$$

or in matrix notation

$$\begin{bmatrix} \hat{y}_{1i} \\ \hat{y}_{2i} \\ \vdots \\ \hat{y}_{mi} \end{bmatrix} = \begin{bmatrix} x_{11i} & x_{12i} & \cdots & x_{1,p-1,i} & 1 \\ x_{21i} & x_{22i} & \cdots & x_{2,p-1,i} & 1 \\ \vdots & \vdots & & \ddots & \vdots \\ x_{m1i} & x_{m2i} & \cdots & x_{m,p-1,i} & 1 \end{bmatrix} \begin{bmatrix} k_1 \\ k_2 \\ \vdots \\ k_p \end{bmatrix} + \begin{bmatrix} \varepsilon_{1i} \\ \varepsilon_{2i} \\ \vdots \\ \varepsilon_{mi} \end{bmatrix} \tag{3.5b}$$

The above equation can be written more compactly as,

$$\hat{\mathbf{y}}_i = \mathbf{X}_i \mathbf{k} + \boldsymbol{\varepsilon}_i \tag{3.5c}$$

where the matrix $\mathbf{X}_i$ is defined as

$$\mathbf{X}_i = \begin{bmatrix} x_{11i} & x_{12i} & \cdots & x_{1,p-1,i} & 1 \\ x_{21i} & x_{22i} & \cdots & x_{2,p-1,i} & 1 \\ \vdots & \vdots & & \ddots & \vdots \\ x_{m1i} & x_{m2i} & \cdots & x_{m,p-1,i} & 1 \end{bmatrix} \tag{3.6}$$

It should be noted that the above definition of $\mathbf{X}_i$ is different from the one often found in linear regression books. There $\mathbf{X}$ is defined for the simple or multiple linear regression model and it contains *all* the measurements. In our case, index i explicitly denotes the ith measurement and we do not group our measurements. Matrix $\mathbf{X}_i$ represents the values of the independent variables from the ith experiment.

3.2 THE LINEAR LEAST SQUARES OBJECTIVE FUNCTION

Given N measurements of the response variables (output vector), the parameters are obtained by minimizing the *Linear Least Squares* (LS) objective function which is given below as the *weighted sum of squares of the residuals*, namely,

$$S_{LS}(\mathbf{k}) = \sum_{i=1}^{N} \mathbf{e}_i^T \mathbf{Q}_i \mathbf{e}_i \tag{3.7}$$

where $\mathbf{e}_i$ is the *m-dimensional* vector of residuals from the ith experiment. It is noted that the residuals $\mathbf{e}_i$ are obtained from Equation 3.2 using the estimated parameter values instead of their true values that yield the error terms $\boldsymbol{\varepsilon}_i$. The LS objective function takes the form,

$$S_{LS}(\mathbf{k}) = \sum_{i=1}^{N} [\hat{\mathbf{y}}_i - \mathbf{X}_i\mathbf{k}]^T \mathbf{Q}_i [\hat{\mathbf{y}}_i - \mathbf{X}_i\mathbf{k}] \tag{3.8}$$

where $\mathbf{Q}_i$ is an $m \times m$ user-supplied weighting matrix. Depending on our choice of $\mathbf{Q}_i$, we have the following cases:

Simple Linear Least Squares

In this case we minimize the sum of squares of errors (SSE) without any weighting factor, i.e., we use $\mathbf{Q}_i = \mathbf{I}$ and Equation 3.6 reduces to

$$S_{LS}(\mathbf{k}) = \sum_{i=1}^{N} \mathbf{e}_i^T \mathbf{e}_i \tag{3.9}$$

This choice of $\mathbf{Q}_i$ yields maximum likelihood estimates of the parameters if the error terms in each response variable and for each experiment (ε_{ij}, i=1,...N; j=1,...,m) are all identically and independently distributed (i.i.d) normally with zero mean and variance, σ_e^2. Namely, $E(\boldsymbol{\varepsilon}_i) = \mathbf{0}$ and $COV(\boldsymbol{\varepsilon}_i) = \sigma_e^2 \mathbf{I}$ where $\mathbf{I}$ is the $m \times m$ identity matrix.

Weighted Least Squares (WLS) Estimation

In this case we minimize a weighted sum of squares of residuals with constant weights, i.e., the user-supplied weighting matrix is kept the same for all experiments, $\mathbf{Q}_i = \mathbf{Q}$ for all i=1,...,N and Equation 3.7 reduces to

$$S_{LS}(\mathbf{k}) = \sum_{i=1}^{N} \mathbf{e}_i^T \mathbf{Q} \mathbf{e}_i \tag{3.10}$$

This choice of $\mathbf{Q}_i$ yields ML estimates of the parameters if the error terms in each response variable and for each experiment (ε_{ij}, i=1,...N; j=1,...,m) are independently distributed normally with zero mean and constant variance. Namely, the variance of a particular response variable is constant from experiment to experiment; however, different response variables have different variances, i.e.,

$$COV(\varepsilon_i) = \begin{bmatrix} \sigma_{e1}^2 & 0 & \cdots & 0 \\ 0 & \sigma_{e2}^2 & \cdots & 0 \\ \vdots & \vdots & \ddots & \vdots \\ 0 & 0 & \cdots & \sigma_{em}^2 \end{bmatrix} \quad ; \ i=1,2,\ldots,N \tag{3.11}$$

which can be written as

$$COV(\varepsilon_i) = \sigma^2 \, diag(v_1, v_2, \ldots, v_m) \quad ; \ i=1,2,\ldots,N \tag{3.12}$$

where σ^2 is an unknown scaling factor and $v_1, v_2, \ldots, v_m$ are known constants. ML estimates are obtained if the constant weighting matrix $\mathbf{Q}$ have been chosen as

$$\mathbf{Q} = diag(v_1^{-1}, v_2^{-1}, \ldots, v_m^{-1}) \quad ; \ i=1,2,\ldots,N \tag{3.13}$$

Generalized Least Squares (GLS) Estimation

In this case we minimize a weighted SSE with non-constant weights. The user-supplied weighting matrices differ from experiment to experiment. ML estimates of the parameters are obtained if we choose

$$\mathbf{Q}_i = [COV(\varepsilon_i)]^{-1} \quad ; \ i=1,2,\ldots,N \tag{3.14}$$

3.3 LINEAR LEAST SQUARES ESTIMATION

Let us consider first the most general case of the *multiresponse linear regression model* represented by Equation 3.2. Namely, we assume that we have N measurements of the m-dimensional output vector (response variables), $\hat{\mathbf{y}}_i$, i=1,...,N.

The computation of the parameter estimates is accomplished by minimizing the *least squares* (LS) objective function given by Equation 3.8 which is shown next

$$S_{LS}(\mathbf{k}) = \sum_{i=1}^{N} [\hat{\mathbf{y}}_i - \mathbf{X}_i\mathbf{k}]^T \mathbf{Q}_i [\hat{\mathbf{y}}_i - \mathbf{X}_i\mathbf{k}]$$

Use of the stationary criterion

$$\frac{\partial S_{LS}(\mathbf{k})}{\partial \mathbf{k}} = \mathbf{0} \tag{3.15}$$

yields a linear equation of the form

$$\mathbf{A}\,\mathbf{k} = \mathbf{b} \tag{3.16}$$

where the *(p×p) dimensional* matrix **A** is given by

$$\mathbf{A} = \sum_{i=1}^{N} \mathbf{X}_i^T \mathbf{Q}_i \mathbf{X}_i \tag{3.17a}$$

and the *p-dimensional* vector **b** is given by

$$\mathbf{b} = \sum_{i=1}^{N} \mathbf{X}_i^T \mathbf{Q}_i \hat{\mathbf{y}}_i \tag{3.17b}$$

Solution of the above linear equation yields the least squares estimates of the parameter vector, **k***,

$$\mathbf{k}^* = \left[\sum_{i=1}^{N} \mathbf{X}_i^T \mathbf{Q}_i \mathbf{X}_i\right]^{-1} \left[\sum_{i=1}^{N} \mathbf{X}_i^T \mathbf{Q}_i \hat{\mathbf{y}}_i\right] \tag{3.18}$$

For the *single response* linear regression model (m=1), Equations (3.17a) and (3.17b) reduce to

$$\mathbf{A} = \sum_{i=1}^{N} \mathbf{x}_i \mathbf{x}_i^T Q_i \tag{3.19a}$$

and

$$\mathbf{b} = \sum_{i=1}^{N} \mathbf{x}_i \hat{y}_i Q_i \tag{3.19b}$$

where Q_i is a scalar weighting factor and $\mathbf{x}_i$ is the augmented *p-dimensional* vector of independent variables $[x_{1i}, x_{2i}, \ldots, x_{p-1,i}, 1]^T$. The optimal parameter estimates are obtained from

$$\mathbf{k}^* = \left[\sum_{i=1}^{N} \mathbf{x}_i \mathbf{x}_i^T Q_i\right]^{-1} \left[\sum_{i=1}^{N} \mathbf{x}_i \hat{y}_i Q_i\right] \tag{3.20}$$

In practice, the solution of Equation 3.16 for the estimation of the parameters is not done by computing the inverse of matrix **A**. Instead, any good linear equation solver should be employed. Our preference is to perform first an eigenvalue decomposition of the real symmetric matrix **A** which provides significant additional information about potential ill-conditioning of the parameter estimation problem (see Chapter 8).

3.4 POLYNOMIAL CURVE FITTING

In engineering practice we are often faced with the task of fitting a low-order polynomial curve to a set of data. Namely, given a set of N pair data, (y_i, x_i), i=1,...,N, we are interested in the following cases,

(i) Fitting a *straight line* to the data,

$$y = k_1 x + k_2 \tag{3.21}$$

This problem corresponds to the simple linear regression model (m=1, n=1, p=2). Taking as Q_i=1 (all data points are weighed equally) Equations 3.19a and 3.19b become

$$\mathbf{A} = \sum_{i=1}^{N} \mathbf{x}_i \mathbf{x}_i^T = \sum_{i=1}^{N} \begin{bmatrix} x_i \\ 1 \end{bmatrix} \begin{bmatrix} x_i & 1 \end{bmatrix} = \begin{bmatrix} \sum_{i=1}^{N} x_i^2 & \sum_{i=1}^{N} x_i \\ \sum_{i=1}^{N} x_i & N \end{bmatrix} \tag{3.22a}$$

and

$$\mathbf{b} = \sum_{i=1}^{N} \mathbf{x}_i \hat{y}_i = \sum_{i=1}^{N} \begin{bmatrix} x_i \\ 1 \end{bmatrix} \hat{y}_i = \begin{bmatrix} \sum_{i=1}^{N} x_i \hat{y}_i \\ \sum_{i=1}^{N} \hat{y}_i \end{bmatrix} \tag{3.22b}$$

The parameters estimates are now obtained from

$$\begin{bmatrix} k_1 \\ k_2 \end{bmatrix} = \begin{bmatrix} \sum_{i=1}^{N} x_i^2 & \sum_{i=1}^{N} x_i \\ \sum_{i=1}^{N} x_i & N \end{bmatrix}^{-1} \begin{bmatrix} \sum_{i=1}^{N} x_i \hat{y}_i \\ \sum_{i=1}^{N} \hat{y}_i \end{bmatrix} \tag{3.23}$$

(ii) Fitting a *quadratic polynomial* to the data,

$$y = k_1 x^2 + k_2 x + k_3 \tag{3.24}$$

This problem corresponds to the multiple linear regression model with m=1, n=2 and p=3. In this case we take $x_1 = x^2$, $x_2 = x$ and with $Q_i = 1$ (all data points are weighed equally) Equations 3.19a and 3.19b become

$$\mathbf{A} = \sum_{i=1}^{N} \begin{bmatrix} x_i^2 \\ x_i \\ 1 \end{bmatrix} \begin{bmatrix} x_i^2 & x_i & 1 \end{bmatrix} = \begin{bmatrix} \sum_{i=1}^{N} x_i^4 & \sum_{i=1}^{N} x_i^3 & \sum_{i=1}^{N} x_i^2 \\ \sum_{i=1}^{N} x_i^3 & \sum_{i=1}^{N} x_i^2 & \sum_{i=1}^{N} x_i \\ \sum_{i=1}^{N} x_i^2 & \sum_{i=1}^{N} x_i & N \end{bmatrix} \tag{3.25a}$$

and

$$\mathbf{b} = \sum_{i=1}^{N} \begin{bmatrix} x_i^2 \\ x_i \\ 1 \end{bmatrix} \hat{y}_i = \begin{bmatrix} \sum_{i=1}^{N} x_i^2 \hat{y}_i \\ \sum_{i=1}^{N} x_i \hat{y}_i \\ \sum_{i=1}^{N} \hat{y}_i \end{bmatrix} \tag{3.25b}$$

The parameters estimates are now obtained from

$$\begin{bmatrix} k_1 \\ k_2 \\ k_3 \end{bmatrix} = \begin{bmatrix} \sum_{i=1}^{N} x_i^4 & \sum_{i=1}^{N} x_i^3 & \sum_{i=1}^{N} x_i^2 \\ \sum_{i=1}^{N} x_i^3 & \sum_{i=1}^{N} x_i^2 & \sum_{i=1}^{N} x_i \\ \sum_{i=1}^{N} x_i^2 & \sum_{i=1}^{N} x_i & N \end{bmatrix}^{-1} \begin{bmatrix} \sum_{i=1}^{N} x_i^2 \hat{y}_i \\ \sum_{i=1}^{N} x_i \hat{y}_i \\ \sum_{i=1}^{N} \hat{y}_i \end{bmatrix} \tag{3.26}$$

(iii) Fitting a *cubic polynomial* to the data

$$y = k_1 x^3 + k_2 x^2 + k_3 x + k_4 \tag{3.27}$$

This problem corresponds to the multiple linear regression model with m=1, n=3 and p=4. In this case we take $x_1=x^3$, $x_2=x^2$, $x_3=x$ and with $Q_i=1$ (all data points are weighed equally) Equations 3.19a and 3.19b become

$$\mathbf{A} = \sum_{i=1}^{N} \begin{bmatrix} x_i^3 \\ x_i^2 \\ x_i \\ 1 \end{bmatrix} \begin{bmatrix} x_i^3 & x_i^2 & x_i & 1 \end{bmatrix} = \begin{bmatrix} \sum_{i=1}^{N} x_i^6 & \sum_{i=1}^{N} x_i^5 & \sum_{i=1}^{N} x_i^4 & \sum_{i=1}^{N} x_i^3 \\ \sum_{i=1}^{N} x_i^5 & \sum_{i=1}^{N} x_i^4 & \sum_{i=1}^{N} x_i^3 & \sum_{i=1}^{N} x_i^2 \\ \sum_{i=1}^{N} x_i^4 & \sum_{i=1}^{N} x_i^3 & \sum_{i=1}^{N} x_i^2 & \sum_{i=1}^{N} x_i \\ \sum_{i=1}^{N} x_i^3 & \sum_{i=1}^{N} x_i^2 & \sum_{i=1}^{N} x_i & N \end{bmatrix} \tag{3.28a}$$

and

$$\mathbf{b} = \sum_{i=1}^{N} \begin{bmatrix} x_i^3 \\ x_i^2 \\ x_i \\ 1 \end{bmatrix} \hat{y}_i = \begin{bmatrix} \sum_{i=1}^{N} x_i^3 \hat{y}_i \\ \sum_{i=1}^{N} x_i^2 \hat{y}_i \\ \sum_{i=1}^{N} x_i \hat{y}_i \\ \sum_{i=1}^{N} \hat{y}_i \end{bmatrix} \tag{3.28b}$$

The parameters estimates are now obtained from

$$\begin{bmatrix} k_1 \\ k_2 \\ k_3 \\ k_4 \end{bmatrix} = \begin{bmatrix} \sum_{i=1}^{N} x_i^6 & \sum_{i=1}^{N} x_i^5 & \sum_{i=1}^{N} x_i^4 & \sum_{i=1}^{N} x_i^3 \\ \sum_{i=1}^{N} x_i^5 & \sum_{i=1}^{N} x_i^4 & \sum_{i=1}^{N} x_i^3 & \sum_{i=1}^{N} x_i^2 \\ \sum_{i=1}^{N} x_i^4 & \sum_{i=1}^{N} x_i^3 & \sum_{i=1}^{N} x_i^2 & \sum_{i=1}^{N} x_i \\ \sum_{i=1}^{N} x_i^3 & \sum_{i=1}^{N} x_i^2 & \sum_{i=1}^{N} x_i & N \end{bmatrix}^{-1} \begin{bmatrix} \sum_{i=1}^{N} x_i^3 \hat{y}_i \\ \sum_{i=1}^{N} x_i^2 \hat{y}_i \\ \sum_{i=1}^{N} x_i \hat{y}_i \\ \sum_{i=1}^{N} \hat{y}_i \end{bmatrix} \tag{3.29}$$

3.5 STATISTICAL INFERENCES

Once we have estimated the unknown parameter values in a linear regression model and the underlying assumptions appear to be reasonable, we can proceed and make statistical inferences about the parameter estimates and the response variables.

3.5.1 Inference on the Parameters

The least squares estimator has several desirable properties. Namely, the parameter estimates are normally distributed, unbiased (i.e., $E(\mathbf{k}^*)=\mathbf{k}$) and their covariance matrix is given by

$$COV(\mathbf{k}^*) = \sigma_\varepsilon^2 \mathbf{A}^{-1} \tag{3.30}$$

where matrix **A** is given by Equation 3.17a or 3.19a. An estimate, $\hat{\sigma}_\varepsilon^2$, of the variance σ_ε^2 is given by

$$\hat{\sigma}_\varepsilon^2 = \frac{S_{LS}(\mathbf{k}^*)}{(\text{d.f.})} \tag{3.31}$$

where (d.f.)=($Nm-p$) are the *degrees of freedom*, namely the total number of measurements minus the number of unknown parameters.

Given all the above it can be shown that *the (1−α)100% joint confidence region* for the parameter vector **k** is an ellipsoid given by the equation:

$$[\mathbf{k}-\mathbf{k}^*]^T\ \mathbf{A}^{-1}[\mathbf{k}-\mathbf{k}^*] = p\hat{\sigma}_\varepsilon^2\, F^{\alpha}_{p,Nm-p} \tag{3.32a}$$

or

$$[\mathbf{k}-\mathbf{k}^*]^T\ \mathbf{A}^{-1}[\mathbf{k}-\mathbf{k}^*] = \frac{pS_{LS}(\mathbf{k}^*)}{Nm-p}\, F^{\alpha}_{p,Nm-p} \tag{3.32b}$$

where α is the selected probability level in Fisher's F-distribution and $F^{\alpha}_{p,Nm-p}$ is obtained from the F-distribution tables with $\nu_1=p$ and $\nu_2=(Nm-p)$ degrees of freedom.

The corresponding *(1−α)100% marginal confidence interval* for each parameter, k_i, i=1,2,...,p, is

$$k_i^* - t^{\nu}_{\alpha/2}\hat{\sigma}_{k_i} \le k_i \le k_i^* + t^{\nu}_{\alpha/2}\hat{\sigma}_{k_i} \tag{3.33}$$

where $t^{\nu}_{\alpha/2}$ is obtained from the tables of Student's T-distribution with $\nu=Nm-p$ degrees of freedom.

The standard error of parameter k_i, $\hat{\sigma}_{k_i}$, is obtained as the square root of the corresponding diagonal element of the inverse of matrix **A** multiplied by $\hat{\sigma}_\varepsilon$, i.e.,

$$\hat{\sigma}_{k_i} = \hat{\sigma}_\varepsilon \sqrt{\left\{\mathbf{A}^{-1}\right\}_{ii}} \tag{3.34}$$

Practically, for $\nu \ge 30$ we can use the approximation $t^{\nu}_{\alpha/2} \approx z_{\alpha/2}$ where $z_{\alpha/2}$ is obtained from the tables of the standard normal distribution. That is why when the degrees of freedom are high, the 95% confidence intervals are simply taken as twice the standard error (recall that $z_{0.025}=1.96$ and $t^{30}_{0.025}=2.042$).

3.5.2 Inference on the Expected Response Variables

A valuable inference that can be made to infer the quality of the model predictions is the (1−α)100% confidence interval of the predicted mean response at $\mathbf{x}_0$. It should be noted that the *predicted mean response* of the linear regression model at $\mathbf{x}_0$ is $\mathbf{y}_0 = \mathbf{F}(\mathbf{x}_0)\mathbf{k}^*$ or simply $\mathbf{y}_0 = \mathbf{X}_0\mathbf{k}^*$. Although the error term ε_0 is not included, there is some uncertainty in the predicted mean response due to the uncertainty in $\mathbf{k}^*$. Under the usual assumptions of normality and independence, the covariance matrix of the predicted mean response is given by

$$COV(\mathbf{y}_0) = \mathbf{F}^T(\mathbf{x}_0)\, COV(\mathbf{k}^*)\, \mathbf{F}(\mathbf{x}_0) \tag{3.35a}$$

or for the standard multiresponse linear regression model where $\mathbf{F}(\mathbf{x}_0) \equiv \mathbf{X}_0$,

$$COV(\mathbf{y}_0) = \mathbf{X}_0^T\, COV(\mathbf{k}^*)\, \mathbf{X}_0 \tag{3.35b}$$

The covariance matrix $COV(\mathbf{k}^*)$ is obtained by Equation 3.30. Let us now concentrate on the expected mean response of a particular response variable. The *(1−α)100% confidence interval* of y_{i0} (i=1,...,m), the ith element of the response vector $\mathbf{y}_0$ at $\mathbf{x}_0$ is given below

$$y_{i0} - t^{\nu}_{\alpha/2}\hat{\sigma}_{y_{i0}} \le \mu_{y_{i0}} \le y_{i0} + t^{\nu}_{\alpha/2}\hat{\sigma}_{y_{i0}} \tag{3.36}$$

The standard error of y_{i0}, $\hat{\sigma}_{y_{i0}}$, is the square root of the ith diagonal element of $COV(\mathbf{y}_0)$, namely,

$$\hat{\sigma}_{y_{i0}} = \hat{\sigma}_{\varepsilon}\sqrt{\left\{\mathbf{F}^T(\mathbf{x}_0)\mathbf{A}^{-1}\mathbf{F}(\mathbf{x}_0)\right\}_{ii}} \tag{3.37a}$$

or for the standard multiresponse linear regression model,

$$\hat{\sigma}_{y_{i0}} = \hat{\sigma}_{\varepsilon}\sqrt{\left\{\mathbf{X}_0^T\mathbf{A}^{-1}\mathbf{X}_0\right\}_{ii}} \tag{3.37b}$$

For the *single* response y_0 in the case of simple or multiple linear regression (i.e., m=1), the *(1−α)100% confidence interval* of y_0 is,

$$y_0 - t^{\nu}_{\alpha/2}\hat{\sigma}_{y_0} \le \mu_{y_0} \le y_0 + t^{\nu}_{\alpha/2}\hat{\sigma}_{y_0} \tag{3.38a}$$

or equivalently

$$\mathbf{x}_0^T\mathbf{k}^* - t^{\nu}_{\alpha/2}\hat{\sigma}_{y_0} \le \mu_{y_0} \le \mathbf{x}_0^T\mathbf{k}^* + t^{\nu}_{\alpha/2}\hat{\sigma}_{y_0} \tag{3.38b}$$

where $t^{\nu}_{\alpha/2}$ is obtained from the tables of Student's T-distribution with ν=(N−p) degrees of freedom and $\hat{\sigma}_{y_0}$ is the *standard error of prediction* at $\mathbf{x}_0$. This quantity usually appears in the standard output of many regression computer packages. It is computed by

$$\hat{\sigma}_{y_0} = \hat{\sigma}_\varepsilon \sqrt{\mathbf{x}_0^T \mathbf{A}^{-1} \mathbf{x}_0} \tag{3.39}$$

In all the above cases we presented confidence intervals for the *mean* expected response rather than a *future observation (future measurement)* of the response variable, $\hat{\mathbf{y}}_0$. In this case, besides the uncertainty in the estimated parameters, we must include the uncertainty due to the measurement error (ε_0).

The corresponding *(1–α)100% confidence interval* for the multiresponse linear model is

$$y_{i0} - t^{\nu}_{\alpha/2}\hat{\sigma}_{\hat{y}_{i0}} \le \hat{y}_{i0} \le y_{i0} + t^{\nu}_{\alpha/2}\hat{\sigma}_{\hat{y}_{i0}} \quad ; i=1,\ldots,m \tag{3.40}$$

where the corresponding standard error of $\hat{y}_{i0}$ is given by

$$\hat{\sigma}_{\hat{y}_{i0}} = \hat{\sigma}_\varepsilon \sqrt{1 + \left\{\mathbf{X}_0^T \mathbf{A}^{-1} \mathbf{X}_0\right\}_{ii}} \tag{3.41}$$

For the case of a single response model (i.e., m=1), the *(1–α)100% confidence interval* of $\hat{y}_0$ is,

$$\mathbf{x}_0^T \mathbf{k}^* - t^{\nu}_{\alpha/2}\hat{\sigma}_{\hat{y}_0} \le \hat{y}_0 \le \mathbf{x}_0^T \mathbf{k}^* + t^{\nu}_{\alpha/2}\hat{\sigma}_{\hat{y}_0} \tag{3.42}$$

where the corresponding standard error of $\hat{y}_0$ is given by

$$\hat{\sigma}_{\hat{y}_0} = \hat{\sigma}_\varepsilon \sqrt{1 + \mathbf{x}_0^T \mathbf{A}^{-1} \mathbf{x}_0} \tag{3.43}$$

3.6 SOLUTION OF MULTIPLE LINEAR REGRESSION PROBLEMS

Problems that can be described by a multiple linear regression model (i.e., they have a single response variable, m=1) can be readily solved by available software. We will demonstrate such problems can be solved by using Microsoft Excel™ and SigmaPlot™.

3.6.1 Procedure for Using Microsoft Excel™ for Windows

Step 1. First the data are entered in columns. The single dependent variable is designated by y whereas the independent ones by x_1, x_2, x_3 etc.

Step 2. Select cells below the entered data to form a rectangle [5 × p] where p is the number of parameters sought after; e.g. in the equation $y=k_1x_1+k_2x_2+k_3$ you are looking for k_1, k_2 and k_3. Therefore p would be equal to 3.
Note: Excel casts the p-parameter model in the following form: $y=m_1x_1+m_2x_2+\ldots+m_{p-1}x_{p-1}+b$.

Step 3. Now that you have selected an area [5 × p] on the spreadsheet, go to the f_x (Paste Function button) and click.

Step 4. Click on Statistical on the left scroll menu and click on LINEST on the right scroll menu; then hit OK.

A box will now appear asking for the following
Known Y`s
Known X`s
Const
Stats

Step 5. Click in the text box for known values for the singe response variable y; then go to the Excel sheet and highlight the y values.

Step 6 Repeat Step 5 for the known values for the independent variables x_1, x_2, etc. by clicking on the box containing these values. This the program lets you highlight the area that encloses all the x values (x_1, x_2, x_3, etc....).

Step 7. Set the logical value *Const=true* if you wish to calculate a y intercept value.

Step 8. Set the logical value *Stats=true* if you wish the program to return additional regression statistics.

Step 9. Now that you have entered all the data you do not hit the OK button but instead press *Control-Shift-Enter*.
This command allows all the elements in the array to be displayed. If you hit OK you will only see one element in the array.

Once the above steps have been followed, the program returns the following information on the worksheet. The information is displayed in a [5 × p] table where p is the number of parameters

1st row:	parameter values					
	m_{p-1}	m_{p-2}	...	m_2	m_1	b
or	k_{p-1}	k_{p-2}	...	k_2	k_1	k_p

2nd row: Standard errors for the estimated parameter values

$se\,(k_{p-1})$ $se(k_{p-2})$ ... $se(k_2)$ $se(k_1)$ $se(k_p)$

3rd row: Coefficient of determination and standard error of the y value

R^2 sev

Note: The coefficient of determination ranges in value from 0 to 1. If it is equal to one then there is a perfect correlation in the sample. If on the other hand it has a value of zero then the model is not useful in calculating a y-value.

4th row: F statistic and the number of degrees of freedom (d.f.)

5th row: Information about the regression

ssreg (regression sum of squares) ssresid (residual sum of squares)

The above procedure will be followed in the following three examples with two, three and four parameter linear single response models

Example 1

Table 3.1 gives a set of pressure versus temperature data of equilibrium CO_2 hydrate formation conditions that were obtained from a series of experiments in a 20 wt % aqueous glycerol solution. The objective is to fit a function of the form

$$\ln P = A + BT + CT^{-2} \tag{3.44a}$$

or

$$\ln P = A + BT \tag{3.44b}$$

Solution of Example 1

Following our notation, Equations 3.44a and 3.44b are written in the following form

$$y = k_1 x_1 + k_2 x_2 + k_3 \tag{3.45a}$$

$$y = k_1 x_1 + k_2 \tag{3.45b}$$

where $y=lnP$, $k_1=B$, $k_2=C$, $k_3=A$, $x_1=T$ and $x_2=T^{-2}$ in Equation 3.34a and $y=lnP$, $k_1=B$, $k_2=A$ and $x_1=T$ in Equation 3.45b.

Excel casts the model in the following forms

$$y=m_1x_1+m_2x_2+b \tag{3.46a}$$

$$y=m_1x_1+b \tag{3.46b}$$

where $y=lnP$, $m_1=k_1=B$, $m_2=k_2=C$, $m_3=k_3=A$, $x_1=T$ and $x_2=T^{-2}$ in Equation 3.46a and $y=lnP$, $m_1=k_1=B$, $m_2=k_2=A$ and $x_1=T$ in Equation 3.46b.

Excel calculates the following parameters for the three-parameter model

m_2	m_1	b
13087246	1.402919	-557.928
2168476.8	0.211819	86.92737
0.9984569	0.014561	#N/A
1617.6016	5	#N/A
0.685894	0.00106	#N/A

and the following parameters for the two-parameter model

m_1	b
0.12461	-33.3
0.00579	1.585
0.98722	0.038
463.323	6
0.67817	0.009

In Table 3.1 the experimentally determined pressures together with the calculated ones using the two models are shown. As seen the three-parameter model represents the data better than the two-parameter one.

Table 3.1 Incipient Equilibrium Data on CO_2 Hydrate Formation in 20 (wt%) Aqueous Glycerol Solutions

Temperature *(K)*	Experimental Pressure *(MPa)*	Calculated Pressure with three-parameter model *(MPa)*	Calculated Pressure with two-parameter model *(MPa)*
270.4	1.502	1.513	1.812
270.6	1.556	1.537	1.858
272.3	1.776	1.804	2.296
273.6	2.096	2.097	2.700
274.1	2.281	2.236	2.874
275.5	2.721	2.727	3.421
276.2	3.001	3.041	3.733
277.1	3.556	3.534	4.176

Source: Brelland and Englezos (1996)

Example 2

In Table 3.2 a set of data that relate the pH with the charge on wood fibers are provided. They were obtained from potentiometric titrations in wood fiber suspensions.

We seek to establish a correlation of the form given by Equation 3.47 by fitting the equation to the charge (Q) versus pH data given in Table 3.2.

$$Q=C_1+C_2(pH)+C_3(pH)^2+C_4(pH)^3 \tag{3.47}$$

Solution of Example 2

Following our notation, Equation 3.47 is written in the following form

$$y=k_1x_1+k_2x_2+k_3x_3+k_4 \tag{3.48a}$$

where $y=Q$, $k_1=C_2$, $k_2=C_3$, $k_3=C_4$, $k_4=C_1$, $x_1=pH$, $x_2=(pH)^2$ and $x_3=(pH)^3$.

Table 3.2 Charge on Wood Fibers

pH	Charge (Q)	Calculated Charge (Q)
2.8535	19.0	16.9
3.2003	32.6	34.5
3.6347	52.8	54.0
4.0910	71.4	71.6
4.5283	86.2	86.3
5.0390	99.6	100.7
5.6107	115.4	114.0
6.3183	130.7	127.2
7.0748	138.4	138.4
7.7353	144.1	147.0
8.2385	151.6	153.3
8.8961	159.9	162.1
9.5342	172.2	171.9
10.0733	183.9	181.9
10.4700	193.1	190.5
10.9921	200.7	203.9

Source: Bygrave (1997).

Excel casts the model in the following form

$$y = m_1x_1 + m_2x_2 + m_3x_3 + b \tag{3.48b}$$

where $y=Q$, $m_1=k_1=C_2$, $m_2=k_2=C_3$, $m_3=k_3=C_4$, $x_1=pH$ $x_2=(pH)^2$ and $x_3=(pH)^3$.

The program returns the following results

m_3 k_3 C_4	m_2 k_2 C_3	m_1 k_1 C_2	b k_4 C_1
0.540448	-12.793	113.4188	-215.213
0.047329	0.9852	6.380755	12.61603
0.998721	2.27446	#N/A	#N/A
3122.612	12	#N/A	#N/A
48461.41	62.07804	#N/A	#N/A

In Table 3.2 the calculated charge values by the model are also shown.

Example 3

Table 3.3 gives another set of equilibrium CO_2 hydrate formation pressure versus temperature data that were obtained from a series of experiments in pure water. Again, the objective is to fit a function of the form of Equation 3.44a.

Table 3.3 Incipient Equilibrium Data on CO_2 Hydrate Formation in Pure Water

Temperature *(K)*	Experimental Pressure *(MPa)*	Calculated Pressure *(MPa)*
275.1	1.542	1.486
275.5	1.651	1.562
276.8	1.936	1.863
277.1	1.954	1.939
277.7	2.126	2.100
279.2	2.601	2.564
280.2	3.021	2.929
281.6	3.488	3.529
282.7	4.155	4.084

Source: Brelland and Englezos (1996), Englezos and Hull (1994).

Following the notation of Example 1 we have the following results from Excel. The calculated values are also shown in Table 3.3.

m_2	m_1	b
0.097679	0.133029	-36.199
0.193309	0.009196	2.600791
0.998128	0.017101	#N/A
1599.801	6	#N/A
0.935707	0.001755	#N/A

3.6.2 Procedure for Using SigmaPlot™ for Windows

The use of this program is illustrated by solving Examples 1 and 2.

Example 1

Step 1. A file called EX-1.FIT that contains the data is created. The data are given in a table form with column 1 for temperature (T), column 2 for pressure (P) and column 3 for *ln*P.

Step 2. Open application SigmaPlot V2.0 for Windows.
A window at the menu bar (Menu → Math → Curve Fit) is edited as follows

```
[Parameters]
        A=1
        B=1
        C=1
[Variables]
        T=col(1)
        lnP=col(3)
[Equations]
        f=A+B*T+C/T**2
fit f to lnP
```

The program returns the following results

Parameter	*Value*	*Standard Error*	*CV(%)*	*Dependencies*
A	5.578e+2	8.693e+1	1.558e+1	1.0000000
B	1.403e+0	2.118e-1	1.510e+1	1.0000000
C	1.308e+7	2.168e+6	1.657e+1	1.0000000

Step 3. Use estimated parameters to calculate formation pressures and input the results to file EX-1.XFM).

Step 4. A window at the menu bar (Menu → Math → Transforms) is edited as follows:

```
T=data(270.3,277.1,0.1)
col(5)=T
A=-557.8
B=1.403
C=1.308E+7
LnP=A+B*T+C*(T^(-2))
col(7)=lnP
col(6)=exp(lnP)
```

Step 5. A file to generate a graph showing the experimental data and the calculated values is then created (File name : EX-1.SPW).

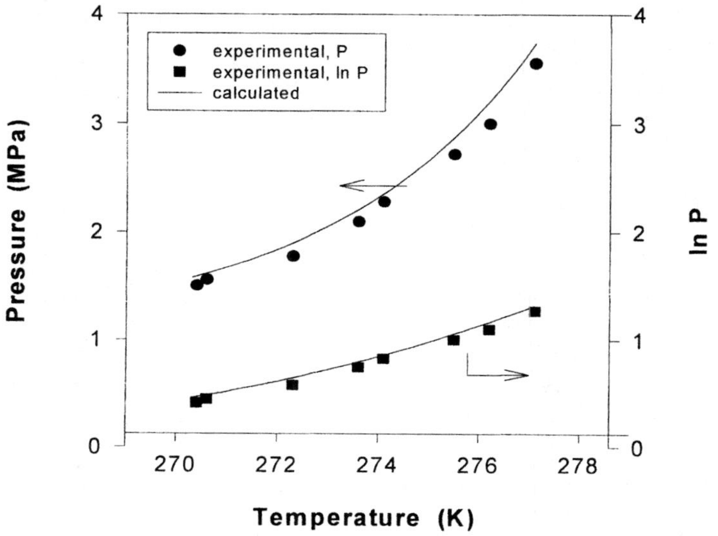

Figure 3.1 Experimental and calculated hydrate formation pressures.

Example 2

Step 1. A file called EX-2.FIT that contains the data given in Table 3.3 is created. The data are given in a table form with column 1 for pH, column 2 for charge (Q)

Step 2. Open application SigmaPlot V2.0 for Windows. A window at the menu bar (Menu → Math → Curve Fit) is edited as follows

```
[Parameters]
        C1=1
        C2=1
        C3=1
        C4=1
[Variables]
        pH=col(1)
        Q=col(2)
[Equations]
     f=C1+C2*pH+C3*(pH^2)+C4*(pH^3)

fit f to Q
```

The program returns the following results

Parameter	*Value*	*Standard Error*	*CV(%)*	*Dependencies*
C_1	-215.2	12.6	5.862	0.9979686
C_2	113.4	6.4	5.626	0.9998497
C_3	-12.8	0.98	7.701	0.9999197
C_4	0.54	0.05	8.757	0.9996247

Step 3. Use estimated parameters to calculate the charge and input the results to file EX-2.XFM.

Step 4. A window at the menu bar (Menu → Math → Transforms) is edited as follows.

```
pH=data(2.8,11.0,0.1)
col(5)=pH
    C1=-215.2
    C2=113.4
    C3=-12.8
    C4=0.54
    Q=C1+C2*pH+C3*(pH^2)+C4*(pH^3)
col(6)=Q
```

Step 5. A file to generate a graph showing the experimental data and the calculated values is then created (File name : EX-2.SPW).

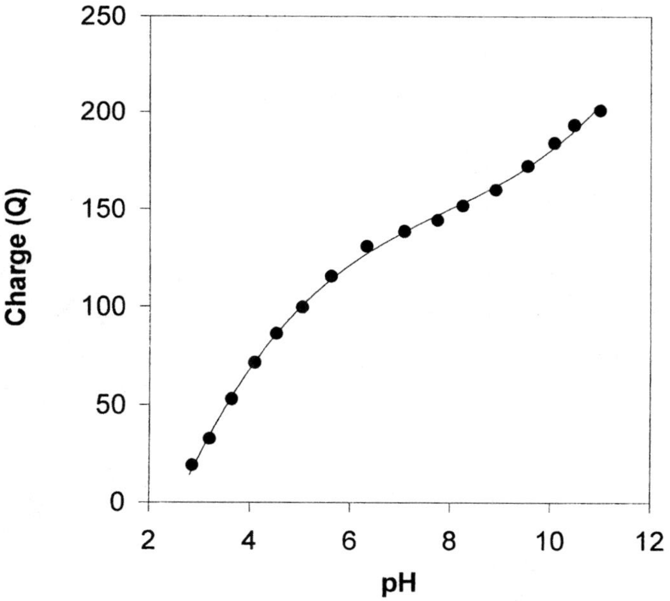

Figure 3.2 *Experimental and calculated charge. The line represents calculated values.*

3.7 SOLUTION OF MULTIRESPONSE LINEAR REGRESSION PROBLEMS

These problems refer to models that have more than one (m>1) response variables, ($m \times n$) independent variables and p (=n+1) unknown parameters. These problems cannot be solved with the readily available software that was used in the previous three examples. These problems can be solved by using Equation 3.18. We often use our nonlinear parameter estimation computer program. Obviously, since it is a linear estimation problem, convergence occurs in one iteration.

3.8 PROBLEMS ON LINEAR REGRESSION

3.8.1 Vapor Pressure Data for Pyridine and Piperidine

Blanco et al. (1994) reported measurements of the vapor pressure (P^{sat}) for p-xylene, γ-picoline, piperidine, pyridine and tetralin. The data for piperidine and pyridine are given in Table 3.4. A suitable equation to correlate these data is Antoine's relationship given next

$$log\left(P^{sat}\right) = A - \frac{B}{t + C} \tag{3.49}$$

where P^{sat} is the vapor pressure in $mmHg$ and t is the temperature in ℃. Determine the values for the parameters A,B and C in Antoine's equation for piperidine and pyridine.

Table 3.4 Vapor Pressure Data for Piperidine and Pyridine

Temperature *(K)*	P^{sat} (piperidine) *(kPa)*	Temperature *(K)*	P^{sat} (Pyridine) *(kPa)*
339.35	26.44	345.45	23.89
339.70	26:66	347.20	25.54
344.50	32.17	348.20	26.66
350.25	39.57	349.60	28.02
352.50	42.84	351.65	30.09
354.90	46.61	353.55	32.35
357.50	50.75	355.05	34.14
359.05	53.55	356.75	36.27
360.55	56.22	359.40	39.87
		362.50	44.26

Source: Blanco et al. (1994).

3.8.2 Vapor Pressure Data for R142b and R152a

Silva and Weber (1993) reported vapor pressure measurements for the 1-chloro-1,1-Difluoroethane (R142b) and 1,1-Difluoroethane (R152a) refrigerants. The data are given in Tables 3.5 and 3.6 respectively. Use Antoine's equation to correlate the data for R142b and the following equation for R152a (Silva and Weber, 1993)

$$log\left(\frac{P^{sat}}{P_c}\right) = \frac{T_c}{T}\left(k_1R + k_2R^{1.5} + k_3R^{2.5} + k_4R^5\right) \tag{3.50}$$

where

$$R = 1 - \frac{T}{T_c} \tag{3.51}$$

and where P_c=4514.73 *kPa* and T_c=386.41*K* are the critical pressure and temperature (for R152a) respectively. You are asked to estimate Antoine's parameters A, B and C for R142b and parameters k_1, k_2, k_3 and k_4 for R152a by using the data given in Table 3.6.

Table 3.5 Vapor Pressure Data for R142b

Temperature (*K*)	P^{sat} for R142b (*kPa*)	Temperature (*K*)	P^{sat} for R142b (*kPa*)
224.804	14.870	257.176	76.073
227.067	16.965	258.755	81.384
229.130	19.088	258.100	79.133
231.623	21.931	260.635	88.059
233.213	23.919	262.374	94.650
234.821	26.070	264.090	101.506
238.184	31.105	265.820	108.809
237.420	29.902	267.684	117.147
240.139	34.354	269.467	125.584
241.542	36.874	271.343	134.991
243.423	40.453	273.095	144.120
245.147	43.982	274.940	154.403
246.740	47.460	276.899	165.907
248.564	51.703	279.206	180.222
249.763	54.665	280.677	189.867
251.993	60.525	280.776	190.553
253.837	65.744	282.535	202.653
255.544	70.871	284.688	218.216

Source: Silva and Weber (1993).

Table 3.6 Vapor Pressure Data for R152a

Temperature (*K*)	P^{sat} for R152a (*kPa*)	Temperature (*K*)	P^{sat} for R152a (*kPa*)
219.921	22.723	247.489	94.123
223.082	27.322	249.262	101.864
224.670	29.885	250.770	108.836
225.959	32,125	250.126	105.788
227.850	35.645	251.534	112.496
229.278	38.501	253.219	120.958
230.924	42.018	254.768	129.162
232.567	45.792	256.519	138.939
234.081	49.497	258.110	148.355
235.540	53.303	259.745	158.511
234.902	51.588	261.310	168.744
236.496	55.911	262.679	178.124
236.059	54.700	264.757	193.084
237.938	60.052	266.603	207.226
239.528	64.912	268.321	221.084
241.295	70.666	268.206	220.076
242.822	75.955	269.918	234.549
244.149	80.807	271.573	249.218
245.844	87.351	273.139	263.726

Source: Silva and Weber (1993).

4

Gauss-Newton Method for Algebraic Models

As seen in Chapter 2 a suitable measure of the discrepancy between a model and a set of data is the objective function, S(**k**), and hence, the parameter values are obtained by minimizing this function. Therefore, the estimation of the parameters can be viewed as an optimization problem whereby any of the available general purpose optimization methods can be utilized. In particular, it was found that the Gauss-Newton method is the most efficient method for estimating parameters in nonlinear models (Bard, 1970). As we strongly believe that this is indeed the best method to use for nonlinear regression problems, the Gauss-Newton method is presented in detail in this chapter. It is assumed that the parameters are free to take any values.

4.1 FORMULATION OF THE PROBLEM

In this chapter we are focusing on a particular technique, the Gauss-Newton method, for the estimation of the unknown parameters that appear in a model described by a set of algebraic equations. Namely, it is assumed that both the structure of the mathematical model and the objective function to be minimized are known. In mathematical terms, we are given the model

$$\mathbf{y} = \mathbf{f}(\mathbf{x}, \mathbf{k}) \tag{4.1}$$

where $\mathbf{k}=[k_1,k_2,\ldots,k_p]^T$ is a *p-dimensional* vector of parameters whose numerical values are unknown, $\mathbf{x}=[x_1,x_2,\ldots,x_n]^T$ is an *n-dimensional* vector of independent variables (which are often set by the experimentalist and their numerical values are either known precisely or have been measured), $\mathbf{f}$ is a *m-dimensional* vector function of known form (the algebraic equations) and $\mathbf{y}=[y_1,y_2,\ldots,y_m]^T$ is the *m-dimensional* vector of depended variables which are measured experimentally (output vector).

Furthermore, we are given a set of experimental data, $[\hat{\mathbf{y}}_i, \mathbf{x}_i]$, i=1,...,N that we need to match to the values calculated by the model in some optimal fashion. Based on the statistical properties of the experimental error involved in the measurement of the output vector **y** (and possibly in the measurement of some of the independent variables **x**) we generate the objective function to be minimized as mentioned in detail in Chapter 2. In most cases the objective function can be written as

$$S(\mathbf{k}) = \sum_{i=1}^{N} \mathbf{e}_i^T \mathbf{Q}_i \mathbf{e}_i \tag{4.2a}$$

where $\mathbf{e}_i = [\hat{\mathbf{y}}_i - \mathbf{f}(\mathbf{x}_i, \mathbf{k})]$ are the residuals and the weighting matrices $\mathbf{Q}_i$, i=1,...,N are chosen as described in Chapter 2. Equation 4.2a can also be written as

$$S(\mathbf{k}) = \sum_{i=1}^{N} [\hat{\mathbf{y}}_i - \mathbf{f}(\mathbf{x}_i, \mathbf{k})]^T \mathbf{Q}_i [\hat{\mathbf{y}}_i - \mathbf{f}(\mathbf{x}_i, \mathbf{k})] \tag{4.2b}$$

Finally, the above equation can also be written as follows

$$S(\mathbf{k}) = \sum_{i=1}^{N} \left(\sum_{k=1}^{m} \sum_{l=1}^{m} e_k Q_{kl} e_l \right)_i \tag{4.2c}$$

Minimization of S(**k**) can be accomplished by using almost any technique available from optimization theory. Next we shall present the Gauss-Newton method as we have found it to be overall the best one (Bard, 1970).

4.2 THE GAUSS-NEWTON METHOD

Let us assume that an estimate $\mathbf{k}^{(j)}$ is available at the jth iteration. We shall try to obtain a better estimate, $\mathbf{k}^{(j+1)}$. Linearization of the model equations around $\mathbf{k}^{(j)}$ yields,

$$\mathbf{f}(\mathbf{x}_i, \mathbf{k}^{(j+1)}) = \mathbf{f}(\mathbf{x}_i, \mathbf{k}^{(j)}) + \left(\frac{\partial \mathbf{f}^T}{\partial \mathbf{k}}\right)_i^T \Delta\mathbf{k}^{(j+1)} + \text{H.O.T.} \quad ; \; i=1,\dots,N \tag{4.3}$$

Neglecting all higher order terms (H.O.T.), the model output at $\mathbf{k}^{(j+1)}$ can be approximated by

$$\mathbf{y}(\mathbf{x}_i,\mathbf{k}^{(j+1)}) = \mathbf{y}(\mathbf{x}_i,\mathbf{k}^{(j)}) + \mathbf{G}_i\Delta\mathbf{k}^{(j+1)} \quad ; \; i=1,\dots,N \tag{4.4}$$

where $\mathbf{G}_i$ is the *(m×p)*-sensitivity matrix $\left(\partial\mathbf{f}^T / \partial\mathbf{k}\right)_i^T \equiv \left(\nabla\mathbf{f}^T\right)_i^T$ evaluated at $\mathbf{x}_i$ and $\mathbf{k}^{(j)}$. It is noted that $\mathbf{G}$ is also the Jacobean matrix of the vector function $\mathbf{f}(\mathbf{x},\mathbf{k})$. Substitution of $\mathbf{y}(\mathbf{x}_i,\mathbf{k}^{(j+1)})$ as approximated by Equation 4.4, into the LS objective function and use of the critical point criterion

$$\frac{\partial S(\mathbf{k}^{(j+1)})}{\partial \mathbf{k}^{(j+1)}} = \mathbf{0} \tag{4.5}$$

yields a linear equation of the form

$$\mathbf{A}\,\Delta\mathbf{k}^{(j+1)} = \mathbf{b} \tag{4.6}$$

where

$$\mathbf{A} = \sum_{i=1}^{N} \mathbf{G}_i^T \mathbf{Q}_i \mathbf{G}_i \tag{4.7}$$

and

$$\mathbf{b} = \sum_{i=1}^{N} \mathbf{G}_i^T \mathbf{Q}_i \left[\, \hat{\mathbf{y}}_i - \mathbf{f}(\mathbf{x}_i,\mathbf{k}^{(j)})\right] \tag{4.8}$$

Solution of the above equation using any standard linear equation solver yields $\Delta\mathbf{k}^{(j+1)}$. The next estimate of the parameter vector, $\mathbf{k}^{(j+1)}$, is obtained as

$$\mathbf{k}^{(j+1)} = \mathbf{k}^{(j)} + \mu\Delta\mathbf{k}^{(j+1)} \tag{4.9}$$

where a stepping parameter, μ ($0<\mu\leq 1$), has been introduced to avoid the problem of overstepping. There are several techniques to arrive at an optimal value for μ; however, the simplest and most widely used is the bisection rule described below.

4.2.1 Bisection Rule

The bisection rule constitutes the simplest and most robust way available to determine an acceptable value for the stepping parameter μ. Normally, one starts with μ=1 and keeps on halving μ until the objective function becomes less than that obtained in the previous iteration (Hartley, 1961). Namely we "accept" the first value of μ that satisfies the inequality

$$S(\mathbf{k}^{(j)} + \mu\Delta\mathbf{k}^{(j+1)}) < S(\mathbf{k}^{(j)}) \tag{4.10}$$

More elaborate techniques have been published in the literature to obtain optimal or near optimal stepping parameter values. Essentially one performs a univariate search to determine the minimum value of the objective function along the chosen direction ($\Delta\mathbf{k}^{(j+1)}$) by the Gauss-Newton method.

4.2.2 Convergence Criteria

A typical test for convergence is $\|\Delta\mathbf{k}^{(j+1)}\|$ < TOL where TOL is a user-specified tolerance. This test is suitable only when the unknown parameters are of the same order of magnitude. A more general convergence criterion is

$$\frac{1}{p}\sum_{i=1}^{p}\left|\frac{\Delta k_i^{(j+1)}}{k_i^{(j)}}\right| \le 10^{-NSIG} \tag{4.11}$$

where p is the number of parameters and NSIG is the number of significant digits desired in the parameter estimates. Although this is not guaranteed, the above convergence criterion yields consistent results assuming of course that no parameter converges to zero!

Algorithm - Implementation Steps:

1. Input the initial guess for the parameters, $\mathbf{k}^{(0)}$ and NSIG
2. For j=0,1, 2,..., repeat the following
3. Compute $\mathbf{y}(\mathbf{x}_i,\mathbf{k}^{(j)})$ and $\mathbf{G}_i$ for each i=1,...,N, and set up matrix **A** & vector **b**
4. Solve the linear equation $\mathbf{A}\Delta\mathbf{k}^{(j+1)}=\mathbf{b}$ and obtain $\Delta\mathbf{k}^{(j+1)}$.
5. Determine μ using the bisection rule and obtain $\mathbf{k}^{(j+1)}=\mathbf{k}^{(j)}+\mu\Delta\mathbf{k}^{(j+1)}$

6. Continue until the maximum number of iterations is reached or convergence is achieved (i.e., $\frac{1}{p}\sum_{i=1}^{p}\left|\frac{\Delta k_i^{(j+1)}}{k_i^{(j)}}\right| \le 10^{-NSIG}$).

7. Compute statistical properties of parameter estimates (see Chapter 11).

In summary, at each iteration of the estimation method we compute the model output, $\mathbf{y}(\mathbf{x}_i,\mathbf{k}^{(j)})$, and the sensitivity coefficients, $\mathbf{G}_i$, for each data point i=1,...,N which are used to set up matrix **A** and vector **b**. Subsequent solution of the linear equation yields $\Delta\mathbf{k}^{(j+1)}$ and hence $\mathbf{k}^{(j+1)}$ is obtained.

The converged parameter values represent the Least Squares (LS), Weighted LS or Generalized LS estimates depending on the choice of the weighting matrices $\mathbf{Q}_i$. Furthermore, if certain assumptions regarding the statistical distribution of the residuals hold, these parameter values could also be the Maximum Likelihood (ML) estimates.

4.2.3 Formulation of the Solution Steps for the Gauss-Newton Method: Two Consecutive Chemical Reactions

Let us consider a batch reactor where the following consecutive reactions take place (Smith, 1981)

$$A \xrightarrow{k_1} B \xrightarrow{k_2} D \tag{4.12}$$

Taking into account the concentration invariant $C_A+C_B+C_D= C_{A0}$, i.e. that there is no change in the total number of moles, the integrated forms of the isothermal rate equations are

$$C_A(t) = C_{A0}e^{-k_1 t} \tag{4.13a}$$

$$C_B(t) = C_{A0}k_1\left(\frac{e^{-k_1 t}}{k_2 - k_1} + \frac{e^{-k_2 t}}{k_2 - k_1}\right) \tag{4.13b}$$

$$C_D(t) = C_{A0} - C_A(t) - C_B(t) \tag{4.13c}$$

where C_A, C_B and C_D are the concentrations of A, B and D respectively, t is the reaction time, and k_1, k_2 are the unknown rate constants. During a typical experiment, the concentrations of A and B are only measured as a function of time. Namely, a typical dataset is of the form $[t_i, C_{Ai}, C_{Bi}]$, i=1,...,N.

The variables, the parameters and the governing equations for this problem can be rewritten in our standard notation as follows:

Parameter vector:	$\mathbf{k} = [k_1, k_2]^T$	
Vector of independent variables:	$\mathbf{x} = [x_1]$	where $x_1 = t$
Output vector (dependent variables):	$\mathbf{y} = [y_1, y_2]^T$	where $y_1 = C_A$, $y_2 = C_B$
Model equations:	$\mathbf{f} = [f_1, f_2]^T$	

where

$$f_1(x_1, k_1, k_2) = C_{A0} e^{-k_1 x_1} \tag{4.14a}$$

$$f_2(x_1, k_1, k_2) = C_{A0} k_1 \left(\frac{e^{-k_1 x_1}}{k_2 - k_1} + \frac{e^{-k_2 x_1}}{k_2 - k_1} \right) \tag{4.14b}$$

The elements of the *(2×2)-sensitivity coefficient matrix* **G** are obtained as follows:

$$G_{11} = \left(\frac{\partial f_1}{\partial k_1} \right) = -x_1 f_1 \tag{4.15a}$$

$$G_{12} = \left(\frac{\partial f_1}{\partial k_2} \right) = 0 \tag{4.15b}$$

$$G_{21} = \left(\frac{\partial f_2}{\partial k_1} \right) = \frac{f_1 \left(k_2 \left(1 + e^{(k_2 - k_1) x_1} \right) + k_1 x_1 (k_1 - k_2) \right)}{(k_2 - k_1)^2} \tag{4.15c}$$

$$G_{22} = \left(\frac{\partial f_2}{\partial k_2} \right) = -\frac{k_1 f_1 \left(1 + (1 + (k_2 - k_1) x_1) e^{(k_2 - k_1) x_1} \right)}{(k_2 - k_1)^2} \tag{4.15d}$$

Equations 4.14 and 4.15 are used to evaluate the model response and the sensitivity coefficients that are required for setting up matrix **A** and vector **b** at each iteration of the Gauss-Newton method.

4.2.4 Notes on the Gauss-Newton Method

This is the well-known Gauss-Newton method which exhibits quadratic convergence to the optimum parameter values when the initial guess is sufficiently close. The Gauss-Newton method can also be looked at as a procedure that converts the nonlinear regression problem into a series of linear regressions by linearizing the nonlinear algebraic equations. It is worth noting that when the model equations are linear with respect to the parameters, there are no higher order terms (HOT) and the linearization is exact. As expected, the optimum solution is obtained in a single iteration since the sensitivity coefficients do not depend on **k**.

In order to enlarge the region of convergence of the Gauss-Newton method and at the same time make it much more robust, a stepping parameter is used to avoid the problem of overstepping particularly when the parameter estimates are away from the optimum. This modification makes the convergence to the optimum monotonic (i.e., the objective function is always reduced from one iteration to the next) and the overall estimation method becomes essentially globally convergent. When the parameters are close to the optimum the bisection rule can be omitted without any problem.

Finally, an important advantage of the Gauss-Newton method is also that at the end of the estimation besides the best parameter estimates their covariance matrix is also readily available without any additional computations. Details will be given in Chapter 11.

4.3 EXAMPLES

4.3.1 Chemical Kinetics: Catalytic Oxidation of 3-Hexanol

Gallot et al. (1998) studied the catalytic oxidation of 3-hexanol with hydrogen peroxide. The data on the effect of the solvent (CH_3OH) on the partial conversion, y, of hydrogen peroxide were read from Figure 1a of the paper by Gallot et al. (1998) and are also given here in Table 4.1. They proposed a model which is given by Equation 4.16.

$$y = k_1[1 - exp(-k_2 t)] \tag{4.16}$$

In this case, the unknown parameter vector **k** is the 2-dimensional vector $[k_1,k_2]^T$. There is only one independent variable (x_1=t) and only one output variable. Therefore, the model in our standard notation is

$$y_1 = f_1(x_1,k_1,k_2) = k_1[1 - exp(-k_2 x_1)] \tag{4.17}$$

Table 4.1 Catalytic Oxidation of 3-Hexanol

Modified Reaction Time (t) *(h kg/kmol)*	Partial Conversion (y)	
	0.75 *g* of CH_3OH	1.30 *g* of CH_3OH
3	0.055	0.040
6	0.090	0.070
13	0.120	0.100
18	0.150	0.130
26	0.165	0.150
28	0.175	0.160

Source: Gallot et al. (1998)

The *(1×2) dimensional* sensitivity coefficient matrix **G** = $[G_{11}, G_{12}]$ is given by

$$G_{11} = \left(\frac{\partial f_1}{\partial k_1}\right) = 1 - exp(-k_2 x_1) \qquad (4.18a)$$

$$G_{12} = \left(\frac{\partial f_1}{\partial k_2}\right) = k_1 x_1\, exp(-k_2 x_1) \qquad (4.18b)$$

Equations 4.17 and 4.18 are used to evaluate the model response and the sensitivity coefficients that are required for setting up matrix **A** and vector **b** at each iteration of the Gauss Newton method.

4.3.2 Biological Oxygen Demand (BOD)

Data on biological oxygen demand versus time are usually modeled by the following equation

$$y = k_1\left[1 - exp(-k_2 x)\right] \qquad (4.19)$$

where k_1 is the ultimate carbonaceous oxygen demand (*mg/L*) and k_2 is the BOD reaction rate constant (d^{-1}). A set of BOD data were obtained by 3rd year Environmental Engineering students at the Technical University of Crete and are given in Table 4.2.

Table 4.2 A Set of BOD Data

Time (*days*)	BOD (*mg/L*)
1	110
2	180
3	230
4	260
5	280
6	290
7	310
8	330

As seen the model for the BOD is identical in mathematical form with the model given by Equation 4.17.

4.3.3 Numerical Example 1

Let us consider the following nonlinear model (Bard, 1970). Data for the model are given in Table 4.3.

$$y = k_1 + \frac{x_1}{k_2 x_2 + k_3 x_3} \tag{4.20}$$

This model is assumed to be able to fit the data given in Table 4.3. Using our standard notation [**y**=**f**(**x**,**k**)] we have,

Parameter vector:	$\mathbf{k} = [k_1, k_2, k_3]^T$
Vector of independent variables:	$\mathbf{x} = [x_1, x_2, x_3]^T$
Output vector:	$\mathbf{y} = [y]$
Model Equation:	$\mathbf{f} = [f_1]$

where

$$f_1(x_1, x_2, x_3, k_1, k_2, k_3) = k_1 + \frac{x_1}{k_2 x_2 + k_3 x_3} \tag{4.21}$$

The elements of the (*1x3*)-dimensional sensitivity coefficient matrix **G** are obtained by evaluating the partial derivatives:

$$G_{11} = \left(\frac{\partial f_1}{\partial k_1}\right) = 1. \tag{4.22a}$$

Table 4.3 Data for Numerical Example 1

Run	y	x_1	x_2	x_3	y_{calc}
1	0.14	1	15	1	0.1341
2	0.18	2	14	2	0.1797
3	0.22	3	13	3	0.2203
4	0.25	4	12	4	0.2565
5	0.29	5	11	5	0.2892
6	0.32	6	10	6	0.3187
7	0.35	7	9	7	0.3455
8	0.39	8	8	8	0.3700
9	0.37	9	7	7	0.4522
10	0.58	10	6	6	0.5618
11	0.73	11	5	5	0.7152
12	0.96	12	4	4	0.9453
13	1.34	13	3	3	1.3288
14	2.10	14	2	2	2.0958
15	4.39	15	1	1	4.3968

Source: Bard (1970).

$$G_{12} = \left(\frac{\partial f_1}{\partial k_2}\right) = -\frac{x_1 x_2}{(k_2 x_2 + k_3 x_3)^2} \tag{4.22b}$$

$$G_{13} = \left(\frac{\partial f_1}{\partial k_3}\right) = -\frac{x_1 x_3}{(k_2 x_2 + k_3 x_3)^2} \tag{4.22c}$$

Equations 4.21 and 4.22 are used to evaluate the model response and the sensitivity coefficients that are required for setting up matrix **A** and vector **b** at each iteration of the Gauss-Newton method.

4.3.4 Chemical Kinetics: Isomerization of Bicyclo [2,1,1] Hexane

Data on the thermal isomerization of bicyclo [2,1,1] hexane were measured by Srinivasan and Levi (1963). The data are given in Table 4.4. The following nonlinear model was proposed to describe the fraction of original material remaining (y) as a function of time (x_1) and temperature (x_2). The model was reproduced from Draper and Smith (1998)

$$y = exp\left\{-k_1 x_1 \, exp\left[-k_2\left(\frac{1}{x_2} - \frac{1}{620}\right)\right]\right\} \tag{4.23}$$

Using our standard notation [**y**=**f**(**x**,**k**)] we have,

Parameter vector:	$\mathbf{k} = [k_1, k_2]^T$
Vector of independent variables:	$\mathbf{x} = [x_1, x_2]^T$
Output vector:	$\mathbf{y} = [y]$
Model Equation:	$\mathbf{f} = [f_1]$

where

$$f_1(x_1, x_2, k_1, k_2) = exp\left\{-k_1 x_1 \, exp\left[-k_2\left(\frac{1}{x_2} - \frac{1}{620}\right)\right]\right\} \tag{4.24}$$

Table 4.4 Isomerization of Bicyclo (2,1,1) Hexane

Run	x_1	x_2	y	Run	x_1	x_2	y
1	120.0	600	0.900	21	120.0	620	0.673
2	60.0	600	0.949	22	60.0	620	0.802
3	60.0	612	0.886	23	60.0	620	0.802
4	120.0	612	0.785	24	60.0	620	0.804
5	120.0	612	0.791	25	60.0	620	0.794
6	60.0	612	0.890	26	60.0	620	0.804
7	60.0	620	0.787	27	60.0	620	0.799
8	30.0	620	0.877	28	30.0	631	0.764
9	15.0	620	0.938	29	45.1	631	0.688
10	60.0	620	0.782	30	30.0	631	0.717
11	45.1	620	0.827	31	30.0	631	0.802
12	90.0	620	0.696	32	45.0	631	0.695
13	150.0	620	0.582	33	15.0	639	0.808
14	60.0	620	0.795	34	30.0	639	0.655
15	60.0	620	0.800	35	90.0	639	0.309
16	60.0	620	0.790	36	25.0	639	0.689
17	30.0	620	0.883	37	60.1	639	0.437
18	90.0	620	0.712	38	60.0	639	0.425
19	150.0	620	0.576	39	30.0	639	0.638
20	90.4	620	0.715	40	30.0	639	0.659
				41	60.0	639	0.449

Source: Srinivasan and Levi (1963).

The elements of the (1×2)-dimensional sensitivity coefficient matrix **G** are obtained by evaluating the partial derivatives:

$$G_{11} = \left(\frac{\partial f_1}{\partial k_1}\right) = -x_1\, exp\left\{-k_2\left(\frac{1}{x_2}-\frac{1}{620}\right)\right\} f_1 \tag{4.25a}$$

$$G_{12} = \left(\frac{\partial f_1}{\partial k_2}\right) = k_1 x_1\left(\frac{1}{x_2}-\frac{1}{620}\right) exp\left\{-k_2\left(\frac{1}{x_2}-\frac{1}{620}\right)\right\} f_1 \tag{4.25b}$$

Equations 4.24 and 4.25 are used to evaluate the model response and the sensitivity coefficients that are required for setting up matrix **A** and vector **b** at each iteration of the Gauss-Newton method.

4.3.5 Enzyme Kinetics

Let us consider the determination of two parameters, the maximum reaction rate (r_{max}) and the saturation constant (K_m) in an enzyme-catalyzed reaction following Michaelis-Menten kinetics. The Michaelis-Menten kinetic rate equation relates the reaction rate (r) to the substrate concentrations (S) by

$$r = \frac{r_{max} S}{K_m + S} \tag{4.26}$$

The parameters are usually obtained from a series of initial rate experiments performed at various substrate concentrations. Data for the hydrolysis of benzoyl-L-tyrosine ethyl ester (BTEE) by trypsin at 30 ℃ and pH 7.5 are given below:

S *(μM)*	20	15	10	5.0	2.5
r *(μM/min)*	330	300	260	220	110

Source: Blanch and Clark (1996)

In this case, the unknown parameter vector **k** is the 2-dimensional vector $[r_{max}, K_m]^T$, the independent variables are only one, **x** = [S] and similarly for the output vector, **y** = [r]. Therefore, the model in our standard notation is

$$y_1 = f_1(x_1, k_1, k_2) = \frac{k_1 x_1}{k_2 + x_1} \tag{4.27}$$

The *(1×2) dimensional* sensitivity coefficient matrix **G** = [G_{11}, G_{12}] is given by

$$G_{11} = \left(\frac{\partial f_1}{\partial k_1}\right) = \frac{x_1}{k_2 + x_1} \tag{4.28a}$$

$$G_{12} = \left(\frac{\partial f_1}{\partial k_2}\right) = \frac{-k_1 x_1}{(k_2 + x_1)^2} \tag{4.28b}$$

Equations 4.27 and 4.28 are used to evaluate the model response and the sensitivity coefficients that are required for setting up matrix **A** and vector **b** at each iteration of the Gauss-Newton method.

4.3.6 Catalytic Reduction of Nitric Oxide

As another example from chemical kinetics, we consider the catalytic reduction of nitric oxide (NO) by hydrogen which was studied using a flow reactor operated differentially at atmospheric pressure (Ayen and Peters, 1962). The following reaction was considered to be important

$$NO + H_2 \longleftrightarrow H_2O + \frac{1}{2}N_2 \tag{4.29}$$

Data were taken at 375, 400 ℃, and 425 ℃ using nitrogen as the diluent. The reaction rate in *gmol/(min·g-catalyst)* and the total NO conversion were measured at different partial pressures for H_2 and NO.

A Langmuir-Hinshelwood reaction rate model for the reaction between an adsorbed nitric oxide molecule and one adjacently adsorbed hydrogen molecule is described by:

$$r = \frac{k K_{H_2} K_{NO} p_{H_2} p_{NO}}{\left(1 + K_{NO} p_{NO} + K_{H_2} p_{H_2}\right)^2} \tag{4.30}$$

where r is the reaction rate in *gmol/(min·g-catalyst)*, p_{H2} is the partial pressure of hydrogen (*atm*), p_{NO} is the partial pressure of NO (*atm*), K_{NO}= $A_2 exp\{-E_2/RT\}$ atm^{-1} is the adsorption equilibrium constant for NO, K_{H2}= $A_3 exp\{-E_3/RT\}$ atm^{-1} is the adsorption equilibrium constant for H_2 and k=$A_1 exp\{-E_1/RT\}$ *gmol/(min·g-catalyst)* is the forward reaction rate constant for surface reaction. The data for the above problem are given in Table 4.5.

The objective of the estimation procedure is to determine the parameters k, K_{H2} and K_{NO} (if data from one isotherm are only considered) or the parameters A_1, A_2, A_3, E_1, E_2, E_3 (when all data are regressed together). The units of E_1, E_2, E_3 are in *cal/mol* and R is the universal gas constant (1.987 *cal/mol K*).

For the isothermal regression of the data, using our standard notation [**y**=**f**(**x**,**k**)] we have,

Parameter vector:	$\mathbf{k} = [k_1, k_2, k_3]^T$	where k_1=k, k_2=K_{H2} & k_3=K_{NO}
Independent variables:	$\mathbf{x} = [x_1, x_2]^T$	where x_1=p_{H2}, x_2=p_{NO}
Output vector	$\mathbf{y} = [y_1]$	where y_1=r
Model Equation	$\mathbf{f} = [f_1]$	

where

$$f_1(x_1, x_2, k_1, k_2, k_3) = \frac{k_1 k_2 k_3 x_1 x_2}{(1 + k_3 x_2 + k_2 x_1)^2} \tag{4.31}$$

The elements of the *(1×3)*-dimensional sensitivity coefficient matrix **G** are obtained by evaluating the partial derivatives:

$$G_{11} = \left(\frac{\partial f_1}{\partial k_1}\right) = \frac{k_2 k_3 x_1 x_2}{(1 + k_3 x_2 + k_2 x_1)^2} \tag{4.32a}$$

$$G_{12} = \left(\frac{\partial f_1}{\partial k_2}\right) = \frac{k_1 k_3 x_1 x_2}{(1 + k_3 x_2 + k_2 x_1)^2} - \frac{2 k_1 k_2 k_3 x_1^2 x_2}{(1 + k_3 x_2 + k_2 x_1)^3} \tag{4.32b}$$

$$G_{13} = \left(\frac{\partial f_1}{\partial k_3}\right) = \frac{k_1 k_2 x_1 x_2}{(1 + k_3 x_2 + k_2 x_1)^2} - \frac{2 k_1 k_2 k_3 x_1 x_2^2}{(1 + k_3 x_2 + k_2 x_1)^3} \tag{4.32c}$$

Equations 4.31 and 4.32 are used to evaluate the model response and the sensitivity coefficients that are required for setting up matrix **A** and vector **b** at each iteration of the Gauss Newton method.

4.3.7 Numerical Example 2

Let us consider the following nonlinear model (Hartley, 1961).

$$y = k_1 + k_2\, exp(k_3 x) \tag{4.33}$$

Table 4.5 Experimental Data for the Catalytic Reduction of Nitric Oxide

p_{H2} *(atm)*	p_{NO} *(atm)*	Reaction Rate, $r\times10^5$ *gmol/(min·g-catalyst)*	Total NO Conversion *(%)*
T=375 ℃, Weight of catalyst=2.39 *g*			
0.00922	0.0500	1.60	1.96
0.0136	0.0500	2.56	2.36
0.0197	0.0500	3.27	2.99
0.0280	0.0500	3.64	3.54
0.0291	0.0500	3.48	3.41
0.0389	0.0500	4.46	4.23
0.0485	0.0500	4.75	4.78
0.0500	0.00918	1.47	14.0
0.0500	0.0184	2.48	9.15
0.0500	0.0298	3.45	6.24
0.0500	0.0378	4.06	5.40
0.0500	0.0491	4.75	4.30
T=400 ℃, Weight of catalyst=1.066 *g*			
0.00659	0.0500	2.52	0.59
0.0113	0.0500	4.21	1.05
0.0228	0.0500	5.41	1.44
0.0311	0.0500	6.61	1.76
0.0402	0.0500	6.86	1.91
0.0500	0.0500	8.79	2.57
0.0500	0.0100	3.64	8.83
0.0500	0.0153	4.77	6.05
0.0500	0.0270	6.61	4.06
0.0500	0.0361	7.94	3.20
0.0500	0.0432	7.82	2.70
T=425 ℃, Weight of catalyst=1.066 *g*			
0.00474	0.0500	5.02	2.62
0.0136	0.0500	7.23	4.17
0.0290	0.0500	11.35	6.84
0.0400	0.0500	13.00	8.19
0.0500	0.0500	13.91	8.53
0.0500	0.0269	9.29	13.3
0.0500	0.0302	9.75	12.3
0.0500	0.0387	11.89	10.4

Source: Ayen and Peters (1962).

Data for the model are given below in Table 4.6. The variable y represents yields of wheat corresponding to six rates of application of fertilizer, x, on a coded scale. The model equation is often called Mitcherlisch's law of diminishing returns.

According to our standard notation the model equation is written as follows

$$y = f_1(x_1, k_1, k_2, k_3) = k_1 + k_2\, exp(k_3 x_1) \tag{4.34}$$

The elements of the (1×3)-dimensional sensitivity coefficient matrix **G** are obtained by evaluating the partial derivatives:

$$G_{11} = \left(\frac{\partial f_1}{\partial k_1}\right) = 1 \tag{4.35a}$$

$$G_{12} = \left(\frac{\partial f_1}{\partial k_2}\right) = exp(k_3 x_1) \tag{4.35b}$$

$$G_{13} = \left(\frac{\partial f_1}{\partial k_3}\right) = k_2 x_1\, exp(k_3 x_1) \tag{4.35c}$$

Table 4.6 Data for Numerical Example 2

x	*y*
-5	*127*
-3	*151*
-1	*379*
1	*421*
3	*460*
5	*426*

Source: Hartley (1961).

4.4 SOLUTIONS

The solutions to the Numerical Examples 1 and 2 will be given here. The rest of the solutions will be presented in Chapter 16 where applications in chemical reaction engineering are illustrated.

4.4.1 Numerical Example 1

Starting with the initial guess $\mathbf{k}^{(0)}=[1, 1, 1]^T$ the Gauss-Newton method easily converged to the parameter estimates within 4 iterations as shown in Table 4.7. In the same table the standard error (%) in the estimation of each parameter is also shown. Bard (1970) also reported the same parameter estimates [0.08241, 1.1330, 2.3437] starting from the same initial guess.

The structure of the model characterizes the shape of the region of convergence. For example if we change the initial guess for k_1 substantially, the algorithm converges very quickly since it enters the model in a linear fashion. This is clearly shown in Table 4.8 where we have used $\mathbf{k}^{(0)}=[100000, 1, 1]^T$. On the other hand, if we use for k_2 a value which is just within one order of magnitude away from the optimum, the Gauss-Newton method fails to converge. For example if $\mathbf{k}^{(0)}=[1, 2, 1]^T$ is used, the method converges within 3 iterations. If however, $\mathbf{k}^{(0)}=[1, 8, 1]^T$ or $\mathbf{k}^{(0)}=[1, 10, 1]^T$ is used, the Gauss-Newton method fails to converge. The actual shape of the region of convergence can be fairly irregular. For example if we use $\mathbf{k}^{(0)}=[1, 14, 1]^T$ or $\mathbf{k}^{(0)}=[1, 15, 1]^T$ the Gauss-Newton method converges within 8 iterations for both cases. But again, when $\mathbf{k}^{(0)}=[1, 16, 1]^T$ is used, the Gauss-Newton method fails to converge.

Table 4.7 Parameter Estimates at Each Iteration of the Gauss-Newton Method for Numerical Example-1with Initial Guess [1, 1, 1]

Iteration	LS Objective function	k_1	k_2	k_3
0	41.6817	1	1	1
1	1.26470	0.08265	1.183	1.666
2	0.03751	0.08249	1.165	2.198
3	0.00824387	0.08243	1.135	2.338
4	0.00824387	0.08241	1.133	2.344
Standard Error (%)		15.02	27.17	12.64

Table 4.8 Parameter Estimates at Each Iteration of the Gauss-Newton Method for Numerical Example 1 with Initial Guess [100000, 1, 1]

Iteration	LS Objective function	k_1	k_2	k_3
0	1.50×10^9	100000	1	1
1	1.26470	0.08265	1.183	1.666
2	0.03751	0.08249	1.165	2.198
3	0.00824387	0.08243	1.135	2.338
4	0.00824387	0.08241	1.133	2.344

4.4.2 Numerical Example 2

Starting with the initial guess $\mathbf{k}^{(0)}$=[100, -200, -1] the Gauss-Newton method converged to the optimal parameter estimates given in Table 4.9 in 12 iterations. The number of iterations depends heavily on the chosen initial guess. If for example we use $\mathbf{k}^{(0)}$=[1000, -200, -0.2] as initial guess, the Gauss-Newton method converges to the optimum within 3 iterations as shown in Table 4.10. At the bottom of Table 4.10 we also report the standard error (%) in the parameter estimates. As expected the uncertainty is quite high since we are estimating 3 parameters from only 6 data points and the structure of the model naturally leads to a high correlation between k_2 and k_3.

Hartley (1961) reported also convergence to the same parameter values $\mathbf{k}^*=[523.3, -156.9, -0.1997]^T$ by using as initial guess $\mathbf{k}^{(0)}=[500, -140, -0.18]^T$.

Table 4.9 Parameter Estimates at Each Iteration of the Gauss-Newton Method for Numerical Example 2. Initial Guess [100, -200, -1]

Iteration	LS Objective Function	k_1	k_2	k_3
0	9.003×10^8	100	-200	-1
1	1.514×10^7	443.6	-32.98	-0.9692
2	1.801×10^7	445.1	-36.00	-0.7660
3	8.722×10^4	457.9	-62.79	-0.4572
4	2.471×10^4	494.1	-127.0	-0.1751
5	1.392×10^4	508.2	-140.0	-0.2253
6	1.346×10^4	528.9	-164.1	-0.1897
7	1.340×10^4	518.7	-151.7	-0.2041
8	1.339×10^4	524.8	-158.8	-0.1977
9	1.339×10^4	522.5	-156.1	-0.2005
10	1.339×10^4	523.6	-157.3	-0.1993
11	1.339×10^4	523.2	-156.8	-0.1998
12	1.339×10^4	523.3	-156.9	-0.1997

Table 4.10 Parameter Estimates at Each Iteration of the Gauss-Newton Method for Numerical Example 2. Initial Guess [10^3, -200, -0.2]

Iteration	LS Objective Function	k_1	k_2	k_3
0	1.826×10^7	1000	-200	-0.2
1	1.339×10^4	523.4	-157.1	-0.1993
2	1.339×10^4	523.1	-156.8	-0.1998
3	1.339×10^4	523.3	-156.9	-0.1997
Standard Error (%)		30.4	115.2	85.2

5

Other Nonlinear Regression Methods for Algebraic Models

There is a variety of general purpose unconstrained optimization methods that can be used to estimate unknown parameters. These methods are broadly classified into two categories: *direct search methods* and *gradient methods* (Edgar and Himmelblau, 1988; Gill et al. 1981; Kowalik and Osborne, 1968; Sargent, 1980; Reklaitis, 1983; Scales, 1985).

A brief overview of this relatively vast subject is presented and several of these methods are briefly discussed in the following sections. Over the years many comparisons of the performance of many methods have been carried out and reported in the literature. For example, Box (1966) evaluated eight unconstrained optimization methods using a set of problems with up to twenty variables.

5.1 GRADIENT MINIMIZATION METHODS

The gradient search methods require derivatives of the objective functions whereas the direct methods are derivative-free. The derivatives may be available analytically or otherwise they are approximated in some way. It is assumed that the objective function has continuous second derivatives, whether or not these are explicitly available. Gradient methods are still efficient if there are some discontinuities in the derivatives. On the other hand, direct search techniques, which use function values, are more efficient for highly discontinuous functions.

The basic problem is to search for the parameter vector **k** that minimizes S(**k**) by following an iterative scheme, i.e.,

$$\textit{Minimize} \qquad S(\mathbf{k}) = \sum_{i=1}^{N} \mathbf{e}_i^T \mathbf{Q}_i \mathbf{e}_i \tag{5.1}$$

where $\mathbf{k}=[k_1, k_2,\ldots,k_p]^T$ the *p-dimensional* vector of parameters, $\mathbf{e}=[e_1, e_2,\ldots,e_m]^T$ the *m-dimensional* vector of residuals where $\mathbf{e}_i = [\hat{\mathbf{y}}_i - \mathbf{f}(\mathbf{x}_i, \mathbf{k})]$ and $\mathbf{Q}_i$ is a user specified positive definite weighting matrix.

The need to utilize an iterative scheme stems from the fact that it is usually impossible to find the exact solutions of the equation that gives the stationary points of S(**k**) (Peressini et al. 1988),

$$\nabla S(\mathbf{k}) \equiv \frac{\partial S(\mathbf{k})}{\partial \mathbf{k}} = \mathbf{0} \tag{5.2a}$$

where the operator $\nabla \equiv \left(\frac{\partial}{\partial k_1}, \frac{\partial}{\partial k_2}, \ldots, \frac{\partial}{\partial k_p} \right)^T$ is applied to the scalar function S(**k**) yielding the column vector

$$\nabla S(\mathbf{k}) \equiv \begin{bmatrix} \frac{\partial S(\mathbf{k})}{\partial k_1} \\ \frac{\partial S(\mathbf{k})}{\partial k_2} \\ \vdots \\ \frac{\partial S(\mathbf{k})}{\partial k_p} \end{bmatrix} \tag{5.2b}$$

Vector $\nabla S(\mathbf{k})$ contains the first partial derivatives of the objective function S(**k**) with respect to **k** and it is often called the *gradient vector*. For simplicity, we denoted it as **g**(**k**) in this chapter.

In order to find the solution, one starts from an initial guess for the parameters, $\mathbf{k}^{(0)}=[k_1^{(0)}, k_2^{(0)}, \ldots, k_p^{(0)}]^T$. There is no general rule to follow in order to obtain an initial guess. However, some heuristic rules can be used and they are discussed in Chapter 8.

At the start of the jth iteration we denote by $\mathbf{k}^{(j)}$ the current estimate of the parameters. The jth iteration consists of the computation of a search vector $\Delta\mathbf{k}^{(j+1)}$ from which we obtain the new estimate $\mathbf{k}^{(j+1)}$ according to the following equation

$$\mathbf{k}^{(j+1)} = \mathbf{k}^{(j)} + \mu^{(j)} \Delta\mathbf{k}^{(j+1)} \tag{5.3}$$

where $\mu^{(j)}$ is the step-size, also known as dumping or relaxation factor. It is obtained by univariate minimization or prior knowledge based upon the theory of the method (Edgar and Himmelblau, 1988). As seen from Equation 5.3 our main concern is to determine the search vector $\Delta\mathbf{k}^{(j+1)}$. Based on the chosen method to calculate the search vector, different solution methods to the minimization problem arise.

The iterative procedure stops when the convergence criterion for termination is satisfied. When the unknown parameters are of the same order of magnitude then a typical test for convergence is $\|\Delta\mathbf{k}^{(j+1)}\| < \text{TOL}$ where TOL is a user-specified tolerance. A more general convergence criterion is

$$\frac{1}{p}\sum_{i=1}^{p}\left|\frac{\Delta k_i^{(j+1)}}{k_i^{(j)}}\right| \leq 10^{-NSIG} \tag{5.4}$$

where p is the number of parameters and NSIG is the number of desired significant digits in the parameter values. It is assumed that no parameter converges to zero.

The minimization method must be *computationally efficient* and *robust* (Edgar and Himmelblau, 1988). *Robustness* refers to the ability to arrive at a solution. *Computationally efficiency* is important since iterative procedures are employed. The speed with which convergence to the optimal parameter values, $\mathbf{k}^*$, is reached is defined with the *asymptotic rate of convergence* (Scales, 1985). An algorithm is considered to have a θth order rate of convergence when θ is the largest integer for which the following limit exists.

$$0 \leq \lim_{j\to\infty}\frac{\left\|\mathbf{k}^{(j+1)}-\mathbf{k}^*\right\|}{\left\|\mathbf{k}^{(j)}-\mathbf{k}^*\right\|^{\theta}} < 1 \tag{5.5}$$

In the above equation, the norm $\|\cdot\|$ is usually the Euclidean norm. We have a linear convergence rate when θ is equal to 1. Superlinear convergence rate refers to the case where θ=1 and the limit is equal to zero. When θ=2 the convergence rate is called quadratic. In general, the value of θ depends on the algorithm while the value of the limit depends upon the function that is being minimized.

5.1.1 Steepest Descent Method

In this method the search vector is the negative of the gradient of the objective function and is given by the next equation

$$\Delta \mathbf{k}^{(j+1)} = -\nabla S(\mathbf{k}^{(j)}) \equiv -\mathbf{g}(\mathbf{k}^{(j)}) \tag{5.6a}$$

Based on Equation 5.1 the search vector is related to the residuals $\mathbf{e}_i = [\hat{\mathbf{y}}_i - \mathbf{f}(\mathbf{x}_i, \mathbf{k})]$ as follows

$$\Delta \mathbf{k}^{(j+1)} = -2\sum_{i=1}^{N} \left(\nabla \mathbf{e}_i^T\right) \mathbf{Q}_i \, \mathbf{e}_i = 2\sum_{i=1}^{N} \left(\nabla \mathbf{f}_i^T\right) \mathbf{Q}_i \left[\hat{\mathbf{y}}_i - \mathbf{f}(\mathbf{x}_i, \mathbf{k})\right] \tag{5.6b}$$

where

$$\nabla \mathbf{e}_i^T = \begin{bmatrix} \dfrac{\partial}{\partial k_1} \\ \dfrac{\partial}{\partial k_2} \\ \cdot \\ \cdot \\ \dfrac{\partial}{\partial k_p} \end{bmatrix} \begin{bmatrix} e_1 & e_2 & \cdot & e_m \end{bmatrix}_i = \begin{bmatrix} \dfrac{\partial e_1}{\partial k_1} & \dfrac{\partial e_2}{\partial k_1} & \cdot\ \cdot & \dfrac{\partial e_m}{\partial k_1} \\ \dfrac{\partial e_1}{\partial k_2} & \dfrac{\partial e_2}{\partial k_2} & \cdot\ \cdot & \dfrac{\partial e_m}{\partial k_2} \\ \cdot & \cdot & \cdot\ \cdot & \cdot \\ \cdot & \cdot & \cdot\ \cdot & \cdot \\ \dfrac{\partial e_1}{\partial k_p} & \dfrac{\partial e_2}{\partial k_p} & \cdot\ \cdot & \dfrac{\partial e_m}{\partial k_p} \end{bmatrix}_i \tag{5.7}$$

and

$$\nabla \mathbf{f}_i^T = \begin{bmatrix} \dfrac{\partial}{\partial k_1} \\ \dfrac{\partial}{\partial k_2} \\ \cdot \\ \cdot \\ \dfrac{\partial}{\partial k_p} \end{bmatrix} \begin{bmatrix} f_1 & f_2 & \cdot\ \cdot & f_m \end{bmatrix}_i = \begin{bmatrix} \dfrac{\partial f_1}{\partial k_1} & \dfrac{\partial f_2}{\partial k_1} & \cdot\ \cdot & \dfrac{\partial f_m}{\partial k_1} \\ \dfrac{\partial f_1}{\partial k_2} & \dfrac{\partial f_2}{\partial k_2} & \cdot\ \cdot & \dfrac{\partial f_m}{\partial k_2} \\ \cdot & \cdot & \cdot\ \cdot & \cdot \\ \cdot & \cdot & \cdot\ \cdot & \cdot \\ \dfrac{\partial f_1}{\partial k_p} & \dfrac{\partial f_2}{\partial k_p} & \cdot\ \cdot & \dfrac{\partial f_m}{\partial k_p} \end{bmatrix}_i \tag{5.8}$$

As seen from the above equations, the *(m×p)* matrix $(\nabla \mathbf{e}^T)^T$ is the Jacobean matrix, **J**, of the vector function **e**, and the *(m×p)* matrix$(\nabla \mathbf{f}^T)^T$ is the Jacobean matrix, **G**, of the vector function **f(x,k)**. The srth element of the Jacobean matrix **J**, is given by

$$J_{sr} = \left(\frac{\partial e_s}{\partial k_r}\right) \quad ; \qquad s=1,2,\ldots,m\,, \quad r=1,2,\ldots,p \tag{5.9a}$$

Similarly, the sr[th] element of the Jacobean matrix **G**, is given by

$$G_{sr} = \left(\frac{\partial f_s}{\partial k_r}\right) \quad ; \quad s=1,2,\ldots,m\ , \quad r=1,2,\ldots,p. \tag{5.9b}$$

The rate of convergence of the Steepest Descent method is first order. The basic difficulty with steepest descent is that the method is too sensitive to the scaling of S(**k**), so that convergence is very slow and oscillations in the **k**-space can easily occur. In general a *well scaled* problem is one in which similar changes in the variables lead to similar changes in the objective function (Kowalik and Osborne, 1968). For these reasons, steepest descent/ascent is not a viable method for the general purpose minimization of nonlinear functions. It is of interest only for historical and theoretical reasons.

Algorithm - Implementation Steps

1. Input the initial guess for the parameters, $\mathbf{k}^{(0)}$ and NSIG or TOL

2. Specify weighting matrix $\mathbf{Q}_i$ for i=1,2,...N.

3. For j=0,1, 2,..., repeat

4. Compute $\Delta\mathbf{k}^{(j+1)}$ using Equation 5.6

5. Determine μ using the bisection rule and obtain $\mathbf{k}^{(j+1)}=\mathbf{k}^{(j)}+\mu^{(j)}\Delta\mathbf{k}^{(j+1)}$

6. Continue until the maximum number of iterations is reached or convergence is achieved (i.e., $\frac{1}{p}\sum_{i=1}^{p}\left|\frac{\Delta k_i^{(j+1)}}{k_i^{(j)}}\right| \le 10^{-NSIG}$ or $\left\|\Delta\mathbf{k}^{(j+1)}\right\| < TOL$)

5.1.2 Newton's Method

By this method the step-size parameter $\mu^{(j)}$ is taken equal to 1 and the search vector is obtained from

$$\Delta\mathbf{k}^{(j+1)} = -\left[\nabla^2 S(\mathbf{k})\right]^{-1}\nabla S(\mathbf{k}^{(j+1)}) \tag{5.10}$$

where $\nabla^2 S(\mathbf{k})$ is the Hessian matrix of S(**k**) evaluated at $\mathbf{k}^{(j)}$. It is denoted as **H**(**k**) and thus, the Equation 5.10 can be written as

$$\Delta \mathbf{k}^{(j+1)} = -\left[\mathbf{H}(\mathbf{k}^{(j)})\right]^{-1} \mathbf{g}(\mathbf{k}^{(j)}) \tag{5.11}$$

The above formula is obtained by differentiating the quadratic approximation of S(**k**) with respect to each of the components of **k** and equating the resulting expression to zero (Edgar and Himmelblau, 1988; Gill et al. 1981; Scales, 1985). It should be noted that in practice there is no need to obtain the inverse of the Hessian matrix because it is better to solve the following linear system of equations (Peressini et al. 1988)

$$\left[\nabla^2 S(\mathbf{k})\right] \Delta \mathbf{k}^{(j+1)} = -\nabla S(\mathbf{k}) \tag{5.12a}$$

or equivalently

$$\mathbf{H}(\mathbf{k}^{(j+1)})\, \Delta \mathbf{k}^{(j+1)} = -\,\mathbf{g}(\mathbf{k}^{(j)}) \tag{5.12b}$$

As seen by comparing Equations 5.6 and 5.12 the steepest-descent method arises from Newton's method if we assume that the Hessian matrix of S(**k**) is approximated by the identity matrix.

Newton's method is not a satisfactory general-purpose algorithm for function minimization, even when a stepping parameter μ is introduced. Fortunately, it can be modified to provide extremely reliable algorithms with the same asymptotic rate of convergence. There is an extensive literature on the Newton's method and the main points of interest are summarized below (Edgar and Himmelblau, 1988; Gill et al. 1981; Pertessini et al. 1988; Scales, 1985):

(i) It is the most rapidly convergent method when the Hessian matrix of S(**k**) is available.
(ii) There is no guarantee that it will converge to a minimum from an arbitrary starting point.
(iii) Problems arise when the Hessian matrix is indefinite or singular.
(iv) The method requires analytical first and second order derivatives which may not be practical to obtain. In that case, finite difference techniques may be employed.

The ratio of the largest to the smallest eigenvalue of the Hessian matrix at the minimum is defined as the condition number. For most algorithms the larger the condition number, the larger the limit in Equation 5.5 and the more difficult it is for the minimization to converge (Scales, 1985).

One approach to solve the linear Equations 5.12 is the method of Gill and Murray that uses the Cholesky factorization of **H** as in the following (Gill and Murray, 1974; Scales, 1985):

$$\mathbf{H} = \mathbf{L}\,\mathbf{D}\,\mathbf{L}^T \tag{5.13}$$

In the above equation, **D** is a diagonal matrix, **L** is a lower triagonal matrix with diagonal elements of unity.

As shown previously the negative of the gradient of the objective function is related to the residuals by Equation 5.6. Therefore, the gradient of the objective function is given by

$$\mathbf{g}(\mathbf{k}) \equiv \nabla S(\mathbf{k}) = 2\sum_{i=1}^{N}\left(\nabla \mathbf{e}_i^T\right)\mathbf{Q}_i\,\mathbf{e}_i \tag{5.14}$$

where $\mathbf{e}_i = \left[\hat{\mathbf{y}}_i - \mathbf{f}(\mathbf{x}_i, \mathbf{k})\right]$ and $\nabla \mathbf{e}_i^T = \left(\mathbf{J}_e^T\right)_i$ is the transpose of the Jacobean matrix of the vector function $\mathbf{e}$. The srth element of this Jacobean was defined by Equation 5.9a. Equation 5.14 can now be written in an expanded form as follows

$$\nabla S(\mathbf{k}) = 2\sum_{i=1}^{N}\begin{bmatrix} \frac{\partial e_1}{\partial k_1} & \frac{\partial e_2}{\partial k_1} & \cdots & \frac{\partial e_m}{\partial k_1} \\ \frac{\partial e_1}{\partial k_2} & \frac{\partial e_2}{\partial k_2} & \cdots & \frac{\partial e_m}{\partial k_2} \\ \vdots & \vdots & & \vdots \\ \frac{\partial e_1}{\partial k_p} & \frac{\partial e_2}{\partial k_p} & \cdots & \frac{\partial e_m}{\partial k_p} \end{bmatrix}_i \begin{bmatrix} Q_{11} & Q_{12} & \cdots & Q_{1m} \\ Q_{21} & Q_{22} & \cdots & Q_{2m} \\ \vdots & \vdots & & \vdots \\ Q_{m1} & Q_{m2} & \cdots & Q_{mm} \end{bmatrix}_i \begin{bmatrix} e_1 \\ e_2 \\ \vdots \\ e_m \end{bmatrix}_i \tag{5.15}$$

After completing the matrix multiplication operations, we obtain

$$\mathbf{g}(\mathbf{k}) \equiv \frac{\partial S(\mathbf{k})}{\partial \mathbf{k}} = \nabla S(\mathbf{k}) = 2\sum_{i=1}^{N}\begin{bmatrix} \sum_{l=1}^{m}\sum_{r=1}^{m}\frac{\partial e_l}{\partial k_1}Q_{lr}e_r \\ \sum_{l=1}^{m}\sum_{r=1}^{m}\frac{\partial e_l}{\partial k_2}Q_{lr}e_r \\ \vdots \\ \sum_{l=1}^{m}\sum_{r=1}^{m}\frac{\partial e_l}{\partial k_p}Q_{lr}e_r \end{bmatrix}_i \tag{5.16}$$

Thus, the sth element of the gradient vector $\mathbf{g}(\mathbf{k})$ of the objective function $S(\mathbf{k})$ is given by the following equation

$$g_s = 2\sum_{i=1}^{N}\sum_{l=1}^{m}\sum_{r=1}^{m}\frac{\partial e_l}{\partial k_s}Q_{lr}e_r \quad ; \; s = 1,2,\ldots,p \tag{5.17}$$

We are now able to obtain the Hessian matrix of the objective function S(**k**) which is denoted by **H** and is given by the following equation

$$\nabla^2 S(\mathbf{k}) \equiv \nabla\nabla^T S(\mathbf{k}) = \begin{bmatrix} \frac{\partial}{\partial k_1} \\ \frac{\partial}{\partial k_2} \\ \vdots \\ \frac{\partial}{\partial k_p} \end{bmatrix} \begin{bmatrix} \frac{\partial S(\mathbf{k})}{\partial k_1} & \frac{\partial S(\mathbf{k})}{\partial k_2} & \cdots & \frac{\partial S(\mathbf{k})}{\partial k_p} \end{bmatrix} =$$

$$= \begin{bmatrix} \frac{\partial^2 S(\mathbf{k})}{\partial k_1^2} & \frac{\partial^2 S(\mathbf{k})}{\partial k_1 \partial k_2} & \cdots & \frac{\partial^2 S(\mathbf{k})}{\partial k_1 \partial k_p} \\ \frac{\partial^2 S(\mathbf{k})}{\partial k_2 \partial k_1} & \frac{\partial^2 S(\mathbf{k})}{\partial k_2^2} & \cdots & \frac{\partial^2 S(\mathbf{k})}{\partial k_2 \partial k_p} \\ \vdots & \vdots & \ddots & \vdots \\ \frac{\partial^2 S(\mathbf{k})}{\partial k_p \partial k_1} & \frac{\partial^2 S(\mathbf{k})}{\partial k_p \partial k_2} & \cdots & \frac{\partial^2 S(\mathbf{k})}{\partial k_p^2} \end{bmatrix} \equiv \begin{bmatrix} \frac{\partial g_1}{\partial k_1} & \frac{\partial g_2}{\partial k_1} & \cdots & \frac{\partial g_m}{\partial k_1} \\ \frac{\partial g_1}{\partial k_2} & \frac{\partial g_2}{\partial k_2} & \cdots & \frac{\partial g_m}{\partial k_2} \\ \vdots & \vdots & \ddots & \vdots \\ \frac{\partial g_1}{\partial k_p} & \frac{\partial g_2}{\partial k_p} & \cdots & \frac{\partial g_m}{\partial k_p} \end{bmatrix} \tag{5.18}$$

where we use the notation $g_s \equiv \frac{\partial S(\mathbf{k})}{\partial k_s}$.

Thus, the $\xi\mu^{th}$ element of the Hessian matrix is defined by

$$G_{\xi\mu} = \frac{\partial g_\xi}{\partial k_\mu} \quad \xi = 1,2,\ldots,p \quad \text{and} \quad \mu = 1,2,\ldots,p \tag{5.19}$$

and this element can be calculated by taking into account Equation 5.16 as follows

$$G_{\xi\mu} = 2\sum_{i=1}^{N}\sum_{l=1}^{m}\sum_{r=1}^{m}\left(\frac{\partial e_l}{\partial k_\xi} Q_{lr} \frac{\partial e_r}{\partial k_\mu}\right)_i + 2\sum_{i=1}^{N}\sum_{l=1}^{m}\sum_{r=1}^{m}\left(\frac{\partial^2 e_l}{\partial k_\xi \partial k_\mu} Q_{lr} e_r\right)_i \tag{5.20}$$

where ξ=1,2,...,p and μ=1,2,...,p.

The Gauss-Newton method arises when the second order terms on the right hand side of Equation 5.20 are ignored. As seen, the Hessian matrix used in Equation 5.11 contains only first derivatives of the model equations **f(x,k)**. Leaving out the second derivative containing terms may be justified by the fact that these terms contain the residuals e_r as factors. These residuals are expected to be small quantities.

The Gauss-Newton method is directly related to Newton's method. The main difference between the two is that Newton's method requires the computation of second order derivatives as they arise from the direct differentiation of the objective function with respect to **k**. These second order terms are avoided when the Gauss-Newton method is used since the model equations are first linearized and then substituted into the objective function. The latter constitutes a key advantage of the Gauss-Newton method compared to Newton's method, which also exhibits quadratic convergence.

Algorithm - Implementation Steps

1. Input the initial guess for the parameters, $\mathbf{k}^{(0)}$ and NSIG or TOL.

2. Specify weighting matrix $\mathbf{Q}_i$ for I=1,2,...N.

3. For j=0,1, 2,..., repeat.

4. Compute $\Delta\mathbf{k}^{(j+1)}$ by solving Equation 5.12b.

5. Determine μ using the bisection rule and obtain $\mathbf{k}^{(j+1)} = \mathbf{k}^{(j)} + \mu\Delta\mathbf{k}^{(j+1)}$.

6. Continue until the maximum number of iterations is reached or convergence is achieved (i.e., $\frac{1}{p}\sum_{i=1}^{p}\left|\frac{\Delta k_i^{(j+1)}}{k_i^{(j)}}\right| \le 10^{-NSIG}$ or $\left\|\Delta\mathbf{k}^{(j+1)}\right\| < TOL$).

According to Scales (1985) the best way to solve Equation 5.12b is by performing a Cholesky factorization of the Hessian matrix. One may also perform a Gauss-Jordan elimination method (Press et al., 1992). An excellent user-oriented presentation of solution methods is provided by Lawson and Hanson (1974). We prefer to perform an eigenvalue decomposition as discussed in Chapter 8.

5.1.3 Modified Newton's Method

Modified Newton methods attempt to alleviate the deficiencies of Newton's method (Bard, 1970). The basic problem arises if the Hesssian matrix, **G**, is not positive definite. That can be checked by examining if all the eigenvalues of **G** are positive numbers. If any of the eigenvalues are not positive then a procedure proposed by Marquardt (1963) based on earlier work by Levenberg (1944) should be followed. A positive value γ can be added to all the eigenvalues such that the resulting poitive quantities, $\lambda_i+\gamma$, $i=1,2,\ldots,m$ are the eigenvalues of a positive matrix $\mathbf{H}_{LM}$, given by

$$\mathbf{H}_{LM} = \mathbf{G} + \gamma\,\mathbf{I} \tag{5.21}$$

where **I** is the identity matrix.

Algorithm – Implementation Steps

1. Input the initial guess for the parameters, $\mathbf{k}^{(0)}$, γ and NSIG or TOL.

2. Specify weighting matrix $\mathbf{Q}_i$ for $i=1,2,\ldots N$.

3. For $j=0,1,2,\ldots$, repeat.

4. Compute $\Delta\mathbf{k}^{(j+1)}$ by solving Equation 5.12b but in this case the Hessian matrix $\mathbf{H}(\mathbf{k})$ has been replaced by that given by Equation 5.22.

5. Determine μ using the bisection rule and obtain $\mathbf{k}^{(j+1)} = \mathbf{k}^{(j)} + \mu\Delta\mathbf{k}^{(j+1)}$.

6. Continue until the maximum number of iterations is reached or convergence is achieved (i.e., $\frac{1}{p}\sum_{i=1}^{p}\left|\frac{\Delta k_i^{(j+1)}}{k_i^{(j)}}\right| \le 10^{-NSIG}$ or $\left\|\Delta\mathbf{k}^{(j+1)}\right\| < TOL$).

The Gill-Murray modified Newton's method uses a Cholesky factorization of the Hessian matrix (Gill and Murray, 1974). The method is described in detail by Scales (1985).

5.1.4 Conjugate Gradient Methods

Modified Newton methods require calculation of second derivatives. There might be cases where these derivatives are not available analytically. One may then calculate them by finite differences (Edgar and Himmelblau, 1988; Gill et al. 1981; Press et al. 1992). The latter, however, requires a considerable number of gradient evaluations if the number of parameters, p, is large. In addition, finite difference approximations of derivatives are prone to truncation and round-off errors (Bard, 1974; Edgar and Himmelblau, 1988; Gill et al. 1981).

Conjugate gradient-type methods form a class of minimization procedures that accomplish two objectives:

(a) There is no need for calculation of second order derivatives.
(b) They have relatively small computer storage requirements.

Thus, these methods are suitable for problems with a very large number of parameters. They are essential in circumstances when methods based on matrix factorization are not viable because the relevant matrix is too large or too dense (Gill et al. 1981).

Two versions of the method have been formulated (Scales, 1986):

(a) Fletcher-Reeves version;
(b) Polak-Ribiere version

Scales (1986) recommends the Polak Ribiere version because it has slightly better convergence properties. Scales also gives an algorithm which is used for both methods that differ only in the formula for the updating of the search vector.

It is noted that the Rosenbrock function given by the next equation has been used to test the performance of various algorithms including modified Newton's and conjugate gradient methods (Scales, 1986)

$$f(\mathbf{x}) = 100(x_1^2 - x_2)^2 + (1 - x_1)^2 \tag{5.22}$$

5.1.5 Quasi-Newton or Variable Metric or Secant Methods

These methods utilize only values of the objective function, S(**k**), and values of the first derivatives of the objective function. Thus, they avoid calculation of the elements of the (*p*×*p*) Hessian matrix. The quasi-Newton methods rely on formulas that approximate the Hessian and its inverse. Two algorithms have been developed:

(a) The Davidon-Fletcher-Powell Formula (DFP)
(b) The Broyden-Fletcher-Goldfard-Shanno Formula (BFGS)

The DFP and BFGS methods exhibit superlinear convergence on suitably smooth functions. They are in general more rapidly convergent, robust and economical than conjugate gradient methods. However, they require much more storage and are not suitable for large problems i.e., problems with many parameters. Their storage requirements are equivalent to Newton's method.

The BFGS method is considered to be superior to DFP in most cases because (a) it is less prone to loss of positive definiteness or to singularity problems through round off errors and (b) it has better theoretical convergence properties (Scales, 1985; Gill et al. 1981; Edgar and Himmelblau, 1988).

Algorithms are not given here because they are readily available elsewhere (Gill and Murray, 1972, 1975; Goldfard, 1976; Scales, 1985; Edgar and Himmelblau, 1988; Gill et al. 1981).

5.2 DIRECT SEARCH OR DERIVATIVE FREE METHODS

Direct search methods use only function evaluations. They search for the minimum of an objective function without calculating derivatives analytically or numerically. Direct methods are based upon heuristic rules which make no *a priori* assumptions about the objective function. They tend to have much poorer convergence rates than gradient methods when applied to smooth functions. Several authors claim that direct search methods are not as efficient and robust as the indirect or gradient search methods (Bard, 1974; Edgar and Himmelblau, 1988; Scales, 1986). However, in many instances direct search methods have proved to be robust and reliable particularly for systems that exhibit local minima or have complex nonlinear constraints (Wang and Luus, 1978).

The Simplex algorithm and that of Powell's are examples of derivative-free methods (Edgar and Himmelblau, 1988; Seber and Wild, 1989, Powell, 1965). In this chapter only two algorithms will be presented (1) the LJ optimization procedure and (2) the simplex method. The well known golden section and Fibonacci methods for minimizing a function along a line will not be presented. Kowalik and Osborne (1968) and Press et al. (1992) among others discuss these methods in detail.

In an effort to address the problem of "combinatorial explosion" in optimization, several "new" global optimization methods have been introduced. These methods include: (i) neural networks (Bishop, 1995), (ii) genetic algorithms (Holland, 1975), (iii) simulated annealing techniques (Kirkpatrick et al., 1983, Černy, 1985; Otten and Ginneken, 1989), (iv) target analysis (Glover 1986) and (v) threshold accepting (Dueck and Scheuer, 1990) to name a few. These methods have attracted significant attention as being the most appropriate ones for *large scale optimization problems* whose objective functions exhibit a plethora of local optima.

The simulated annealing technique is probably the most popular one. It tries to mimic the physical process of annealing whereby a material starts in a melted state and its temperature is gradually lowered until it reaches its minimum energy state. In the physical system the temperature should not be rapidly lowered because a sub-optimal structure may be formed in the crystallized system and lead to quenching. In an analogous fashion, we consider the minimization of the objective function in a series of steps. A slow reduction in temperature corresponds to allowing non-improving steps to be taken with a certain probability which is higher in the beginning of the algorithm when the temperature is high. Simulated annealing is essentially a *probabilistic hill climbing algorithm* and hence, the method has the capability to move away from a local minimum. The probability used in simulated annealing algorithms is the Gibbs-Boltzmann distribution encountered in statistical mechanics. One of its characteristics is that for very high temperatures each state has almost an equal chance of being chosen to be the current state. For low temperatures only states with low energies have a high probability of becoming the current state. In practice, simulated annealing is implemented using the Metropolis et al. (1953) algorithm. Simulated annealing has solved the famous *travelling salesman problem* of finding the shortest itinerary for a salesman who visits N cities. The method has also been successfully used to determine the arrangement of several hundred thousand circuit elements on a small silicon wafer by minimizing the interference between their connecting wires.

Usually the space over which the objective function is minimized is not defined as the *p-dimensional* space of p continuously variable parameters. Instead it is a discrete configuration space of *very high dimensionality*. In general the number of elements in the configuration space is exceptionally large so that they cannot be fully explored with a reasonable computation time.

For parameter estimation purposes, simulated annealing can be implemented by discretizing the parameter space. Alternatively, we can specify minimum and maximum values for each unknown parameter, and by using a random number uniformly distributed in the range [0,1], we can specify randomly the potential parameter values as

$$k_i = k_{min,i} + R[k_{max,i} - k_{min,i}] \quad ; \quad i=1,\ldots,p \tag{5.23}$$

where R is a random number.

Another interesting implementation of simulated annealing for continuous minimization (like a typical parameter estimation problem) utilizes a modification of the downhill simplex method. Press et al. (1992) provide a brief overview of simulated annealing techniques accompanied with listings of computer programs that cover all the above cases.

A detailed presentation of the simulated annealing techniques can be found in *The Annealing Algorithm* by Otten and Ginneken (1989).

5.2.1 LJ Optimization Procedure

One of the most reliable direct search methods is the LJ optimization procedure (Luus and Jaakola, 1973). This procedure uses random search points and systematic contraction of the search region. The method is easy to program and handles the problem of multiple optima with high reliability (Wang and Luus, 1977, 1978). A important advantage of the method is its ability to handle multiple nonlinear constraints.

The adaptation of the original LJ optimization procedure to parameter estimation problems for algebraic equation models is given next.

(i) Choose an initial guess for the p-dimensional unknown parameter vector, $\mathbf{k}^{(0)}$; the region contraction coefficient, δ (typically δ=0.95 is used); the number of random evaluations of the objective function, N_R (typically N_R=100 is used) within an iteration; the maximum number of iterations, j_{max} (typically j_{max}=200 is used) and an initial search region, $\mathbf{r}^{(0)}$ (a typical choice is $\mathbf{r}^{(0)} = \mathbf{k}_{max} - \mathbf{k}_{min}$).

(ii) Set the iteration index j=1 and $\mathbf{k}^{(j-1)} = \mathbf{k}^{(0)}$ and $\mathbf{r}^{(j-1)} = \mathbf{r}^{(0)}$.

(iii) Generate or read from a file, $N_R \times p$ random numbers (R_{ni}) uniformly distributed in [-0.5, 0.5]

(iv) For n=1,2,...,N_R, generate the corresponding random trial parameter vectors from

$$\mathbf{k}_n = \mathbf{k}^{(j-1)} + \boldsymbol{R}_n \mathbf{r}^{(j-1)} \tag{5.24}$$

where $\boldsymbol{R}_n = diag(R_{n1}, R_{n2}, ..., R_{np})$.

(v) Find the parameter vector among the N_R trial ones that minimizes the LS Objective function

$$S_{LS}(\mathbf{k}) = \sum_{i=1}^{N} [\hat{\mathbf{y}}_i - \mathbf{f}(\mathbf{x}_i, \mathbf{k})]^T \mathbf{Q}_i [\hat{\mathbf{y}}_i - \mathbf{f}(\mathbf{x}_i, \mathbf{k})] \tag{5.25}$$

(vi) Keep the best trial parameter vector, $\mathbf{k}^*$, up to now and the corresponding minimum value of the objective function, S*.

(vii) Set $\mathbf{k}^{(j)} = \mathbf{k}^*$ and compute the search region for the next iteration as

$$\mathbf{r}^{(j)} = \delta \times \mathbf{r}^{(j-1)} \tag{5.26}$$

(viii) If $j < j_{max}$, increment j by 1 go to Step (iii); else STOP.

Given the fact that in parameter estimation we normally have a relatively smooth LS objective function, we do not need to be exceptionally concerned about local optima (although this may not be the case for ill-conditioned estimation problems). This is particularly true if we have a good idea of the range where the parameter values should be. As a result, it may be more *efficient* to consider using a value for N_R which is a function of the number of unknown parameters. For example, we may consider

$$N_R = 50 + 10 \times p \tag{5.27}$$

Typical values would be N_R=60 when p=1, N_R=110 when p=5 and N_R=160 when p=10.

At the same time we may wish to consider a slower contraction of the search region as the dimensionality of the parameter space increases. For example we could use a *constant reduction of the volume* (say 10%) of the search region rather than a constant reduction in the search region of *each* parameter. Namely we could use,

$$\delta = (0.90)^{1/p} \tag{5.28}$$

Typical values would be δ=0.90 when p=1, δ=0.949 when p=2, δ=0.974 when p=4, δ=0.987 when p=8 and δ=0.993 when p=16.

Since we have a minimization problem, significant computational savings can be realized by noting in the implementation of the LJ optimization procedure that for each trial parameter vector, we do not need to complete the summation in Equation 5.23. Once the LS Objective function exceeds the smallest value found up to that point (S*), a new trial parameter vector can be selected.

Finally, we may wish to consider a multi-pass approach whereby the search region for each unknown parameter is determined by the maximum change of the parameter during the last pass (Luus, 1998).

5.2.2 Simplex Method

The "Sequential Simplex" or simply Simplex method relies on geometry to create a heuristic rule for finding the minimum of a function. It is noted that the Simplex method of linear programming is a different method.

Kowalik and Osborn (1968) define simplex as following

A set of N+1 points in the N-dimensional space forms a simplex. When the points are equidistant the simplex is said to be regular.

For a function of N variables one needs a (N+1)-dimensional geometric figure or simplex to use and select points on the vertices to evaluate the function to be minimized. Thus, for a function of two variables an equilateral triangle is used whereas for a function of three variables a regular tetrahedron.

Edgar and Himmelblau (1988) demonstrate the use of the method for a function of two variables. Nelder and Mead (1965) presented the method for a function of N variables as a flow diagram. They demonstrated its use by applying it to minimize *Rosenbrock's function* (Equation 5.22) as well as to the following functions:

Powell's quartic function

$$f(\mathbf{x}) = (x_1 + 10x_2)^2 + 5(x_3 - x_4)^2 + (x_2 - 2x_3)^4 + 10(x_1 - x_4) \qquad (5.29)$$

Fletcher and Powell's function

$$f(\mathbf{x}) = 100[x_3 - 100(x_1, x_2)]^2 + \left[\sqrt{(x_1^2 + x_2^2)} - 1\right]^2 + x_3^2 \qquad (5.30)$$

where

$$2\pi\theta(x_1, x_2) = \begin{cases} arctan(x_2 / x_1) ; & x_1 > 0 \\ \pi + arctan(x_2 / x_1) ; & x_1 < 0 \end{cases} \qquad (5.31)$$

In general, for a function of N variables the Simplex method proceeds as follows:

Step 1. Form an initial simplex e.g. an equidistant triangle for a function of two variables.

Step 2. Evaluate the function at each of the vertices.

Step 3. Reject the vertex where the function has the largest value. This point is replaced by another one that is found in the direction away from the rejected vertex and through the centroid of the simplex. The distance from the rejected vertex is always constant at each search step.
In the case of a function of two variables the direction is from the rejected vertex through the middle of the line of the triangle that is opposite to this point . The new point together with the previous two points define a new equilateral triangle.

Step 4. Proceed until a simplex that encloses the minimum is found. Stop when the difference between two consecutive function evaluations is less than a preset value (tolerance).

It is noted that Press et al. (1992) give a subroutine that implements the simplex method of Nelder and Mead. They also recommend to restart the minimization routine at a point where it claims to have found a minimum

The Simplex optimization method can also be used in the search for optimal experimental conditions (Walters et al. 1991). A starting simplex is usually formed from existing experimental information. Subsequently, the response that plays the role of the objective function is evaluated and a new vertex is found to replace the worst one. The experimenter then performs an experiment at this new vertex to determine the response. A new vertex then is found as before. Thus, sequentially one forms a new simplex and stops when the response remains practically the same. At that point the experimenter may switch to a factorial experimental design to further optimize the experimental conditions.

For example, Kurniawan (1998) investigated the in-situ electrochemical brightening of thermo-mechanical pulp using sodium carbonate as the brightening chemical. The electrochemical brightening process was optimized by performing a simplex optimization. In particular, she performed two sequential simplex optimizations. The objective of the first was to maximize the brightness gain and minimize the yellowness (or maximize the absolute yellowness gain) whereas that of the second was to maximize the brightness gain only. Four factors were considered: current (*Amp*), anode area (cm^2), temperature (*K*) and pH. Thus, the simplex was a pentahedron.

Kurniawan noticed that the first vertex was the same in both optimizations. This was due to the fact that in both cases the worse vertex was the same. Kurniawan also noticed that the search for the optimal conditions was more effective when two responses were optimized. Finally, she noticed that for the Simplex method to perform well, the initial vertices should define extreme ranges of the factors.

5.3 EXERCISES

You are asked to estimate the unknown parameters in the examples given in Chapter 4 by employing methods presented in this chapter.

6

Gauss-Newton Method for Ordinary Differential Equation (ODE) Models

In this chapter we are concentrating on the Gauss-Newton method for the estimation of unknown parameters in models described by a set of ordinary differential equations (ODEs).

6.1 FORMULATION OF THE PROBLEM

As it was mentioned in Chapter 2, the mathematical models are of the form

$$\frac{d\mathbf{x}(t)}{dt} = \mathbf{f}(\mathbf{x}(t), \mathbf{u}, \mathbf{k}) \quad ; \quad \mathbf{x}(t_0) = \mathbf{x}_0 \tag{6.1}$$

$$\mathbf{y}(t) = \mathbf{C}\mathbf{x}(t) \tag{6.2}$$

or more generally

$$\mathbf{y}(t) = \mathbf{h}(\mathbf{x}(t),\mathbf{k}) \tag{6.3}$$

where

$\mathbf{k}=[k_1,k_2,\ldots,k_p]^T$ is a *p-dimensional* vector of parameters whose numerical values are unknown;

$\mathbf{x}=[x_1,x_2,\ldots,x_n]^T$ is an *n-dimensional* vector of state variables;

$\mathbf{x}_0$ is an *n-dimensional* vector of initial conditions for state variables which are assumed to be known precisely;

$\mathbf{u}=[u_1,u_2,\ldots,u_r]^T$ is an *r-dimensional* vector of manipulated variables which are either set by the experimentalist or they have been measured and it is assumed that their numerical values are precisely known;

$\mathbf{f}=[f_1,f_2,\ldots,f_n]^T$ is a *n-dimensional* vector function of known form (the differential equations);

$\mathbf{y}=[y_1,y_2,\ldots,y_m]^T$ is the *m-dimensional* output vector i.e., the set of variables that are measured experimentally; and

$\mathbf{C}$ is the $m \times n$ observation matrix, which indicates the state variables (or linear combinations of state variables) that are measured experimentally.

Experimental data are available as measurements of the output vector as a function of time, i.e., $[\hat{\mathbf{y}}_i, t_i]$, i=1,...,N where with $\hat{\mathbf{y}}_i$ we denote the measurement of the output vector at time t_i. These are to be matched to the values calculated by the model at the same time, $\mathbf{y}(t_i)$, in some optimal fashion. Based on the statistical properties of the experimental error involved in the measurement of the output vector, we determine the weighting matrices $\mathbf{Q}_i$ (i=1,...,N) that should be used in the objective function to be minimized as mentioned earlier in Chapter 2. The objective function is of the form,

$$S(\mathbf{k})=\sum_{i=1}^{N}\left[\hat{\mathbf{y}}_i-\mathbf{y}(t_i,\mathbf{k})\right]^T \mathbf{Q}_i\left[\hat{\mathbf{y}}_i-\mathbf{y}(t_i,\mathbf{k})\right] \tag{6.4}$$

Minimization of $S(\mathbf{k})$ can be accomplished by using almost any technique available from optimization theory, however since each objective function evaluation requires the integration of the state equations, the use of quadratically convergent algorithms is highly recommended. The Gauss-Newton method is the most appropriate one for ODE models (Bard, 1970) and it presented in detail below.

6.2 THE GAUSS-NEWTON METHOD

Again, let us assume that an estimate $\mathbf{k}^{(j)}$ of the unknown parameters is available at the jth iteration. Linearization of the output vector around $\mathbf{k}^{(j)}$ and retaining first order terms yields

$$\mathbf{y}(t_i,\mathbf{k}^{(j+1)}) = \mathbf{y}(t_i,\mathbf{k}^{(j)}) + \left(\frac{\partial \mathbf{y}^T}{\partial \mathbf{k}}\right)_i^T \Delta\mathbf{k}^{(j+1)} \tag{6.5}$$

Assuming a linear relationship between the output vector and the state variables ($\mathbf{y} = \mathbf{C}\mathbf{x}$), the above equation becomes

$$\mathbf{y}(t_i,\mathbf{k}^{(j+1)}) = \mathbf{C}\mathbf{x}(t_i,\mathbf{k}^{(j)}) + \mathbf{C}\left(\frac{\partial \mathbf{x}^T}{\partial \mathbf{k}}\right)_i^T \Delta\mathbf{k}^{(j+1)} \tag{6.6}$$

In the case of ODE models, the sensitivity matrix $\mathbf{G}(t_i) \equiv (\partial\mathbf{x}^T/\partial\mathbf{k})^T$ cannot be obtained by a simple differentiation. However, we can find a differential equation that $\mathbf{G}(t)$ satisfies and hence, the sensitivity matrix $\mathbf{G}(t)$ can be determined as a function of time by solving simultaneously with the state ODEs another set of differential equations. This set of ODEs is obtained by differentiating both sides of Equation 6.1 (the state equations) with respect to $\mathbf{k}$, namely

$$\frac{\partial}{\partial \mathbf{k}}\left(\frac{d\mathbf{x}}{dt}\right) = \frac{\partial}{\partial \mathbf{k}}\left(\mathbf{f}(\mathbf{x},\mathbf{u},\mathbf{k})\right) \tag{6.7}$$

Reversing the order of differentiation on the left-hand side of Equation 6.7 and performing the implicit differentiation of the right-hand side, we obtain

$$\frac{d}{dt}\left[\left(\frac{\partial \mathbf{x}^T}{\partial \mathbf{k}}\right)^T\right] = \left(\frac{\partial \mathbf{f}^T}{\partial \mathbf{x}}\right)^T\left(\frac{\partial \mathbf{x}^T}{\partial \mathbf{k}}\right)^T + \left(\frac{\partial \mathbf{f}^T}{\partial \mathbf{k}}\right)^T \tag{6.8}$$

or better

$$\frac{d\mathbf{G}(t)}{dt} = \left(\frac{\partial \mathbf{f}^T}{\partial \mathbf{x}}\right)^T \mathbf{G}(t) + \left(\frac{\partial \mathbf{f}^T}{\partial \mathbf{k}}\right)^T \tag{6.9}$$

The initial condition $\mathbf{G}(t_0)$ is obtained by differentiating the initial condition, $\mathbf{x}(t_0)=\mathbf{x}_0$, with respect to $\mathbf{k}$ and since the initial state is independent of the parameters, we have:

$$\mathbf{G}(t_0) = \mathbf{0}. \tag{6.10}$$

Equation 6.9 is a matrix differential equation and represents a set of $n \times p$ ODEs. Once the sensitivity coefficients are obtained by solving numerically the above ODEs, the output vector, $\mathbf{y}(t_i,\mathbf{k}^{(j+1)})$, can be computed.

Substitution of the latter into the objective function and use of the stationary condition $\partial S(\mathbf{k}^{(j+1)})/\partial \mathbf{k}^{(j+1)} = \mathbf{0}$, yields a linear equation for $\Delta\mathbf{k}^{(j+1)}$

$$\mathbf{A}\,\Delta\mathbf{k}^{(j+1)} = \mathbf{b} \tag{6.11}$$

where

$$\mathbf{A} = \sum_{i=1}^{N} \mathbf{G}^T(t_i)\mathbf{C}^T\mathbf{Q}_i\mathbf{C}\mathbf{G}(t_i) \tag{6.12}$$

and

$$\mathbf{b} = \sum_{i=1}^{N} \mathbf{G}^T(t_i)\mathbf{C}^T\mathbf{Q}_i\,[\,\hat{\mathbf{y}}_i - \mathbf{C}\mathbf{x}(t_i,\mathbf{k}^{(j)})] \tag{6.13}$$

Solution of the above equation yields $\Delta\mathbf{k}^{(j+1)}$ and hence, $\mathbf{k}^{(j+1)}$ is obtained from

$$\mathbf{k}^{(j+1)} = \mathbf{k}^{(j)} + \mu\Delta\mathbf{k}^{(j+1)} \tag{6.14}$$

where μ is a stepping parameter ($0<\mu \le 1$) to be determined by the bisection rule. The simple bisection rule is presented later in this chapter whereas optimal step-size determination procedures are presented in detail in Chapter 8.

In summary, at each iteration given the current estimate of the parameters, $\mathbf{k}^{(j)}$, we obtain $\mathbf{x}(t)$ and $\mathbf{G}(t)$ by integrating the state and sensitivity differential equations. Using these values we compute the model output, $\mathbf{y}(t_i,\mathbf{k}^{(j)})$, and the sensitivity coefficients, $\mathbf{G}(t_i)$, for each data point i=1,...,N which are subsequently used to set up matrix **A** and vector **b**. Solution of the linear equation yields $\Delta\mathbf{k}^{(j+1)}$ and hence $\mathbf{k}^{(j+1)}$ is obtained.

Thus, a sequence of parameter estimates is generated, $\mathbf{k}^{(1)}$, $\mathbf{k}^{(2)}$,... which often converges to the optimum, $\mathbf{k}^*$, if the initial guess, $\mathbf{k}^{(0)}$, is sufficiently close. The converged parameter values represent the Least Squares (LS), Weighted Least Squares (WLS) or Generalized Least Squares (GLS) estimates depending on the choice of the weighting matrices $\mathbf{Q}_i$. Furthermore, if certain assumptions regarding the statistical distribution of the residuals hold, these parameter values could also be the Maximum Likelihood (ML) estimates.

6.2.1 Gauss-Newton Algorithm for ODE Models

1. Input the initial guess for the parameters, $\mathbf{k}^{(0)}$ and NSIG.
2. For j=0,1, 2,..., repeat.
3. Integrate state and sensitivity equations to obtain $\mathbf{x}(t)$ and $\mathbf{G}(t)$. At each sampling period, t_i, i=1,...,N compute $\mathbf{y}(t_i,\mathbf{k}^{(j)})$, and $\mathbf{G}(t_i)$ to set up matrix **A** and vector **b**.
4. Solve the linear equation $\mathbf{A}\Delta\mathbf{k}^{(j+1)}=\mathbf{b}$ and obtain $\Delta\mathbf{k}^{(j+1)}$.
5. Determine μ using the bisection rule and obtain $\mathbf{k}^{(j+1)}=\mathbf{k}^{(j)}+\mu\Delta\mathbf{k}^{(j+1)}$.
6. Continue until the maximum number of iterations is reached or convergence is achieved (i.e., $\frac{1}{p}\sum_{i=1}^{p}\left|\frac{\Delta k_i^{(j+1)}}{k_i^{(j)}}\right| \le 10^{-NSIG}$).
7. Compute statistical properties of parameter estimates (see Chapter 11).

The above method is the well-known Gauss-Newton method for differential equation systems and it exhibits quadratic convergence to the optimum. Computational modifications to the above algorithm for the incorporation of prior knowledge about the parameters (Bayessian estimation) are discussed in detail in Chapter 8.

6.2.2 Implementation Guidelines for ODE Models

1. Use of a Differential Equation Solver

If the dimensionality of the problem is not excessively high, simultaneous integration of the state and sensitivity equations is the easiest approach to implement the Gauss-Newton method without the need to store $\mathbf{x}(t)$ as a function of time. The latter is required in the evaluation of the Jacobeans in Equation 6.9 during the solution of this differential equation to obtain $\mathbf{G}(t)$.

Let us rewrite $\mathbf{G}(t)$ as

$$\mathbf{G}(t) \equiv \left(\frac{\partial \mathbf{x}^T}{\partial \mathbf{k}}\right)^T = \left[\left(\frac{\partial \mathbf{x}}{\partial k_1}\right), \left(\frac{\partial \mathbf{x}}{\partial k_2}\right), \ldots, \left(\frac{\partial \mathbf{x}}{\partial k_p}\right)\right] = [\mathbf{g}_1, \mathbf{g}_2, \ldots, \mathbf{g}_p] \tag{6.15}$$

In this case the *n-dimensional* vector $\mathbf{g}_l$ represents the sensitivity coefficients of the state variables with respect to parameter k_l and satisfies the following ODE,

$$\frac{d\mathbf{g}_1(t)}{dt} = \left(\frac{\partial \mathbf{f}^T}{\partial \mathbf{x}}\right)^T \mathbf{g}_1(t) + \left(\frac{\partial \mathbf{f}}{\partial k_1}\right) \quad ; \quad \mathbf{g}_1(t_0)=\mathbf{0} \tag{6.16a}$$

Similarly, the *n-dimensional* vector $\mathbf{g}_2$ represents the sensitivity coefficients of the state variables with respect to parameter k_2 and satisfies the following ODE,

$$\frac{d\mathbf{g}_2(t)}{dt} = \left(\frac{\partial \mathbf{f}^T}{\partial \mathbf{x}}\right)^T \mathbf{g}_2(t) + \left(\frac{\partial \mathbf{f}}{\partial k_2}\right) \quad ; \quad \mathbf{g}_2(t_0)=\mathbf{0} \tag{6.16b}$$

Finally for the last parameter, k_p, we have the corresponding sensitivity vector $\mathbf{g}_p$

$$\frac{d\mathbf{g}_p(t)}{dt} = \left(\frac{\partial \mathbf{f}^T}{\partial \mathbf{x}}\right)^T \mathbf{g}_p(t) + \left(\frac{\partial \mathbf{f}}{\partial k_p}\right) \quad ; \quad \mathbf{g}_p(t_0)=\mathbf{0} \tag{6.16c}$$

Since most of the numerical differential equation solvers require the equations to be integrated to be of the form

$$\frac{d\mathbf{z}}{dt} = \boldsymbol{\varphi}(\mathbf{z}) \quad ; \quad \mathbf{z}(t_0) = \text{given} \tag{6.17}$$

we generate the following *n×(p+1)-dimensional* vector **z**

$$\mathbf{z} = \begin{bmatrix} \mathbf{x}(t) \\ \hdashline \left(\dfrac{\partial \mathbf{x}}{\partial k_1}\right) \\ \hdashline \left(\dfrac{\partial \mathbf{x}}{\partial k_2}\right) \\ \hdashline \vdots \\ \hdashline \left(\dfrac{\partial \mathbf{x}}{\partial k_p}\right) \end{bmatrix} = \begin{bmatrix} \mathbf{x}(t) \\ \mathbf{g}_1(t) \\ \mathbf{g}_2(t) \\ \vdots \\ \mathbf{g}_p(t) \end{bmatrix} \tag{6.18}$$

and the corresponding *n×(p+1)-dimensional* vector function $\boldsymbol{\varphi}(\mathbf{z})$

$$\boldsymbol{\varphi}(\mathbf{z}) = \begin{bmatrix} \mathbf{f}(\mathbf{x},\mathbf{u},\mathbf{k}) \\ \hdashline \left(\dfrac{\partial \mathbf{f}^T}{\partial \mathbf{x}}\right)^T \mathbf{g}_1(t) + \left(\dfrac{\partial \mathbf{f}}{\partial k_1}\right) \\ \hdashline \left(\dfrac{\partial \mathbf{f}^T}{\partial \mathbf{x}}\right)^T \mathbf{g}_2(t) + \left(\dfrac{\partial \mathbf{f}}{\partial k_2}\right) \\ \hdashline \vdots \\ \hdashline \left(\dfrac{\partial \mathbf{f}^T}{\partial \mathbf{x}}\right)^T \mathbf{g}_p(t) + \left(\dfrac{\partial \mathbf{f}}{\partial k_p}\right) \end{bmatrix} \tag{6.19}$$

If the equation solver permits it, information can also be provided about the Jacobean of $\boldsymbol{\varphi}(\mathbf{z})$, particularly when we are dealing with stiff differential equations. The Jacobean is of the form

$$\boldsymbol{\Phi}_z = \left(\frac{\partial \boldsymbol{\varphi}^T}{\partial \mathbf{z}}\right)^T = \begin{bmatrix} \left(\dfrac{\partial \mathbf{f}^T}{\partial \mathbf{x}}\right)^T & \mathbf{0} & \mathbf{0} & \cdots & \mathbf{0} \\ * & \left(\dfrac{\partial \mathbf{f}^T}{\partial \mathbf{x}}\right)^T & \mathbf{0} & \cdots & \mathbf{0} \\ * & \mathbf{0} & \left(\dfrac{\partial \mathbf{f}^T}{\partial \mathbf{x}}\right)^T & \cdots & \mathbf{0} \\ \vdots & \vdots & \vdots & \ddots & \vdots \\ * & \mathbf{0} & \mathbf{0} & \cdots & \left(\dfrac{\partial \mathbf{f}^T}{\partial \mathbf{x}}\right)^T \end{bmatrix} \tag{6.20}$$

where the "*" in the first column represents terms that have second order derivatives of $\mathbf{f}$ with respect to $\mathbf{x}$. In most practical situations these terms can be neglected and hence, this Jacobean can be considered as a block diagonal matrix as far as the ODE solver is concerned. This results in significant savings in terms of memory requirements and robustness of the numerical integration.

In a typical implementation, the numerical integration routine is requested to provide $\mathbf{z}(t)$ at each sampling point, t_i, $i=1,\ldots,N$ and hence, $\mathbf{x}(t_i)$ and $\mathbf{G}(t_i)$ become available for the computation of $\mathbf{y}(t_i,\mathbf{k}^{(j)})$ as well as for adding the appropriate terms in matrix **A** and **b**.

2. *Implementation of the Bisection Rule*

As mentioned in Chapter 4, an acceptable value for the stepping parameter μ is obtained by starting with $\mu=1$ and halving μ until the objective function becomes less than that obtained in the previous iteration, namely, the first value of μ that satisfies the following inequality is accepted.

$$S(\mathbf{k}^{(j)} + \mu\Delta\mathbf{k}^{(j+1)}) < S(\mathbf{k}^{(j)}) \tag{6.21}$$

In the case of ODE models, evaluation of the objective function, $S(\mathbf{k}^{(j)}+\mu\Delta\mathbf{k}^{(j+1)})$, for a particular value of μ implies the integration of the state equations. It should be emphasized here that it is unnecessary to integrate the state equations for the entire data length $[t_0, t_N]$ for each trial value of μ. Once the objective function becomes greater than $S(\mathbf{k}^{(j)})$, a smaller value of μ can be chosen. By this procedure, besides the savings in computation time, numerical instability is also avoided since the objective function becomes large quickly and the integration is often stopped well before computer overflow is threatened (Kalogerakis and Luus, 1983a).

The importance of using a good integration routine should also be emphasized. When $\Delta\mathbf{k}^{(j+1)}$ is excessively large (severe overstepping) during the determination of an acceptable value for μ numerical instability may cause computer overflow well before we have a chance to compute the output vector at the first data point and compare the objective functions. In this case, the use of a good integration routine is of great importance to provide a message indicating that the tolerence requirements cannot be met. At that moment we can stop the integration and simply halve μ and start integration of the state equations again. Several restarts may be necessary before an acceptable value for μ is obtained.

Furthermore, when $\mathbf{k}^{(j)}+\mu\Delta\mathbf{k}^{(j+1)}$ is used at the next iteration as the current estimate, we do not anticipate any problems in the integration of both the state and sensitivity equations. This is simply due to the fact that the eigenvalues of the Jacobean of the sensitivity equations (inversely related to the governing time constants) are the same as those in the state equations where the integration was performed successfully. These considerations are of particular importance when the model is described by a set of stiff differential equations where the wide range of the prevailing time constants creates additional numerical difficulties that tend to shrink the region of convergence (Kalogerakis and Luus, 1983a).

6.3 THE GAUSS-NEWTON METHOD – NONLINEAR OUTPUT RELATIONSHIP

When the output vector (measured variables) are related to the state variables (and possibly to the parameters) through a nonlinear relationship of the form $\mathbf{y}(t) = \mathbf{h}(\mathbf{x}(t),\mathbf{k})$, we need to make some additional minor modifications. The sensitivity of the output vector to the parameters can be obtained by performing the implicit differentiation to yield:

$$\left(\frac{\partial \mathbf{y}^T}{\partial \mathbf{k}}\right)^T = \left(\frac{\partial \mathbf{h}^T}{\partial \mathbf{x}}\right)^T \left(\frac{\partial \mathbf{x}^T}{\partial \mathbf{k}}\right)^T + \left(\frac{\partial \mathbf{h}^T}{\partial \mathbf{k}}\right)^T \tag{6.22}$$

Substitution into the linearized output vector (Equation 6.5) yields

$$\mathbf{y}(t_i,\mathbf{k}^{(j+1)}) = \mathbf{h}(\mathbf{x}(t_i,\mathbf{k}^{(j)})) + \mathbf{W}(t_i)\,\Delta\mathbf{k}^{(j+1)} \tag{6.23}$$

where

$$\mathbf{W}(t_i) = \left(\frac{\partial \mathbf{h}^T}{\partial \mathbf{x}}\right)_i^T \mathbf{G}(t_i) + \left(\frac{\partial \mathbf{h}^T}{\partial \mathbf{k}}\right)_i^T \tag{6.24}$$

and hence the corresponding normal equations are obtained, i.e.,

$$\mathbf{A}\,\Delta\mathbf{k}^{(j+1)} = \mathbf{b} \tag{6.25}$$

where

$$\mathbf{A} = \sum_{i=1}^{N} \mathbf{W}^T(t_i)\mathbf{Q}_i\mathbf{W}(t_i) \tag{6.26}$$

and

$$\mathbf{b} = \sum_{i=1}^{N} \mathbf{W}^T(t_i)\mathbf{Q}_i\left[\hat{\mathbf{y}}_i - \mathbf{h}\mathbf{x}(t_i,\mathbf{k}^{(j)})\right] \tag{6.27}$$

If the nonlinear output relationship is independent of the parameters, i.e., it is of the form

$$\mathbf{y}(t) = \mathbf{h}(\mathbf{x}(t)) \tag{6.28}$$

then $\mathbf{W}(t_i)$ simplifies to

$$\mathbf{W}(t_i) = \left(\frac{\partial \mathbf{h}^T}{\partial \mathbf{x}}\right)_i^T \mathbf{G}(t_i) \tag{6.29}$$

and the corresponding matrix **A** and vector **b** become

$$\mathbf{A} = \sum_{i=1}^{N} \mathbf{G}^T(t_i)\left(\frac{\partial \mathbf{h}^T}{\partial \mathbf{x}}\right)_i \mathbf{Q}_i \left(\frac{\partial \mathbf{h}^T}{\partial \mathbf{x}}\right)_i^T \mathbf{G}(t_i) \tag{6.30}$$

and

$$\mathbf{b} = \sum_{i=1}^{N} \mathbf{G}^T(t_i)\left(\frac{\partial \mathbf{h}^T}{\partial \mathbf{x}}\right)_i \mathbf{Q}_i \left[\hat{\mathbf{y}}_i - \mathbf{h}(\mathbf{x}(t_i,\mathbf{k}^{(j)}))\right] \tag{6.31}$$

In other words, the observation matrix **C** from the case of a linear output relationship is substituted with the Jacobean matrix $(\partial \mathbf{h}^T/\partial \mathbf{x})^T$ in setting up matrix **A** and vector **b**.

6.4 THE GAUSS-NEWTON METHOD – SYSTEMS WITH UNKNOWN INITIAL CONDITIONS

Let us consider a system described by a set of ODEs as in Section 6.1.

$$\frac{d\mathbf{x}(t)}{dt} = \mathbf{f}(\mathbf{x}(t), \mathbf{u}, \mathbf{k}) \quad ; \quad \mathbf{x}(t_0) = \mathbf{x}_0 \tag{6.32}$$

$$\mathbf{y}(t) = \mathbf{C}\mathbf{x}(t) \tag{6.33}$$

The only difference here is that it is further assumed that some or all of the components of the initial state vector $\mathbf{x}_0$ are unknown. Let the *q-dimensional* vector **p** ($0 < q \le n$) denote the unknown components of the vector $\mathbf{x}_0$. In this class of parameter estimation problems, the objective is to determine not only the parameter vector **k** but also the unknown vector **p** containing the unknown elements of the initial state vector $\mathbf{x}(t_0)$.

Again, we assume that experimental data are available as measurements of the output vector at various points in time, i.e., $[\hat{\mathbf{y}}_i, t_i]$, i=1,...,N. The objective function that should be minimized is the same as before. Tthe only difference is that the minimization is carried out over **k** and **p**, namely the objective function is viewed as

$$S(\mathbf{k},\mathbf{p}) = \sum_{i=1}^{N} [\hat{\mathbf{y}}_i - \mathbf{y}(t_i,\mathbf{k},\mathbf{p})]^T \mathbf{Q}_i [\hat{\mathbf{y}}_i - \mathbf{y}(t_i,\mathbf{k},\mathbf{p})] \tag{6.34}$$

Let us suppose that an estimate $\mathbf{k}^{(j)}$ and $\mathbf{p}^{(j)}$ of the unknown parameter and initial state vectors is available at the jth iteration. Linearization of the output vector around $\mathbf{k}^{(j)}$ and $\mathbf{p}^{(j)}$ yields,

$$\mathbf{y}(t_i,\mathbf{k}^{(j+1)},\mathbf{p}^{(j+1)}) = \mathbf{y}(t_i,\mathbf{k}^{(j)},\mathbf{p}^{(j)}) + \left(\frac{\partial \mathbf{y}^T}{\partial \mathbf{x}}\right)_i^T \left(\frac{\partial \mathbf{x}^T}{\partial \mathbf{k}}\right)_i^T \Delta\mathbf{k}^{(j+1)}$$

$$+ \left(\frac{\partial \mathbf{y}^T}{\partial \mathbf{x}}\right)_i^T \left(\frac{\partial \mathbf{x}^T}{\partial \mathbf{p}}\right)_i^T \Delta\mathbf{p}^{(j+1)} \tag{6.35}$$

Assuming a linear output relationship (i.e., $\mathbf{y}(t) = \mathbf{C}\mathbf{x}(t)$), the above equation becomes

$$\mathbf{y}(t_i,\mathbf{k}^{(j+1)},\mathbf{p}^{(j+1)}) = \mathbf{C}\mathbf{x}(t_i,\mathbf{k}^{(j)},\mathbf{p}^{(j)}) + \mathbf{C}\mathbf{G}(t_i)\,\Delta\mathbf{k}^{(j+1)} + \mathbf{C}\mathbf{P}(t_i)\,\Delta\mathbf{p}^{(j+1)} \tag{6.36}$$

where $\mathbf{G}(t)$ is the usual $n{\times}p$ parameter sensitivity matrix $(\partial\mathbf{x}^T/\partial\mathbf{k})^T$ and $\mathbf{P}(t)$ is the $n{\times}q$ initial state sensitivity matrix $(\partial\mathbf{x}^T/\partial\mathbf{p})^T$.

The parameter sensitivity matrix $\mathbf{G}(t)$ can be obtained as shown in the previous section by solving the matrix differential equation,

$$\frac{d\mathbf{G}(t)}{dt} = \left(\frac{\partial \mathbf{f}^T}{\partial \mathbf{x}}\right)^T \mathbf{G}(t) + \left(\frac{\partial \mathbf{f}^T}{\partial \mathbf{k}}\right)^T \tag{6.37}$$

with the initial condition,

$$\mathbf{G}(t_0) = \mathbf{0}. \tag{6.38}$$

Similar to the parameter sensitivity matrix, the initial state sensitivity matrix, $\mathbf{P}(t)$, cannot be obtained by a simple differentiation. $\mathbf{P}(t)$ is determined by solving a matrix differential equation that is obtained by differentiating both sides of Equation 6.1 (state equation) with respect to $\mathbf{p}^{(j)}$.

Reversing the order of differentiation and performing implicit differentiation on the right-hand side, we arrive at

$$\frac{d}{dt}\left[\left(\frac{\partial \mathbf{x}^T}{\partial \mathbf{p}}\right)^T\right] = \left(\frac{\partial \mathbf{f}^T}{\partial \mathbf{x}}\right)^T \left(\frac{\partial \mathbf{x}^T}{\partial \mathbf{p}}\right)^T \tag{6.39}$$

or better

$$\frac{d\mathbf{P}(t)}{dt} = \left(\frac{\partial \mathbf{f}^T}{\partial \mathbf{x}}\right)^T \mathbf{P}(t) \tag{6.40}$$

The initial condition is obtained by differentiating both sides of the initial condition, $\mathbf{x}(t_0)=\mathbf{x}_0$, with respect to $\mathbf{p}$, yielding

$$\mathbf{P}(t_0)=\begin{bmatrix} \mathbf{I}_{qxq} \\ \mathbf{0}_{(n-q)xq} \end{bmatrix} \tag{6.41}$$

Without any loss of generality, it has been assumed that the unknown initial states correspond to state variables that are placed as the first elements of the state vector $\mathbf{x}(t)$. Hence, the structure of the initial condition in Equation 6.41.

Thus, integrating the state and sensitivity equations (Equations 6.1, 6.9 and 6.40), a total of $n\times(p+q+1)$ differential equations, the output vector, $\mathbf{y}(t,\mathbf{k}^{(j+1)},\mathbf{p}^{(j+1)})$ is obtained as a linear function of $\mathbf{k}^{(j+1)}$ and $\mathbf{p}^{(j+1)}$. Next, substitution of $\mathbf{y}(t_i,\mathbf{k}^{(j+1)},\mathbf{p}^{(j+1)})$ into the objective function and use of the stationary criteria

$$\frac{\partial S(\mathbf{k}^{(j+1)},\mathbf{p}^{(j+1)})}{\partial \mathbf{k}^{(j+1)}} = \mathbf{0} \tag{6.42a}$$

and

$$\frac{\partial S(\mathbf{k}^{(j+1)},\mathbf{p}^{(j+1)})}{\partial \mathbf{p}^{(j+1)}} = \mathbf{0} \tag{6.42b}$$

yields the following linear equation

$$\begin{bmatrix} \sum_{i=1}^{N} \mathbf{G}^T(t_i)\mathbf{C}^T\mathbf{QCG}(t_i) & \sum_{i=1}^{N} \mathbf{G}^T(t_i)\mathbf{C}^T\mathbf{QCP}(t_i) \\ \sum_{i=1}^{N} \mathbf{P}^T(t_i)\mathbf{C}^T\mathbf{QCG}(t_i) & \sum_{i=1}^{N} \mathbf{P}^T(t_i)\mathbf{C}^T\mathbf{QCP}(t_i) \end{bmatrix} \begin{bmatrix} \Delta\mathbf{k}^{(j+1)} \\ \Delta\mathbf{p}^{(j+1)} \end{bmatrix} = \begin{bmatrix} \sum_{i=1}^{N} \mathbf{G}^T(t_i)\mathbf{C}^T\mathbf{Q}\left(\hat{\mathbf{y}}_i - \mathbf{Cx}(t_i, \mathbf{k}^{(j)}, \mathbf{p}^{(j)})\right) \\ \sum_{i=1}^{N} \mathbf{P}^T(t_i)\mathbf{C}^T\mathbf{Q}\left(\hat{\mathbf{y}}_i - \mathbf{Cx}(t_i, \mathbf{k}^{(j)}, \mathbf{p}^{(j)})\right) \end{bmatrix} \tag{6.43}$$

Solution of the above equation yields $\Delta\mathbf{k}^{(j+1)}$ and $\Delta\mathbf{p}^{(j+1)}$. The estimates $\mathbf{k}^{(j+1)}$ and $\mathbf{p}^{(j+1)}$ are obtained next as

$$\begin{bmatrix} \mathbf{k}^{(j+1)} \\ \mathbf{p}^{(j+1)} \end{bmatrix} = \begin{bmatrix} \mathbf{k}^{(j)} \\ \mathbf{p}^{(j)} \end{bmatrix} + \mu \begin{bmatrix} \Delta\mathbf{k}^{(j+1)} \\ \Delta\mathbf{p}^{(j+1)} \end{bmatrix} \tag{6.44}$$

where a stepping parameter μ (to be determined by the bisection rule) is also used.

If the initial guess $\mathbf{k}^{(0)}$, $\mathbf{p}^{(0)}$ is sufficiently close to the optimum, this procedure yields a quadratic convergence to the optimum. However, the same difficulties, as those discussed earlier arise whenever the initial estimates are far from the optimum.

If we consider the limiting case where $p=0$ and $q\neq0$, i.e., the case where there are no unknown parameters and only some of the initial states are to be estimated, the previously outlined procedure represents a quadratically convergent method for the solution of *two-point boundary value problems*. Obviously in this case, we need to compute only the sensitivity matrix **P**(t). It can be shown that under these conditions the Gauss-Newton method is a typical quadratically convergent "shooting method." As such it can be used to solve *optimal control* problems using the Boundary Condition Iteration approach (Kalogerakis, 1983).

6.5 EXAMPLES

6.5.1 A Homogeneous Gas Phase Reaction

Bellman et al. (1967) have considered the estimation of the two rate constants k_1 and k_2 in the Bodenstein-Linder model for the homogeneous gas phase reaction of NO with O_2:

$$2NO + O_2 \longleftrightarrow 2NO_2$$

The model is described by the following equation

$$\frac{dx}{dt} = k_1(\alpha - x)(\beta - x)^2 - k_2x^2 \quad ; \quad x(0) = 0 \tag{6.45}$$

where α=126.2, β=91.9 and x is the concentration of NO_2. The concentration of NO_2 was measured experimentally as a function of time and the data are given in Table 6.1

The model is of the form dx/dt=f(x,k_1,k_2) where f(x,k_1,k_2)=$k_1(\alpha\text{-}x)(\beta\text{-}x)^2$-$k_2x^2$. The single state variable x is also the measured variable (i.e., y(t)=x(t)). The sensitivity matrix, **G**(t), is a *(1×2)-dimensional* matrix with elements:

$$\mathbf{G}(t) = [G_1(t),\ G_2(t)] = \left[\left(\frac{\partial x}{\partial k_1}\right), \left(\frac{\partial x}{\partial k_2}\right)\right] \tag{6.46}$$

Table 6.1 Data for the Homogeneous Gas Phase Reaction of NO with O_2.

Time	*Concentration of NO_2*
0	0
1	1.4
2	6.3
3	10.5
4	14.2
5	17.6
6	21.4
7	23.0
9	27.0
11	30.5
14	34.4
19	38.8
24	41.6
29	43.5
39	45.3

Source: Bellman et al. (1967).

In this case, Equation 6.16 simply becomes,

$$\frac{dG_1}{dt} = \left(\frac{\partial f}{\partial x}\right)G_1 + \left(\frac{\partial f}{\partial k_1}\right) \quad ; \quad G_1(0) = 0 \tag{6.47a}$$

and similarly for $G_2(t)$,

$$\frac{dG_2}{dt} = \left(\frac{\partial f}{\partial x}\right) G_2 + \left(\frac{\partial f}{\partial k_2}\right) \quad ; \quad G_2(0) = 0 \tag{6.47b}$$

where

$$\left(\frac{\partial f}{\partial x}\right) = -k_1(\beta-x)^2 - 2k_1(\alpha-x)(\beta-x) - 2k_2 x \tag{6.48a}$$

$$\left(\frac{\partial f}{\partial k_1}\right) = (\alpha-x)(\beta-x)^2 \tag{6.48b}$$

$$\left(\frac{\partial f}{\partial k_2}\right) = -x^2 \tag{6.48c}$$

Equations 6.47a and 6.47b should be solved simultaneously with the state equation (Equation 6.45). The three ODEs are put into the standard form ($d\mathbf{z}/dt = \boldsymbol{\varphi}(\mathbf{z})$) used by differential equation solvers by setting

$$\mathbf{z}(t) = \begin{bmatrix} x(t) \\ G_1(t) \\ G_2(t) \end{bmatrix} \tag{6.49a}$$

and

$$\boldsymbol{\varphi}(\mathbf{z}) = \begin{bmatrix} k_1(\alpha-x)(\beta-x)^2 - k_2 x^2 \\ -[k_1(\beta-x)^2 + 2k_1(\alpha-x)(\beta-x) + 2k_2 x]G_1 + (\alpha-x)(\beta-x)^2 \\ -[k_1(\beta-x)^2 + 2k_1(\alpha-x)(\beta-x) + 2k_2 x]G_2 - x^2 \end{bmatrix} \tag{6.49b}$$

Integration of the above equation yields x(t) and **G**(t) which are used in setting up matrix **A** and vector **b** at each iteration of the Gauss-Newton method.

6.5.2 Pyrolytic Dehydrogenation of Benzene to Diphenyl and Triphenyl

Let us now consider the pyrolytic dehydrogenation of benzene to diphenyl and triphenyl (Seinfeld and Gavalas, 1970; Hougen and Watson, 1948):

$$2C_2H_6 \longleftrightarrow C_{12}H_{10} + H_2$$

$$C_6H_6 + C_{12}H_{10} \longleftrightarrow C_{10}H_{14} + H_2$$

The following kinetic model has been proposed

$$\frac{dx_1}{dt} = -r_1 - r_2 \tag{6.50a}$$

$$\frac{dx_2}{dt} = \frac{r_1}{2} - r_2 \tag{6.50b}$$

where

$$r_1 = k_1\left[x_1^2 - x_2(2 - 2x_1 - x_2)/3K_1\right] \tag{6.51a}$$

$$r_2 = k_2\left[x_1 x_2 - (1 - x_1 - 2x_2)(2 - 2x_1 - x_2)/9K_2\right] \tag{6.51b}$$

where x_1 denotes *lb-mole* of benzene per *lb-mole* of pure benzene feed and x_2 denotes *lb-mole* of diphenyl per *lb-mole* of pure benzene feed. The parameters k_1 and k_2 are unknown reaction rate constants whereas K_1 and K_2 are equilibrium constants. The data consist of measurements of x_1 and x_2 in a flow reactor at eight values of the reciprocal space velocity t and are given below: The feed to the reactor was pure benzene.

Table 6.2. Data for the Pyrolytic Dehydrogenation of Benzene

Reciprocal Space Velocity (t) $\times 10^4$	x_1	x_2
5.63	0.828	0.0737
11.32	0.704	0.113
16.97	0.622	0.1322
22.62	0.565	0.1400
34.0	0.499	0.1468
39.7	0.482	0.1477
45.2	0.470	0.1477
169.7	0.443	0.1476

Source: Seinfeld and Gavalas (1970); Hougen and Watson (1948).

As both state variables are measured, the output vector is the same with the state vector, i.e., $y_1=x_1$ and $y_2=x_2$. The feed to the reactor was pure benzene. The equilibrium constants K_1 and K_2 were determined from the run at the lowest space velocity to be 0.242 and 0.428, respectively.

Using our standard notation, the above problem is written as follows:

$$\frac{dx_1}{dt} = f_1(x_1, x_2; k_1, k_2) \tag{6.52a}$$

$$\frac{dx_2}{dt} = f_2(x_1, x_2, k_1, k_2) \tag{6.52b}$$

where $f_1=(-r_1-r_2)$ and $f_2=r_1/2-r_2$.

The sensitivity matrix, $\mathbf{G}(t)$, is a *(2x2)-dimensional* matrix with elements:

$$\mathbf{G}(t) = [\mathbf{g}_1(t), \mathbf{g}_2(t)] = \left[\left(\frac{\partial \mathbf{x}}{\partial k_1}\right), \left(\frac{\partial \mathbf{x}}{\partial k_2}\right)\right] =$$

$$\begin{bmatrix} G_{11}(t) & G_{12}(t) \\ G_{21}(t) & G_{22}(t) \end{bmatrix} = \begin{bmatrix} \left(\frac{\partial x_1}{\partial k_1}\right) & \left(\frac{\partial x_1}{\partial k_2}\right) \\ \left(\frac{\partial x_2}{\partial k_1}\right) & \left(\frac{\partial x_2}{\partial k_2}\right) \end{bmatrix} \tag{6.53}$$

Equations 6.16 then become,

$$\frac{d\mathbf{g}_1(t)}{dt} = \left(\frac{\partial \mathbf{f}^T}{\partial \mathbf{x}}\right)^T \mathbf{g}_1(t) + \left(\frac{\partial \mathbf{f}}{\partial k_1}\right); \qquad \mathbf{g}_1(t_0)=\mathbf{0} \tag{6.54a}$$

and

$$\frac{d\mathbf{g}_2(t)}{dt} = \left(\frac{\partial \mathbf{f}^T}{\partial \mathbf{x}}\right)^T \mathbf{g}_2(t) + \left(\frac{\partial \mathbf{f}}{\partial k_2}\right); \qquad \mathbf{g}_2(t_0)=\mathbf{0} \tag{6.54b}$$

Taking into account Equation 6.53, the above equations can also be written as follows:

$$\begin{bmatrix} \frac{dG_{11}}{dt} \\ \frac{dG_{21}}{dt} \end{bmatrix} = \begin{bmatrix} \frac{\partial f_1}{\partial x_1} & \frac{\partial f_1}{\partial x_2} \\ \frac{\partial f_2}{\partial x_1} & \frac{\partial f_2}{\partial x_2} \end{bmatrix} \begin{bmatrix} G_{11} \\ G_{21} \end{bmatrix} + \begin{bmatrix} \frac{\partial f_1}{\partial k_1} \\ \frac{\partial f_2}{\partial k_1} \end{bmatrix} ; \quad G_{11}(t_0)=0,\ G_{21}(t_0)=0 \tag{6.55a}$$

and

$$\begin{bmatrix} \frac{dG_{12}}{dt} \\ \frac{dG_{22}}{dt} \end{bmatrix} = \begin{bmatrix} \frac{\partial f_1}{\partial x_1} & \frac{\partial f_1}{\partial x_2} \\ \frac{\partial f_2}{\partial x_1} & \frac{\partial f_2}{\partial x_2} \end{bmatrix} \begin{bmatrix} G_{12} \\ G_{22} \end{bmatrix} + \begin{bmatrix} \frac{\partial f_1}{\partial k_2} \\ \frac{\partial f_2}{\partial k_2} \end{bmatrix} ; \quad G_{21}(t_0)=0,\ G_{22}(t_0)=0 \tag{6.55b}$$

Finally, we obtain the following equations

$$\frac{dG_{11}}{dt} = \left(\frac{\partial f_1}{\partial x_1}\right) G_{11} + \left(\frac{\partial f_1}{\partial x_2}\right) G_{21} + \frac{\partial f_1}{\partial k_1} \quad ; \quad G_{11}(0) = 0 \tag{6.56a}$$

$$\frac{dG_{21}}{dt} = \left(\frac{\partial f_2}{\partial x_1}\right) G_{11} + \left(\frac{\partial f_2}{\partial x_2}\right) G_{21} + \frac{\partial f_2}{\partial k_1} \quad ; \quad G_{21}(0) = 0 \tag{6.56b}$$

$$\frac{dG_{12}}{dt} = \left(\frac{\partial f_1}{\partial x_1}\right) G_{12} + \left(\frac{\partial f_1}{\partial x_2}\right) G_{22} + \frac{\partial f_1}{\partial k_2} \quad ; \quad G_{12}(0) = 0 \tag{6.56c}$$

$$\frac{dG_{22}}{dt} = \left(\frac{\partial f_2}{\partial x_1}\right) G_{12} + \left(\frac{\partial f_2}{\partial x_2}\right) G_{22} + \frac{\partial f_2}{\partial k_2} \quad ; \quad G_{22}(0) = 0 \tag{6.56d}$$

where

$$\left(\frac{\partial f_1}{\partial x_1}\right) = -k_1\left(2x_1 + \frac{2x_2}{3K_1}\right) - k_2\left(x_2 - \frac{1}{9K_2}(4x_1 - 4 + 5x_2)\right) \tag{6.57a}$$

$$\left(\frac{\partial f_1}{\partial x_2}\right) = -k_1\left(\frac{1}{3K_1}(2x_2 + 2x_1 - 2)\right) - k_2\left(x_1 - \frac{1}{9K_2}(5x_1 - 5 + 4x_2)\right) \tag{6.57b}$$

$$\left(\frac{\partial f_2}{\partial x_1}\right) = \frac{k_1}{2}\left(2x_1 + \frac{2x_2}{3K_1}\right) - k_2\left(x_2 - \frac{1}{9K_2}(4x_1 - 4 + 5x_2)\right) \tag{6.57c}$$

$$\left(\frac{\partial f_2}{\partial x_2}\right)=\frac{k_1}{2}\left(\frac{1}{3K_1}(2x_2+2x_1-2)\right)-k_2\left(x_1-\frac{1}{9K_2}(5x_1+4x_2-5)\right) \quad (6.57d)$$

$$\left(\frac{\partial f_1}{\partial k_1}\right)=-\left[x_1^2+\frac{1}{3K_1}\left(x_2^2+2x_1x_2-2x_2\right)\right] \quad (6.57e)$$

$$\left(\frac{\partial f_2}{\partial k_1}\right)=\frac{1}{2}\left[x_1^2+\frac{1}{3K_1}\left(x_2^2+2x_1x_2-2x_2\right)\right] \quad (6.57f)$$

$$\left(\frac{\partial f_1}{\partial k_2}\right)=-\left[x_1x_2-\frac{1}{9K_2}\left(2x_1^2-4x_1+5x_1x_2-5x_2+2x_2^2+2\right)\right] \quad (6.57g)$$

$$\left(\frac{\partial f_2}{\partial k_2}\right)=-\left[x_1x_2-\frac{1}{9K_2}\left(2x_1^2-4x_1+5x_1x_2-5x_2+2x_2^2+2\right)\right] \quad (6.57h)$$

The four sensitivity equations (Equations 6.56a-d) should be solved simultaneously with the two state equations (Equation 6.52). Integration of these six [$=n\times(p+1)=2\times(2+1)$] equations yields **x**(t) and **G**(t) which are used in setting up matrix **A** and vector **b** at each iteration of the Gauss-Newton method.

The ordinary differential equation that a particular element, G_{ij}, of the (*nxp*)-*dimensional* sensitivity matrix satisfies, can be written directly using the following expression,

$$\frac{dG_{ij}}{dt}\equiv\frac{d}{dt}\left(\frac{\partial x_i}{\partial k_j}\right)=\sum_{k=1}^{n}\left(\frac{\partial f_i}{\partial x_k}\right)\left(\frac{\partial x_k}{\partial k_j}\right)+\frac{\partial f_i}{\partial k_j}\equiv\sum_{k=1}^{n}\left(\frac{\partial f_i}{\partial x_k}\right)G_{kj}+\frac{\partial f_i}{\partial k_j} \quad (6.58)$$

6.5.3 Catalytic Hydrogenation of 3-Hydroxypropanal (HPA) to 1,3-Propanediol (PD)

The hydrogenation of 3-hydroxypropanal (HPA) to 1,3-propanediol (PD) over $Ni/SiO_2/Al_2O_3$ catalyst powder was studied by Professor Hoffman's group at the Friedrich-Alexander University in Erlagen, Germany (Zhu et al., 1997). PD is a potentially attractive monomer for polymers like polypropylene terephthalate. They used a batch stirred autoclave. The experimental data were kindly provided by Professor Hoffman and consist of measurements of the concentration of HPA and PD (C_{HPA}, C_{PD}) versus time at various operating temperatures and pressures.

The complete data set will be given in the case studies section. In this chapter, we will discuss how we set up the equations for the regression of an isothermal data set given in Tables 6.3 or 6.4.

The same group also proposed a reaction scheme and a mathematical model that describe the rates of HPA consumption, PD formation as well as the formation of acrolein (Ac). The model is as follows

$$\frac{dC_{HPA}}{dt} = -[r_1 + r_2]C_k - [r_3 + r_4 - r_{-3}] \tag{6.59a}$$

$$\frac{dC_{PD}}{dt} = [r_1 - r_2]C_k \tag{6.59b}$$

$$\frac{dC_{Ac}}{dt} = r_3 - r_4 - r_{-3} \tag{6.59c}$$

where C_k is the concentration of the catalyst (10 *g/L*). The reaction rates are given below

$$r_1 = \frac{k_1 P C_{HPA}}{H\left[1 + \left(\frac{K_1 P}{H}\right)^{0.5} + K_2 C_{HPA}\right]^3} \tag{6.60a}$$

$$r_2 = \frac{k_2 C_{PD} C_{HPA}}{1 + \left(\frac{K_1 P}{H}\right)^{0.5} + K_2 C_{HPA}} \tag{6.60b}$$

$$r_3 = k_3 C_{HPA} \tag{6.60c}$$

$$r_{-3} = k_{-3} C_{Ac} \tag{6.60d}$$

$$r_4 = k_4 C_{Ac} C_{HPA} \tag{6.60e}$$

In the above equations, k_j (j=1, 2, 3, -3, 4) are rate constants *(L/(mol min g)*, K_1 and K_2 are the adsorption equilibrium constants *(L/mol)* for H_2 and HPA respectively. P is the hydrogen pressure *(MPa)* in the reactor and H is the Henry's law constant with a value equal to 1379 *(L bar/mol)* at 298 *K*. The seven parameters (k_1, k_2, k_3, k_{-3}, k_4, K_1 and K_2) are to be determined from the measured concentrations of HPA and PD.

Table 6.3 Data for the Catalytic Hydrogenation of 3-Hydroxypropanal (HPA) to 1,3-Propanediol (PD) at 5.15 MPa and 45 ℃

t (*min*)	C_{HPA} (*mol/L*)	C_{PD} (*mol/L*)
0.0	1.34953	0.0
10	1.36324	0.00262812
20	1.25882	0.0700394
30	1.17918	0.184363
40	0.972102	0.354008
50	0.825203	0.469777
60	0.697109	0.607359
80	0.421451	0.852431
100	0.232296	1.03535
120	0.128095	1.16413
140	0.0289817	1.30053
160	0.00962368	1.31971

Source: Zhu et al. (1997).

Table 6.4 Data for the Catalytic Hydrogenation of 3-Hydroxypropanal (HPA) to 1,3-Propanediol (PD) at 5.15 Mpa and 80 ℃

t (*min*)	C_{HPA} (*mol/L*)	C_{PD} (*mol/L*)
0.0	1.34953	0.0
5	0.873513	0.388568
10	0.44727	0.816032
15	0.140925	0.967017
20	0.0350076	1.05125
25	0.0130859	1.08239
30	0.00581597	1.12024

Source: Zhu et al. (1997).

In order to use our standard notation we introduce the following vectors:

$$\mathbf{x} = [x_1, x_2, x_3]^T = [C_{HPA}, C_{PD}, C_{Ac}]^T$$

$$\mathbf{k} = [k_1, k_2, k_3, k_4, k_5, k_6, k_7]^T = [k_1, k_2, k_3, k_{-3}, k_4, K_1, K_2]^T$$

$$\mathbf{y} = [y_1, y_2]^T = [C_{HPA}, C_{PD}]^T$$

Hence, the differential equation model takes the form,

$$\frac{dx_1}{dt} = f_1(x_1, x_2, x_3; k_1, k_2, ..., k_7; u_1, u_2) \tag{6.61a}$$

$$\frac{dx_2}{dt} = f_2(x_1, x_2, x_3; k_1, k_2, ..., k_7; u_1, u_2) \tag{6.61b}$$

$$\frac{dx_3}{dt} = f_3(x_1, x_2, x_3; k_1, k_2, ..., k_7; u_1, u_2) \tag{6.61c}$$

and the observation matrix is simply

$$\mathbf{C} = \begin{bmatrix} 1 & 0 & 0 \\ 0 & 1 & 0 \end{bmatrix} \tag{6.62}$$

In Equations 6.61, u_1 denotes the concentration of catalyst present in the reactor (C_k) and u_2 the hydrogen pressure (P). As far as the estimation problem is concerned, both these variables are assumed to be known precisely. Actually, as it will be discussed later on experimental design (Chapter 12), the value of such variables is chosen by the experimentalist and can have a paramount effect on the quality of the parameter estimates. Equations 6.61 are rewritten as following

$$\frac{dx_1}{dt} = -u_1(r_1+r_2)-(k_3x_1+k_5x_3x_1-k_4x_3) \tag{6.63a}$$

$$\frac{dx_2}{dt} = u_1(r_1-r_2) \tag{6.63b}$$

$$\frac{dx_3}{dt} = (k_3x_1-k_5x_3x_1-k_4x_3) \tag{6.63c}$$

where

$$r_1 = \frac{k_1u_2x_1}{H\left[1+\left(\frac{k_6u_2}{H}\right)^{0.5}+k_7x_1\right]^3} \tag{6.64a}$$

$$r_2 = \frac{k_2x_2x_1}{1+\left(\frac{k_6u_2}{H}\right)^{0.5}+k_7x_1} \tag{6.64b}$$

$$r_3 = k_3 x_1 \tag{6.64c}$$

$$r_{-3} = k_4 x_3 \tag{6.64d}$$

$$r_4 = k_5 x_3 x_1 \tag{6.64e}$$

The sensitivity matrix, $\mathbf{G}(t)$, is a (3×7)-*dimensional* matrix with elements:

$$\mathbf{G}(t) = [\mathbf{g}_1(t), \mathbf{g}_2(t), \ldots, \mathbf{g}_7(t)] = \left[\left(\frac{\partial \mathbf{x}}{\partial k_1}\right), \left(\frac{\partial \mathbf{x}}{\partial k_2}\right), \ldots, \left(\frac{\partial \mathbf{x}}{\partial k_7}\right)\right] \tag{6.65a}$$

$$\mathbf{G}(t) = \begin{bmatrix} G_{11}(t) & \cdot\;\cdot\;\cdot & G_{17}(t) \\ G_{21}(t) & \cdot\;\cdot\;\cdot & G_{27}(t) \\ G_{31}(t) & \cdot\;\cdot\;\cdot & G_{37}(t) \end{bmatrix} = \begin{bmatrix} \left(\frac{\partial x_1}{\partial k_1}\right) & \cdot\;\cdot\;\cdot & \left(\frac{\partial x_1}{\partial k_7}\right) \\ \left(\frac{\partial x_2}{\partial k_1}\right) & \cdot\;\cdot\;\cdot & \left(\frac{\partial x_2}{\partial k_7}\right) \\ \left(\frac{\partial x_3}{\partial k_1}\right) & \cdot\;\cdot\;\cdot & \left(\frac{\partial x_3}{\partial k_7}\right) \end{bmatrix} \tag{6.65b}$$

Equations 6.16 then become,

$$\frac{d\mathbf{g}_1(t)}{dt} = \left(\frac{\partial \mathbf{f}^T}{\partial \mathbf{x}}\right)^T \mathbf{g}_1(t) + \left(\frac{\partial \mathbf{f}}{\partial k_1}\right); \qquad \mathbf{g}_1(t_0)=\mathbf{0} \tag{6.66a}$$

$$\frac{d\mathbf{g}_2(t)}{dt} = \left(\frac{\partial \mathbf{f}^T}{\partial \mathbf{x}}\right)^T \mathbf{g}_2(t) + \left(\frac{\partial \mathbf{f}}{\partial k_2}\right); \qquad \mathbf{g}_2(t_0)=\mathbf{0} \tag{6.66b}$$

$$\vdots$$

$$\frac{d\mathbf{g}_7(t)}{dt} = \left(\frac{\partial \mathbf{f}^T}{\partial \mathbf{x}}\right)^T \mathbf{g}_7(t) + \left(\frac{\partial \mathbf{f}}{\partial k_7}\right); \qquad \mathbf{g}_7(t_0)=\mathbf{0} \tag{6.66c}$$

where

$$\left(\frac{\partial \mathbf{f}^T}{\partial \mathbf{x}}\right)^T = \begin{bmatrix} \left(\frac{\partial f_1}{\partial x_1}\right) & \left(\frac{\partial f_1}{\partial x_2}\right) & \left(\frac{\partial f_1}{\partial x_3}\right) \\ \left(\frac{\partial f_2}{\partial x_1}\right) & \left(\frac{\partial f_2}{\partial x_2}\right) & \left(\frac{\partial f_2}{\partial x_3}\right) \\ \left(\frac{\partial f_3}{\partial x_1}\right) & \left(\frac{\partial f_3}{\partial x_2}\right) & \left(\frac{\partial f_3}{\partial x_3}\right) \end{bmatrix} \tag{6.67a}$$

and

$$\left(\frac{\partial \mathbf{f}}{\partial k_j}\right) = \begin{bmatrix} \left(\frac{\partial f_1}{\partial k_j}\right) \\ \left(\frac{\partial f_2}{\partial k_j}\right) \\ \left(\frac{\partial f_3}{\partial k_j}\right) \end{bmatrix} \quad ; \quad j=1,2,\ldots,7 \tag{6.67b}$$

Taking into account the above equations we obtain

$$\left.\begin{aligned}
\frac{dG_{11}}{dt} &= \left(\frac{\partial f_1}{\partial x_1}\right)G_{11} + \left(\frac{\partial f_1}{\partial x_2}\right)G_{21} + \left(\frac{\partial f_1}{\partial x_3}\right)G_{31} + \frac{\partial f_1}{\partial k_1} \quad ; \quad G_{11}(0) = 0 \\
\frac{dG_{21}}{dt} &= \left(\frac{\partial f_2}{\partial x_1}\right)G_{11} + \left(\frac{\partial f_2}{\partial x_2}\right)G_{21} + \left(\frac{\partial f_2}{\partial x_3}\right)G_{31} + \frac{\partial f_2}{\partial k_1} \quad ; \quad G_{21}(0) = 0 \\
&\vdots \\
\frac{dG_{17}}{dt} &= \left(\frac{\partial f_1}{\partial x_1}\right)G_{17} + \left(\frac{\partial f_1}{\partial x_2}\right)G_{27} + \left(\frac{\partial f_1}{\partial x_3}\right)G_{37} + \frac{\partial f_1}{\partial k_7} \quad ; \quad G_{17}(0) = 0 \\
\frac{dG_{27}}{dt} &= \left(\frac{\partial f_2}{\partial x_1}\right)G_{17} + \left(\frac{\partial f_2}{\partial x_2}\right)G_{27} + \left(\frac{\partial f_2}{\partial x_3}\right)G_{37} + \frac{\partial f_2}{\partial k_7} ; \quad G_{27}(0) = 0 \\
&\vdots \\
\frac{dG_{37}}{dt} &= \left(\frac{\partial f_3}{\partial x_1}\right)G_{17} + \left(\frac{\partial f_3}{\partial x_2}\right)G_{27} + \left(\frac{\partial f_3}{\partial x_3}\right)G_{37} + \frac{\partial f_3}{\partial k_7} ; \quad G_{37}(0) = 0
\end{aligned}\right\} \tag{6.68}$$

The partial derivatives with respect to the state variables in Equation 6.67a that are needed in the above ODEs are given next

$$\left(\frac{\partial f_1}{\partial x_1}\right) = -u_1\left(\frac{\partial r_1}{\partial x_1} + \frac{\partial r_2}{\partial x_1}\right) - k_3 - k_5 x_3 \tag{6.69a}$$

$$\left(\frac{\partial f_1}{\partial x_2}\right) = -u_1\left(\frac{\partial r_2}{\partial x_2}\right) = -u_1 \frac{r_2}{x_2} \tag{6.69b}$$

$$\left(\frac{\partial f_1}{\partial x_3}\right) = k_4 - k_5 x_1 \tag{6.69c}$$

$$\left(\frac{\partial f_2}{\partial x_1}\right) = u_1\left(\frac{\partial r_1}{\partial x_1} - \frac{\partial r_2}{\partial x_1}\right) \tag{6.69d}$$

$$\left(\frac{\partial f_2}{\partial x_2}\right) = -u_1\left(\frac{\partial r_2}{\partial x_2}\right) = -u_1 \frac{r_2}{x_2} \tag{6.69e}$$

$$\left(\frac{\partial f_2}{\partial x_3}\right) = 0 \tag{6.69f}$$

$$\left(\frac{\partial f_3}{\partial x_1}\right) = k_3 - k_5 x_3 \tag{6.69g}$$

$$\left(\frac{\partial f_3}{\partial x_2}\right) = 0 \tag{6.69h}$$

$$\left(\frac{\partial f_3}{\partial x_3}\right) = -k_5 x_1 - k_4 \tag{6.69i}$$

The partial derivatives with respect to the parameters in Equation 6.67b that are needed in the above ODEs are given next

$$\left(\frac{\partial \mathbf{f}}{\partial k_1}\right) = \begin{bmatrix} \left(\dfrac{\partial f_1}{\partial k_1}\right) \\ \left(\dfrac{\partial f_2}{\partial k_1}\right) \\ \left(\dfrac{\partial f_3}{\partial k_1}\right) \end{bmatrix} = \begin{bmatrix} -u_1\left(\dfrac{\partial r_1}{\partial k_1}\right) \\ u_1\left(\dfrac{\partial r_1}{\partial k_1}\right) \\ 0 \end{bmatrix} \tag{6.70a}$$

$$\left(\frac{\partial \mathbf{f}}{\partial k_2}\right) = \begin{bmatrix} \left(\dfrac{\partial f_1}{\partial k_2}\right) \\ \left(\dfrac{\partial f_2}{\partial k_2}\right) \\ \left(\dfrac{\partial f_3}{\partial k_2}\right) \end{bmatrix} = \begin{bmatrix} -u_1\left(\dfrac{\partial r_2}{\partial k_2}\right) \\ -u_1\left(\dfrac{\partial r_2}{\partial k_2}\right) \\ 0 \end{bmatrix} \tag{6.70b}$$

$$\left(\frac{\partial \mathbf{f}}{\partial k_3}\right) = \begin{bmatrix} \left(\dfrac{\partial f_1}{\partial k_3}\right) \\ \left(\dfrac{\partial f_2}{\partial k_3}\right) \\ \left(\dfrac{\partial f_3}{\partial k_3}\right) \end{bmatrix} = \begin{bmatrix} -x_1 \\ 0 \\ x_1 \end{bmatrix} \tag{6.70c}$$

$$\left(\frac{\partial \mathbf{f}}{\partial k_4}\right) = \begin{bmatrix} \left(\dfrac{\partial f_1}{\partial k_4}\right) \\ \left(\dfrac{\partial f_2}{\partial k_4}\right) \\ \left(\dfrac{\partial f_3}{\partial k_4}\right) \end{bmatrix} = \begin{bmatrix} x_3 \\ 0 \\ -x_3 \end{bmatrix} \tag{6.70d}$$

$$\left(\frac{\partial \mathbf{f}}{\partial k_5}\right) = \begin{bmatrix} \left(\frac{\partial f_1}{\partial k_5}\right) \\ \left(\frac{\partial f_2}{\partial k_5}\right) \\ \left(\frac{\partial f_3}{\partial k_5}\right) \end{bmatrix} = \begin{bmatrix} -x_3 x_1 \\ 0 \\ -x_3 x_1 \end{bmatrix} \tag{6.70e}$$

$$\left(\frac{\partial \mathbf{f}}{\partial k_6}\right) = \begin{bmatrix} \left(\frac{\partial f_1}{\partial k_6}\right) \\ \left(\frac{\partial f_2}{\partial k_6}\right) \\ \left(\frac{\partial f_3}{\partial k_6}\right) \end{bmatrix} = \begin{bmatrix} -u_1\left(\frac{\partial r_1}{\partial k_6} + \frac{\partial r_2}{\partial k_6}\right) \\ u_1\left(\frac{\partial r_1}{\partial k_6} - \frac{\partial r_2}{\partial k_6}\right) \\ 0 \end{bmatrix} \tag{6.70f}$$

$$\left(\frac{\partial \mathbf{f}}{\partial k_7}\right) = \begin{bmatrix} \left(\frac{\partial f_1}{\partial k_7}\right) \\ \left(\frac{\partial f_2}{\partial k_7}\right) \\ \left(\frac{\partial f_3}{\partial k_7}\right) \end{bmatrix} = \begin{bmatrix} -u_1\left(\frac{\partial r_1}{\partial k_7} + \frac{\partial r_2}{\partial k_7}\right) \\ u_1\left(\frac{\partial r_1}{\partial k_7} - \frac{\partial r_2}{\partial k_7}\right) \\ 0 \end{bmatrix} \tag{6.70g}$$

The 21 equations (given as Equation 6.68) should be solved simultaneously with the three state equations (Equation 6.64). Integration of these 24 equations yields $\mathbf{x}(t)$ and $\mathbf{G}(t)$ which are used in setting up matrix $\mathbf{A}$ and vector $\mathbf{b}$ at each iteration of the Gauss-Newton method. Given the complexity of the ODEs when the dimensionality of the problem increases, it is quite helpful to have a general purpose computer program that sets up the sensitivity equations automatically.

Furthermore, since analytical derivatives are subject to user input error, numerical evaluation of the derivatives can also be used in a typical computer implementation of the Gauss-Newton method. Details for a successful implementation of the method are given in Chapter 8.

6.6 EQUIVALENCE OF GAUSS-NEWTON WITH THE QUASI-LINEARIZATION METHOD

The quasilinearization method (QM) is another method for solving off-line parameter estimation problems described by Equations 6.1, 6.2 and 6.3 (Bellman and Kalaba, 1965). Quasilinearization converges quadratically to the optimum but has a small region of convergence (Seinfeld and Gavalas, 1970). Kalogerakis and Luus (1983b) presented an alternative development of the QM that enables a more efficient implementation of the algorithm.

Furthermore, they showed that this simplified QM is very similar to the Gauss-Newton method. Next the quasilinearization method as well as the simplified quasilinearization method are described and the equivalence of QM to the Gauss-Newton method is demonstrated.

6.6.1 The Quasilinearization Method and its Simplification

An estimate $\mathbf{k}^{(j)}$ of the unknown parameter vector is available at the jth iteration. Equation 6.1 then becomes

$$\frac{d\mathbf{x}^{(j)}(t)}{dt} = \mathbf{f}\left(\mathbf{x}^{(j)}(t), \mathbf{k}^{(j)}\right) \tag{6.71}$$

Using the parameter estimate $\mathbf{k}^{(j+1)}$ from the next iteration we obtain from Equation 6.1

$$\frac{d\mathbf{x}^{(j+1)}(t)}{dt} = \mathbf{f}\left(\mathbf{x}^{(j+1)}(t), \mathbf{k}^{(j+1)}\right) \tag{6.72}$$

By using a Taylor series expansion on the right hand side of Equation 6.72 and keeping only the linear terms we obtain the following equation

$$\begin{aligned}\frac{d\mathbf{x}^{(j+1)}(t)}{dt} = \mathbf{f}\left(\mathbf{x}^{(j)}(t), \mathbf{k}^{(j)}\right) + \left(\frac{\partial \mathbf{f}}{\partial \mathbf{x}}\right)^T \left[\mathbf{x}^{(j+1)}(t) - \mathbf{x}^{(j)}(t)\right] \\ + \left(\frac{\partial \mathbf{f}}{\partial \mathbf{k}}\right)^T \left[\mathbf{k}^{(j+1)}(t) - \mathbf{k}^{(j)}(t)\right]\end{aligned} \tag{6.73}$$

where the partial derivatives are evaluated at $\mathbf{x}^{(j)}(t)$.

The above equation is linear in $\mathbf{x}^{(j+1)}$ and $\mathbf{k}^{(j+1)}$. Integration of Equation 6.72 will result in the following equation

$$\mathbf{x}^{(j+1)}(t) = \mathbf{g}(t) + \mathbf{G}(t)\mathbf{k}^{(j+1)} \tag{6.74}$$

where $\mathbf{g}(t)$ is an n-dimensional vector and $\mathbf{G}(t)$ is an $n{\times}p$ matrix.

Equation 6.74 is differentiated and the RHS of the resultant equation is equated with the RHS of Equation 6.73 to yield

$$\frac{d\mathbf{g}(t)}{dt} = \mathbf{f}\left(\mathbf{x}^{(j)}(t), \mathbf{k}^{(j)}\right) + \left(\frac{\partial \mathbf{f}}{\partial \mathbf{x}}\right)^T \left[\mathbf{g}(t) - \mathbf{x}^{(j)}(t)\right] - \left(\frac{\partial \mathbf{f}}{\partial \mathbf{k}}\right)^T \mathbf{k}^{(j)} \tag{6.75}$$

and

$$\frac{d\mathbf{G}(t)}{dt} = \left(\frac{\partial \mathbf{f}}{\partial \mathbf{x}}\right)^T \mathbf{G}(t) + \left(\frac{\partial \mathbf{f}}{\partial \mathbf{k}}\right)^T \tag{6.76}$$

The initial conditions for Equations 6.75 and 6.76 are as follows

$$\mathbf{g}(t_0) = \mathbf{x}_0 \tag{6.77a}$$

$$\mathbf{G}(t_0) = \mathbf{0}. \tag{6.77b}$$

Equations 6.71, 6.75 and 6.76 can be solved simultaneously to yield $\mathbf{g}(t)$ and $\mathbf{G}(t)$ when the initial state vector $\mathbf{x}_0$ and the parameter estimate vector $\mathbf{k}^{(j)}$ are given. In order to determine $\mathbf{k}^{(j+1)}$ the output vector (given by Equation 6.2) is inserted into the objective function (Equation 6.4) and the stationary condition yields,

$$\frac{\partial S(\mathbf{k}^{(j+1)})}{\partial \mathbf{k}^{(j+1)}} = \mathbf{0}. \tag{6.78}$$

The case of a nonlinear observational relationship (Equation 6.3) will be examined later. Equation 6.78 yields the following linear equation which is solved by LU decomposition (or any other technique) to obtain $\mathbf{k}^{(j+1)}$

$$\left[\sum_{i=1}^{N} \mathbf{G}^T(t_i)\mathbf{C}^T\mathbf{Q}_i\mathbf{C}\mathbf{G}(t_i)\right]\mathbf{k}^{(j+1)} = \sum_{i=1}^{N} \mathbf{G}^T(t_i)\mathbf{C}^T\mathbf{Q}_i\left[\hat{\mathbf{y}}(t_i) - \mathbf{C}\mathbf{g}(t_i)\right] \tag{6.79}$$

As matrix $\mathbf{Q}_i$ is positive definite, the above equation gives the minimum of the objective function.

Since linearization of the differential Equation 6.1 around the trajectory $\mathbf{x}^{(j)}(t)$, resulting from the choice of $\mathbf{k}^{(j)}$ has been used, the above method gives $\mathbf{k}^{(j+1)}$ which is an approximation to the best parameter vector. Using this value as $\mathbf{k}^{(j)}$ a new $\mathbf{k}^{(j+1)}$ can be obtained and thus a sequence of vectors $\mathbf{k}^{(0)}$, $\mathbf{k}^{(1)}$, $\mathbf{k}^{(2)}$... is obtained. This sequence converges rapidly to the optimum provided that the initial guess is sufficiently good. The above described methodology constitutes the Quasilinearization Method (QM). The total number of differential equations which must be integrated at each iteration step is $n\times(p+2)$.

Kalogerakis and Luus (1983b) noticed that Equation 6.75 is redundant. Since Equation 6.74 is obtained by linearization around the nominal trajectory $\mathbf{x}^{(j)}(t)$ resulting from $\mathbf{k}^{(j)}$, if we let $\mathbf{k}^{(j+1)}$ be $\mathbf{k}^{(j)}$ then Equation 6.74 becomes

$$\mathbf{x}^{(j)}(t) = \mathbf{g}(t) + \mathbf{G}(t)\mathbf{k}^{(j)} \tag{6.80}$$

Equation 6.80 is exact rather than a first order approximation as Equation 6.74 is. This is simply because Equation 6.80 is Equation 6.74 evaluated at the point of linearization, $\mathbf{k}^{(j)}$. Thus Equation 6.80 can be used to compute $\mathbf{g}(t)$ as

$$\mathbf{g}(t) = \mathbf{x}^{(j)}(t) - \mathbf{G}(t)\mathbf{k}^{(j)} \tag{6.81}$$

It is obvious that the use of Equation 6.81 leads to a simplification because the number of differential equations that now need to be integrated is $n\times(p+1)$. Kalogerakis and Luus (1983b) then proposed the following algorithm for the QM.

Step 1. Select an initial guess $\mathbf{k}^{(0)}$. Hence j=0.

Step 2. Integrate Equations 6.71 and 6.76 simultaneously to obtain $\mathbf{x}^{(j)}(t)$ and $\mathbf{G}(t)$.

Step 3. Use equation 6.81 to obtain $\mathbf{g}(t_i)$, i=1,2,...,N and set up matrix **A** and vector **b** in Equation 6.79.

Step 4. Solve equation 6.79 to obtain $\mathbf{k}^{(j+1)}$.

Step 5. Continue until

$$\left\|\mathbf{k}^{(j+1)} - \mathbf{k}^{(j)}\right\| \leq TOL \tag{6.82}$$

where TOL is a preset small number to ensure termination of the iterations. If the above inequality is not satisfied then we set $\mathbf{k}^{(j)}=\mathbf{k}^{(j+1)}$, increase j by one and go to Step 2 to repeat the calculations.

6.6.2 Equivalence to Gauss-Newton Method

If we compare Equations 6.79 and 6.11 we notice that the only difference between the quasilinearization method and the Gauss-Newton method is the nature of the equation that yields the parameter estimate vector $\mathbf{k}^{(j+1)}$. If one substitutes Equation 6.81 into Equation 6.79 obtains the following equation

$$\left[\sum_{i=1}^{N}\mathbf{G}^T(t_i)\mathbf{C}^T\mathbf{Q}_i\mathbf{C}\mathbf{G}(t_i)\right]\mathbf{k}^{(j+1)} = \sum_{i=1}^{N}\mathbf{G}^T(t_i)\mathbf{C}^T\mathbf{Q}_i\left[\hat{\mathbf{y}}(t_i)-\mathbf{C}\mathbf{x}^{(j)}(t_i)+\mathbf{C}\mathbf{G}(t_i)\mathbf{k}^{(j)}\right] \tag{6.83}$$

By taking the last term on the right hand side of Equation 6.83 to the left hand side one obtains Equation 6.11 that is used for the Gauss-Newton method. Hence, when the output vector is linearly related to the state vector (Equation 6.2) then the simplified quasilinearization method is computationally identical to the Gauss-Newton method.

Kalogerakis and Luus (1983b) compared the computational effort required by Gauss-Newton, simplified quasilinearization and standard quasilinearization methods. They found that all methods produced the same new estimates at each iteration as expected. Furthermore, the required computational time for the Gauss-Newton and the simplified quasilinearization was the same and about 90% of that required by the standard quasilinearization method.

6.6.3 Nonlinear Output Relationship

When the output vector is nonlinearly related to the state vector (Equation 6.3) then substitution of $\mathbf{x}^{(j+1)}$ from Equation 6.74 into the Equation 6.3 followed by substitution of the resulting equation into the objective function (Equation 6.4) yields the following equation after application of the stationary condition (Equation 6.78)

$$\sum_{i=1}^{N}\mathbf{G}^T(t_i)\left(\frac{\partial \mathbf{h}^T(t_i,\mathbf{x}^{(j+1)}(t_i))}{\partial \mathbf{x}^{(j+1)}(t_i)}\right)\mathbf{Q}_i\left[\hat{\mathbf{y}}(t_i)-\mathbf{h}(t_i,\mathbf{x}^{(j+1)}(t_i))\right]=\mathbf{0} \tag{6.84}$$

The above equation represents a set of p nonlinear equations which can be solved to obtain $\mathbf{k}^{(j+1)}$. The solution of this set of equations can be accomplished by two methods. First, by employing Newton's method or alternatively by linearizing the output vector around the trajectory $\mathbf{x}^{(j)}(t)$. Kalogerakis and Luus (1983b) showed that when linearization of the output vector is used, the quasilinearization computational algorithm and the Gauss-Newton method yield the same results.

7

Shortcut Estimation Methods for Ordinary Differential Equation (ODE) Models

Whenever the whole state vector is measured, i.e., when $\mathbf{y}(t_i)=\mathbf{x}(t_i)$, $i=1,..,N$, we can employ approximations of the time-derivatives of the state variables or make use of suitable integrals and thus reduce the parameter estimation problem from an ODE system to one for an algebraic equation system. We shall present two approaches, one approximating time derivatives of the state variables (the *derivative approach*) and other approximating suitable integrals of the state variables (the *integral approach*). In addition, in this chapter we present the methodology for estimating average kinetic rates (e.g., specific growth rates, specific uptake rates, or specific secretion rates) in biological systems operating in the batch, fed-batch, continuous or perfusion mode. These estimates are routinely used by analysts to compare the productivity or growth characteristics among different microorganism populations or cell lines.

7.1 ODE MODELS WITH LINEAR DEPENDENCE ON THE PARAMETERS

Let us consider the special class of problems where all state variables are measured and the parameters enter in a linear fashion into the governing differential equations. As usual, we assume that $\mathbf{x}$ is the *n-dimensional* vector of state variables and $\mathbf{k}$ is the *p-dimensional* vector of unknown parameters. The structure of the ODE model is of the form

$$\left.\begin{aligned}\frac{dx_1}{dt} &= k_1\varphi_{11}(\mathbf{x}) + k_2\varphi_{12}(\mathbf{x}) + \dots + k_p\varphi_{1p}(\mathbf{x})\\ \frac{dx_2}{dt} &= k_1\varphi_{21}(\mathbf{x}) + k_2\varphi_{22}(\mathbf{x}) + \dots + k_p\varphi_{2p}(\mathbf{x})\\ &\vdots\\ \frac{dx_n}{dt} &= k_1\varphi_{n1}(\mathbf{x}) + k_2\varphi_{n2}(\mathbf{x}) + \dots + k_p\varphi_{np}(\mathbf{x})\end{aligned}\right\} \tag{7.1}$$

where $\varphi_{ij}(\mathbf{x})$, $i=1,\dots,n$; $j=1,\dots,p$ are known functions of the state variables only. Quite often these functions are suitable products of the state variables especially when the model is that of a homogeneous reacting system. In a more compact form Equation 7.1 can be written as

$$\frac{d\mathbf{x}}{dt} = \mathbf{\Phi}(\mathbf{x})\mathbf{k} \tag{7.2}$$

where the *n×p dimensional* matrix $\mathbf{\Phi}(\mathbf{x})$ has as elements the functions $\varphi_{ij}(\mathbf{x})$, $i=1,\dots,n$; $j=1,\dots,p$. It is also assumed that the initial condition $\mathbf{x}(t_0)=\mathbf{x}_0$ is known precisely.

7.1.1 Derivative approach

In this case we approximate the time derivatives on the left hand side of Equation 7.1 numerically. In order to minimize the potential numerical errors in the evaluation of the derivatives it is strongly recommended to smooth the data first by polynomial fitting. The order of the polynomial should be the lowest possible that fits the measurements *satisfactorily*.

The time-derivatives can be estimated analytically from the smoothed data as

$$\boldsymbol{\eta}_i = \left[\frac{d\hat{\mathbf{x}}(t)}{dt}\right]_{t=t_i} \tag{7.3}$$

where $\hat{\mathbf{x}}(t)$ is the fitted polynomial. If for example a second or third order polynomial has been used, $\hat{\mathbf{x}}(t)$ will be given respectively by

$$\hat{x}_j(t) = a_{j0} + a_{j1}t + a_{j2}t^2 \quad ; \quad j=1,\dots n \tag{7.4}$$

or

$$\hat{x}_j(t) = a_{j0} + a_{j1}t + a_{j2}t^2 + a_{j3}t^3 \quad ; \quad j=1,\dots n \tag{7.5}$$

The best and easiest way to smooth the data and avoid misuse of the polynomial curve fitting is by employing *smooth cubic splines.* IMSL provides two routines for this purpose: CSSCV and CSSMH. The latter is more versatile as it gives the option to the user to apply different levels of smoothing by controlling a single parameter. Furthermore, IMSL routines CSVAL and CSDER can be used once the coefficients of the cubic spines have been computed by CSSMH to calculate the smoothed values of the state variables and their derivatives respectively.

Having the smoothed values of the state variables at each sampling point and the derivatives, $\boldsymbol{\eta}_i$, we have essentially transformed the problem to a "usual" *linear* regression problem. The parameter vector is obtained by minimizing the following LS objective function

$$S_{LS}(\mathbf{k}) = \sum_{i=1}^{N} [\boldsymbol{\eta}_i - \boldsymbol{\Phi}(\hat{\mathbf{x}}_i)\mathbf{k}]^T \mathbf{Q}_i [\boldsymbol{\eta}_i - \boldsymbol{\Phi}(\hat{\mathbf{x}}_i)\mathbf{k}] \tag{7.6}$$

where $\hat{\mathbf{x}}_i$ is the smoothed value of the measured state variables at $t=t_i$. Any good linear regression package can be used to solve this problem.

However, an important question that needs to be answered is "what constitutes a *satisfactory* polynomial fit?" An answer can come from the following simple reasoning. The purpose of the polynomial fit is to smooth the data, namely, to remove *only* the measurement error (noise) from the data. If the mathematical (ODE) model under consideration is indeed the *true* model (or simply an adequate one) then the calculated values of the output vector based on the ODE model should correspond to the error-free measurements. Obviously, these model-calculated values should ideally be the same as the smoothed data assuming that the correct amount of data-filtering has taken place.

Therefore, in our view polynomial fitting for data smoothing should be an iterative process. First we start with the visually best choice of polynomial order or smooth cubic splines. Subsequently, we estimate the unknown parameters in the ODE model and generate model calculated values of the output vector. Plotting of the raw data, the smoothed data and the ODE model calculated values in the same graph enables visual inspection. If the smoothed data are reasonably close to the model calculated values and the residuals appear to be normal, the polynomial fitting was done correctly. If not, we should go back and redo the polynomial fitting. In addition, we should make sure that the ODE model is adequate. If it is not, the above procedure fails. In this case, the data smoothing should be based simply on the requirement that the differences between the raw data and the smoothed values are normally distributed with zero mean.

Finally, the user should always be aware of the danger in getting numerical estimates of the derivatives from the data. Different smoothing cubic splines or polynomials can result in similar values for the state variables and at the same time have widely different estimates of the derivatives. This problem can be controlled by the previously mentioned iterative procedure if we pay attention not only to the values of the state variables but to their derivatives as well.

7.1.2 Integral Approach

In this case instead of approximating the time derivatives on the left hand side of Equation 7.1, we integrate both sides with respect to time. Namely, integration between t_0 and t_i of Equation 7.1 yields,

$$\left.\begin{aligned}
\int_{t_0}^{t_i} \frac{dx_1}{dt}\,dt &= \int_{t_0}^{t_i} \left[k_1\varphi_{11}(\mathbf{x}) + k_2\varphi_{12}(\mathbf{x}) + \ldots + k_p\varphi_{1p}(\mathbf{x})\right]dt \\
\int_{t_0}^{t_i} \frac{dx_2}{dt}\,dt &= \int_{t_0}^{t_i} \left[k_1\varphi_{21}(\mathbf{x}) + k_2\varphi_{22}(\mathbf{x}) + \ldots + k_p\varphi_{2p}(\mathbf{x})\right]dt \\
&\vdots \\
\int_{t_0}^{t_i} \frac{dx_n}{dt}\,dt &= \int_{t_0}^{t_i} \left[k_1\varphi_{n1}(\mathbf{x}) + k_2\varphi_{n2}(\mathbf{x}) + \ldots + k_p\varphi_{np}(\mathbf{x})\right]dt
\end{aligned}\right\} \tag{7.7}$$

which upon expansion of the integrals yields

$$\left.\begin{aligned}
x_1(t_i) - x_1(t_0) &= k_1\int_{t_0}^{t_i}\varphi_{11}(\mathbf{x})dt + k_2\int_{t_0}^{t_i}\varphi_{12}(\mathbf{x})dt + \ldots + k_p\int_{t_0}^{t_i}\varphi_{1p}(\mathbf{x})dt \\
x_2(t_i) - x_2(t_0) &= k_1\int_{t_0}^{t_i}\varphi_{21}(\mathbf{x})dt + k_2\int_{t_0}^{t_i}\varphi_{22}(\mathbf{x})dt + \ldots + k_p\int_{t_0}^{t_i}\varphi_{2p}(\mathbf{x})dt \\
&\vdots \\
x_n(t_i) - x_n(t_0) &= k_1\int_{t_0}^{t_i}\varphi_{n1}(\mathbf{x})dt + k_2\int_{t_0}^{t_i}\varphi_{n2}(\mathbf{x})dt + \ldots + k_p\int_{t_0}^{t_i}\varphi_{np}(\mathbf{x})dt
\end{aligned}\right\} \tag{7.8}$$

Noting that the initial conditions are known ($\mathbf{x}(t_0)=\mathbf{x}_0$), the above equations can be rewritten as

$$\left.\begin{aligned}
x_1(t_i) - x_1(t_0) &= k_1\Psi_{11}(t_i) + k_2\Psi_{12}(t_i) + \ldots + k_p\Psi_{1p}(t_i) \\
x_2(t_i) - x_2(t_0) &= k_1\Psi_{21}(t_i) + k_2\Psi_{22}(t_i) + \ldots + k_p\Psi_{2p}(t_i) \\
&\vdots \\
x_n(t_i) - x_n(t_0) &= k_1\Psi_{n1}(t_i) + k_2\Psi_{n2}(t_i) + \ldots + k_p\Psi_{np}(t_i)
\end{aligned}\right\} \tag{7.9}$$

or more compactly as

$$\mathbf{x}(t_i) = \mathbf{x}_0 + \boldsymbol{\Psi}(t_i)\mathbf{k} \tag{7.10}$$

where

$$\Psi_{jr}(t_i) = \int_{t_0}^{t_i} \varphi_{jr}(\mathbf{x})dt \quad ; \; j=1,\dots,n \;\&\; r=1,\dots,p \tag{7.11}$$

The above integrals can be calculated since we have measurements of all the state variables as a function of time. In particular, to obtain a good estimate of these integrals it is strongly recommended to smooth the data first using a polynomial fit of a suitable order. Therefore, the integrals $\Psi_{jr}(t_i)$ should be calculated as

$$\Psi_{jr}(t_i) = \int_{t_0}^{t_i} \varphi_{jr}(\hat{\mathbf{x}})dt \quad ; \; j=1,\dots,n \;\&\; r=1,\dots,p \tag{7.12}$$

where $\hat{\mathbf{x}}(t)$ is the fitted polynomial. The same guidelines for the selection of the smoothing polynomials apply.

Having the smoothed values of the state variables at each sampling point, $\hat{\mathbf{x}}_i$ and the integrals, $\boldsymbol{\Psi}(t_i)$, we have essentially transformed the problem to a "usual" *linear* regression problem. The parameter vector is obtained by minimizing the following LS objective function

$$S_{LS}(\mathbf{k}) = \sum_{i=1}^{N} \left[\hat{\mathbf{x}}_i - \mathbf{x}_0 - \boldsymbol{\Psi}(t_i)\mathbf{k}\right]^T \mathbf{Q}_i \left[\hat{\mathbf{x}}_i - \mathbf{x}_0 - \boldsymbol{\Psi}(t_i)\mathbf{k}\right] \tag{7.13}$$

The above linear regression problem can be readily solved using any standard linear regression package.

7.2 GENERALIZATION TO ODE MODELS WITH NONLINEAR DEPENDENCE ON THE PARAMETERS

The derivative approach described previously can be readily extended to ODE models where the unknown parameters enter in a nonlinear fashion. The exact same procedure to obtain good estimates of the time derivatives of the state variables at each sampling point, t_i, can be followed. Thus the governing ODE

$$\frac{d\mathbf{x}(t)}{dt} = \mathbf{f}(\mathbf{x}(t), \mathbf{k}) \tag{7.14}$$

can be written at each sampling point as

$$\boldsymbol{\eta}_i = \mathbf{f}(\mathbf{x}(t_i), \mathbf{k}) \tag{7.15}$$

where again

$$\boldsymbol{\eta}_i = \left[\frac{d\hat{\mathbf{x}}(t)}{dt}\right]_{t=t_i} \tag{7.3}$$

Having the smoothed values of the state variables at each sampling point and having estimated analytically the time derivatives, $\boldsymbol{\eta}_i$, we have transformed the problem to a "usual" nonlinear regression problem for algebraic models. The parameter vector is obtained by minimizing the following LS objective function

$$S_{LS}(\mathbf{k}) = \sum_{i=1}^{N} [\boldsymbol{\eta}_i - \mathbf{f}(\hat{\mathbf{x}}_i, \mathbf{k})]^T \mathbf{Q}_i [\boldsymbol{\eta}_i - \mathbf{f}(\hat{\mathbf{x}}_i, \mathbf{k})] \tag{7.16}$$

where $\hat{\mathbf{x}}_i$ is the smoothed value of the measured state variables at $t=t_i$.

The above parameter estimation problem can now be solved with any estimation method for algebraic models. Again, our preference is to use the Gauss-Newton method as described in Chapter 4.

The only drawback in using this method is that any numerical errors introduced in the estimation of the time derivatives of the state variables have a direct effect on the estimated parameter values. Furthermore, by this approach we can not readily calculate confidence intervals for the unknown parameters. This method is the standard procedure used by the General Algebraic Modeling System (GAMS) for the estimation of parameters in ODE models when all state variables are observed.

7.3 ESTIMATION OF APPARENT RATES IN BIOLOGICAL SYSTEMS

In biochemical engineering we are often faced with the problem of estimating average apparent growth or uptake/secretion rates. Such estimates are particularly useful when we compare the productivity of a culture under different operating conditions or modes of operation. Such computations are routinely done by analysts well before any attempt is made to estimate "true" kinetics parameters like those appearing in the Monod growth model for example.

In this section we shall use the standard notation employed by biochemical engineers and industrial microbiologists in presenting the material. Thus if we denote by X_v the viable cell (*cells/L*) or biomass (*mg/L*) concentration, S the limiting substrate concentration (*mmol/L*) and P the product concentration (*mmol/L*) in the bioreactor, the dynamic component mass balances yield the following ODEs for each mode of operation:

Batch Experiments

$$\frac{dX_v}{dt} = \mu X_v \tag{7.17}$$

$$\frac{dS}{dt} = -q_S X_v \tag{7.18}$$

$$\frac{dP}{dt} = q_p X_v \tag{7.19}$$

where μ is the apparent specific growth rate *(1/h)*, q_p is the apparent specific secretion/production rate, *(mmol/(mg·h·L))* and q_s is the apparent specific uptake rate *(mmol/(mg·h·L))*.

Fed-Batch Experiments

In this case there is a continuous addition of nutrients to the culture by the feed stream. There is no effluent stream. The governing equations are:

$$\frac{dX_v}{dt} = (\mu - D)X_v \tag{7.20}$$

$$\frac{dS}{dt} = -q_S X_v + D(S_f - S) \tag{7.21}$$

$$\frac{dP}{dt} = q_p X_v - DP \tag{7.22}$$

where D is the dilution factor *(1/h)* defined as the feed flowrate over volume (F/V) and Sf (*mmol/L*) is the substrate concentration in the feed. For fed-batch cultures the volume V is increasing continuously until the end of the culture period according to the feeding rate as follows

$$\frac{dV}{dt} = F = DV \tag{7.23}$$

Continuous (Chemostat) Experiments

In this case, there is a continuous supply of nutrients and a continuous withdrawal of the culture broth including the submerged free cells. The governing equations for continuous cultures are the same as the ones for fed-batch cultures (Equations 7.20–7.22). The only difference is that feed flowrate is normally equal to the effluent flowrate (F_{in}=F_{out}=F) and hence the volume, V, stays constant throughout the culture.

Perfusion Experiments

Perfusion cultures of submerged free cells are essentially continuous cultures with a cell retention device so that no cells exit in the effluent stream. The governing ODEs are

$$\frac{dX_v}{dt} = \mu X_v \tag{7.24}$$

$$\frac{dS}{dt} = -q_s X_v + D(S_f - S) \tag{7.25}$$

$$\frac{dP}{dt} = q_p X_v - DP \tag{7.26}$$

Formulation of the problem

Having measurements of X_v, S and P over time, determine the apparent specific growth rate (μ), the apparent specific uptake rate (q_s) and the apparent specific secretion rate (q_p) at any particular point during the culture or obtain an average value over a user-specified time period.

7.3.1 Derivative Approach

This approach is based on the rearrangement of the governing ODEs to yield expressions of the different specific rates, namely,

$$\mu = \begin{cases} \dfrac{1}{X_v(t)}\dfrac{dX_v(t)}{dt} & \textit{batch \& perfusion} \\[2ex] D + \dfrac{1}{X_v(t)}\dfrac{dX_v(t)}{dt} & \textit{fedbatch \& continuous} \end{cases} \tag{7.27}$$

which can be reduced to

$$\mu = \begin{cases} \dfrac{d\,ln\{X_v(t)\}}{dt} & \textit{batch \& perfusion} \\[2ex] D + \dfrac{d\,ln\{X_v(t)\}}{dt} & \textit{fedbatch \& continuous} \end{cases} \tag{7.28}$$

The specific uptake and secretion rates are given by

$$q_S = \begin{cases} \dfrac{-1}{X_v(t)}\dfrac{dS(t)}{dt} & \textit{batch} \\[2ex] \dfrac{-1}{X_v(t)}\left[\dfrac{dS(t)}{dt} - D[S_f - S(t)]\right] & \textit{all other} \end{cases} \tag{7.29}$$

$$q_P = \begin{cases} \dfrac{1}{X_v(t)}\dfrac{dP(t)}{dt} & \textit{batch} \\[2ex] \dfrac{1}{X_v(t)}\left[\dfrac{dP(t)}{dt} + DP(t)\right] & \textit{all other} \end{cases} \tag{7.30}$$

Based on the above expressions, it is obvious that by this approach one needs to estimate numerically the time derivatives of X_v, S and P as a function of time. This can be accomplished by following the same steps as described in Section 7.1.1. Namely, we must first smooth the data by a polynomial fitting and then estimate the derivatives. However, cell culture data are often very noisy and hence, the numerical estimation of time derivatives may be subject to large estimation errors.

7.3.2 Integral Approach

A much better approach for the estimation of specific rates in biological systems is through the use of the integral method. Let us first start with the analysis of data from batch experiments.

Batch Experiments

Integration of the ODE for biomass from t_0 to t_i yields

$$\ln X_v(t_i) - \ln X_v(t_0) = \int_{t_0}^{t_i} \mu \, dt \tag{7.31}$$

Similarly the ODEs for S and P yield,

$$S(t_i) - S(t_0) = - \int_{t_0}^{t_i} q_s X_v(t) dt \tag{7.32}$$

and

$$P(t_i) - P(t_0) = \int_{t_0}^{t_i} q_p X_v(t) dt \tag{7.33}$$

Let us assume at this point that the specific rates are constant in the interval $[t_0,t_i]$. Under this assumption the above equations become,

$$\ln X_v(t_i) = \mu[t_i - t_0] + \ln X_v(t_0) \tag{7.34}$$

$$S(t_i) = -q_s \int_{t_0}^{t_i} X_v(t) dt + S(t_0) \tag{7.35}$$

$$P(t_i) = q_p \int_{t_0}^{t_i} X_v(t) dt + P(t_0) \tag{7.36}$$

Normally a series of measurements, $X_v(t_i)$, $S(t_i)$ and $P(t_i)$, $i=1,\ldots,N$, are available. Equation 7.34 suggests that the specific growth rate (μ) can be obtained as the slope in a plot of $\ln X_v(t_i)$ versus t_i. Equation 7.35 suggests that the specific substrate uptake rate (q_s) can also be obtained as the negative slope in a plot of $S(t_i)$ versus $\int_{t_0}^{t_i} X_v(t) dt$. Similarly, Equation 7.36 suggests that the specific secretion rate (q_p) can be obtained as the slope in a plot of $P(t_i)$ versus $\int_{t_0}^{t_i} X_v(t) dt$.

By constructing a plot of $S(t_i)$ versus $\int X_v dt$, we can *visually* identify distinct time periods during the culture where the specific uptake rate (q_s) is "constant" and estimates of q_s are to be determined. Thus, by using the linear least squares estimation capabilities of any spreadsheet calculation program, we can readily estimate the specific uptake rate over any user-specified time period. The estimated specific uptake rate is essentially its average value of the estimation period. This average value is particularly useful to the analyst who can quickly compare the metabolic characteristics of different microbial populations or cell lines.

Similarly we can estimate the specific secretion rate. It is obvious from the previous analysis that an accurate estimation of the average specific rates can only be done if the integral $\int X_v dt$ is estimated accurately. If measurements of biomass or cell concentrations have been taken very frequently, simple use of the trapezoid rule for the computation of $\int X_v dt$ may suffice. If however the measurements are very noisy or they have been infrequently collected, the data must be first smoothed through polynomial fitting and then the integrals can be obtained analytically using the fitted polynomial.

Estimation Procedure for Batch Experiments

(i) Import measurements into a spreadsheet and plot the data versus time to spot any gross errors or outliers.

(ii) Smooth the data $\{X_v(t_i)$, $S(t_i)$ and $P(t_i)$, $i=1,\ldots,N\}$ by performing a polynomial fit to each variable.

(iii) Compute the required integrals $\int_{t_0}^{t_i} X_v(t)dt$, $i=1,\ldots,N$ using the fitted polynomial for $X_v(t_i)$.

(iv) Plot $\ln X_v(t_i)$ versus t_i.

(v) Plot $S(t_i)$ and $P(t_i)$ versus $\int_{t_0}^{t_i} X_v(t)dt$.

(vi) Identify the segments of the plotted curves where the specific growth, uptake and secretion rates are to be estimated (e.g., exclude the lag phase, concentrate on the exponential growth phase or on the stationary phase or on the death phase of the batch culture).

(vi) Using the linear least-squares function of any spreadsheet program compute the specific uptake/secretion rates within the desired time periods as the slope of the corresponding regression lines.

It is noted that the initial time (t_0) where the computation of all integrals begun, does not affect the determination of the slope at any later time segment. It affects only the estimation of the constant in the linear model.

The major disadvantage of the integral method is the difficulty in computing an estimate of the standard error in the estimation of the specific rates. Obviously, all linear least squares estimation routines provide automatically the standard error of estimate and other statistical information. However, the computed statistics are based on the assumption that there is no error present in the independent variable.

This assumption can be relaxed when the experimental error in the independent variable is much smaller compared to the error present in the measurements of the dependent variable. In our case the assumption of simple linear least squares implies that $\int X_v dt$ is known precisely. Although we do know that there are errors in the measurement of X_v, the polynomial fitting and the subsequent integration provides a certain amount of data filtering which could allows us to assume that experimental error in $\int X_v dt$ is negligible compared to that present in $S(t_i)$ or $P(t_i)$.

Nonetheless, the value for the standard error computed by the linear least square computations reveal an estimate of the uncertainty which is valid for comparison purposes among different experiments.

Fed-Batch & Continuous (Chemostat) Experiments

The integral approach presented for the analysis of batch cultures can be readily extended to fed-batch and continuous cultures. Integration of the ODE for biomass from t_0 to t_i yields

$$ln X_v(t_i) - ln X_v(t_0) = \int_{t_0}^{t_i} [\mu - D] dt \tag{7.37}$$

Similarly the ODEs for S and P yield,

$$S(t_i) - S(t_0) = -\int_{t_0}^{t_i} q_s X_v(t) dt + \int_{t_0}^{t_i} D[S_f - S(t)] dt \tag{7.38}$$

and

$$P(t_i) - P(t_0) = \int_{t_0}^{t_i} q_p X_v(t) dt - \int_{t_0}^{t_i} DP(t) dt \tag{7.39}$$

In this case we assume that we know the dilution rate (D=F/V) precisely as a function of time. In a chemostat D is often constant since the feed flowrate and the volume are kept constant. In a fed-batch culture the volume is continuously increasing. The dilution rate generally varies with respect to time although it could also be kept constant if the operator provides an exponentially varying feeding rate.

Again assuming that the specific rates remain constant in the period $[t_0, t_i]$, the above equations become,

$$ln X_v(t_i) + \int_{t_0}^{t_i} D dt = \mu[t_i - t_0] + ln X_v(t_0) \tag{7.40}$$

Similarly the ODEs for S and P yield,

$$S(t_i) - \int_{t_0}^{t_i} D[S_f - S(t)]dt = -q_s \int_{t_0}^{t_i} X_v(t)dt + S(t_0) \tag{7.41}$$

and

$$P(t_i) + \int_{t_0}^{t_i} DP(t)dt = q_p \int_{t_0}^{t_i} X_v(t)dt + P(t_0) \tag{7.42}$$

Equation 7.40 suggests that the specific growth rate (μ) can be obtained as the slope in a plot of $\{lnX_v(t_i) + \int Ddt\}$ versus t_i. Equation 7.41 suggests that the specific substrate uptake rate (q_s) can also be obtained as the negative slope in a plot of $\{ S(t_i) - \int_0^{t_i} D[S_f - S(t)]dt \}$ versus $\int_0^{t_i} X_v(t)dt$. Similarly, Equation 7.42 suggests that the specific secretion rate (q_p) can be obtained as the slope in a plot of $\{ P(t_i) + \int_0^{t_i} DP(t)dt \}$ versus $\int_0^{t_i} X_v(t)dt$.

By constructing the above plots, we can *visually* identify distinct time periods during the culture where the specific rates are to be estimated. If the dilution rate (D) is taken equal to zero, the usual estimation equations for batch cultures are obtained. This leads to the interpretation of the left hand side of Equations 7.41 and 7.42 as "effective batch culture concentrations" of the substrate (S) and the product (P). The computational procedure is given next.

Estimation Procedure for Fed-Batch & Continuous (Chemostat) Experiments

(i) Import measurements (X_v, S and P) and user-specified variables (such as D and S_f as a function of time) into a spreadsheet and plot the data versus time to spot any gross errors or outliers.

(ii) Smooth the data $\{X_v(t_i)$, $S(t_i)$ and $P(t_i)$, i=1,...,N$\}$ by performing a polynomial fit to each variable.

(iii) Compute the required integrals $\int_0^{t_i} X_v(t)dt$, i=1,...,N using the fitted polynomial for $X_v(t_i)$.

(iv) Compute also the integrals $\int_0^{t_i} Ddt$, $\int_0^{t_i} D[S_f - S(t)]dt$ and $\int_0^{t_i} DP(t)dt$, i=1,...,N using the fitted polynomial for $S(t_i)$, $P(t_i)$ and the user-specified feeding variables D and S_f.

(v) Plot $\{lnX_v(t_i) + \int_{t_0}^{t_i} D dt\}$ versus t_i.

(vi) Plot $\{S(t_i) - \int_{t_0}^{t_i} D[S_f - S(t)]dt\}$ and $\{P(t_i) + \int_{t_0}^{t_i} DP(t)dt\}$ versus $\int_{t_0}^{t_i} X_v(t)dt$.

(vii) Identify the segments of the plotted curves where the specific growth, uptake and secretion rates are to be estimated (e.g., exclude any lag phase or concentrate on two different segments of the culture to compare the productivity or growth characteristics, etc.).

(viii) Using the linear least-squares function of any spreadsheet program compute the specific uptake/secretion rates within the desired time periods as the slope of the corresponding regression lines.

The main advantage of the above estimation procedure is that there is no need to assume steady-state operation. Since practice has shown that steady state operation is not easily established for prolonged periods of time, this approach enables the determination of average specific rates taking into account accumulation terms.

Perfusion Experiments

The procedure for the estimation of q_s and q_p is identical to the one presented for fed-batch and continuous cultures. The only difference is in the estimation of the specific growth rate (μ). Since perfusion cultures behave as batch cultures as far as the biomass is concerned, μ can be obtained as described earlier for batch systems. Namely μ is obtained as the slope in the plot of $lnX_v(t_i)$ versus t_i.

The above procedure can be easily modified to account for the case of a small "bleeding rate", i.e., when there is a known (i.e., measured) small withdrawal of cells (much less than what it could be in a chemostat).

We finish this section with a final note on the estimation of the specific death rate (k_d). The procedures presented earlier (for batch, fed-batch, continuous and perfusion cultures) provide an estimate of the apparent growth rate (μ_a) rather than the true specific growth rate (μ). The true growth rate can only be obtained as (μ_a+k_d) where k_d is the specific death rate.

The specific death rate can be obtained by considering the corresponding mass balance for nonviable (dead) cells. Normally nonviable cell concentration is measured at the same time the measurement for viable cell concentration is made. If viability (ζ) data are available, the nonviable cell concentration can be obtained from the viable one as $X_d = X_v(1-\zeta)/\zeta$.

The dynamic mass balance for nonviable cells in a batch or perfusion culture yields,

$$\frac{dX_d}{dt} = k_d X_v \tag{7.43}$$

This leads to the following integral equation for k_d

$$X_d(t_i) = k_d \int_{t_0}^{t_i} X_v(t)dt + X_d(t_0) \tag{7.44}$$

Hence, in a batch or perfusion culture, k_d can be obtained as the slope in a plot of $X_d(t_i)$ versus $\int_{t_0}^{t_i} X_v(t)dt$.

The dynamic mass balance for nonviable cells in a fed-batch or continuous culture is

$$\frac{dX_d}{dt} = k_d X_v - DX_d \tag{7.45}$$

The above equation yields to the following integral equation for k_d,

$$X_d(t_i) + \int_{t_0}^{t_i} DX_d(t)dt = k_d \int_{t_0}^{t_i} X_v(t)dt + X_d(t_0) \tag{7.46}$$

Thus, in a fed-batch or continuous culture, k_d can be obtained as the slope in a plot of $\{X_d(t_i) + \int_{t_0}^{t_i} DX_d(t)dt\}$ versus $\int_{t_0}^{t_i} X_v(t)dt$.

7.4 EXAMPLES

In Chapter 17, where there is a special section with biochemical engineering problems, examples on shortcut methods are presented.

7.4.1 Derivative Approach - Pyrolytic Dehydrogenation of Benzene

In this section we shall only present the derivative approach for the solution of the pyrolytic dehydrogenation of benzene to diphenyl and triphenyl regression problem. This problem, which was already presented in Chapter 6, is also used here to illustrate the use of shortcut methods. As discussed earlier, both state variables are measured and the two unknown parameters appear linearly in the governing ODEs which are also given below for ease of the reader.

$$\frac{dx_1}{dt} = -r_1 - r_2 \tag{7.47a}$$

$$\frac{dx_2}{dt} = \frac{r_1}{2} - r_2 \tag{7.47b}$$

where

$$r_1 = k_1\left[x_1^2 - x_2(2-2x_1-x_2)/3K_1\right] \tag{7.48a}$$

$$r_2 = k_2\left[x_1x_2 - (1-x_1-2x_2)(2-2x_1-x_2)/9K_2\right] \tag{7.48b}$$

Here x_1 denotes *lb-moles* of benzene per *lb-mole* of pure benzene feed and x_2 denotes *lb-moles* of diphenyl per *lb-mole* of pure benzene feed. The parameters k_1 and k_2 are unknown reaction rate constants whereas K_1 and K_2 are known equilibrium constants. The data consist of measurements of x_1 and x_2 in a flow reactor at eight values of the reciprocal space velocity t. The feed to the reactor was pure benzene. The experimental data are given in Table 6.2 (in Chapter 6). The governing ODEs can also be written as

$$\frac{dx_1}{dt} = k_1\varphi_{11}(\mathbf{x}) + k_2\varphi_{12}(\mathbf{x}) \tag{7.49a}$$

$$\frac{dx_2}{dt} = k_1\varphi_{21}(\mathbf{x}) + k_2\varphi_{22}(\mathbf{x}) \tag{7.49b}$$

where

$$\varphi_{11}(\mathbf{x}) = -\left[x_1^2 - x_2(2-2x_1-x_2)/3K_1\right] \tag{7.50a}$$

$$\varphi_{12}(\mathbf{x}) = -\left[x_1x_2 - (1-x_1-2x_2)(2-2x_1-x_2)/9K_2\right] \tag{7.50b}$$

$$\varphi_{21}(\mathbf{x}) = -\varphi_{11}(\mathbf{x})/2 \tag{7.50c}$$

$$\varphi_{22}(\mathbf{x}) = \varphi_{12}(\mathbf{x}) \tag{7.50d}$$

As we mentioned, the first and probably most crucial step is the computation of the time derivatives of the state variables from smoothed data. The best and easiest way to smooth the data is using smooth cubic splines using the IMSL routines CSSMH, CSVAL & CSDER. The latter two are used once the cubic splines coefficients and break points have been computed by CSSMH to generate the values of the smoothed measurements and their derivatives (η_1 and η_2).

Table 7.1: Estimated parameter values with short cut methods for different values of the smoothing parameter (s/N) in IMSL routine CSSMH

Smoothing Parameter*	k_1	k_2
0.01	329.38	415.88
0.1	319.01	403.68
1	293.32	368.21
Estimates based on Gauss-Newton method for ODEs	354.61	400.23

* Used in IMSL routine CSSMH (s/N).

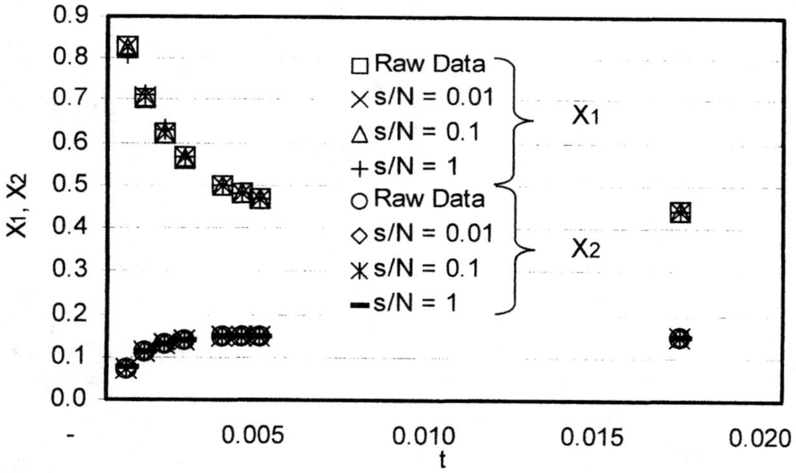

Figure 7.1: Smoothed data for variables x_1 and x_2 using a smooth cubic spline approximation (s/N=0.01, 0.1 and 1).

Once we have the smoothed values of the state variables, we can proceed and compute φ_{11}, φ_{12}, φ_{21} and φ_{22}. All these computed quantities (η_1, η_2, φ_{11}, φ_{12}, φ_{21} and φ_{22}) constitute the data set for the linear least squares regression. In Figure 7.1 the original data and their smoothed values are shown for 3 different values of the smoothing parameter "s" required by CSSMH. An one percent (1%) standard error in the measurements has been assumed and used as weights in data smoothing by CSSMH. IMSL recommends this smoothing parameter to be in the range $[N-(2N)^{0.5}, N+(2N)^{0.5}]$. For this problem this means $0.5 \le s/N \le 1.5$. As seen in Figure 7.1 the smoothed data do not differ significantly for s/N equal to 0.01, 0.1 and 1. However, this is not the case with the estimated derivatives of the state variables. This is clearly seen in Figure 7.2 where the estimated values of $-dx_2/dt$ are quite different in the beginning of the transient.

Subsequent use of linear regression yields the two unknown parameters k_1 and k_2. The results are shown in Table 7.1 for the three different values of the smoothing parameter.

As seen, there is significant variability in the estimates. This is the reason why we should avoid using this technique if possible (unless we wish to generate initial guesses for the Gauss-Newton method for ODE systems). As it was mentioned earlier, the numerical computation of derivatives from noisy data is a risky business!

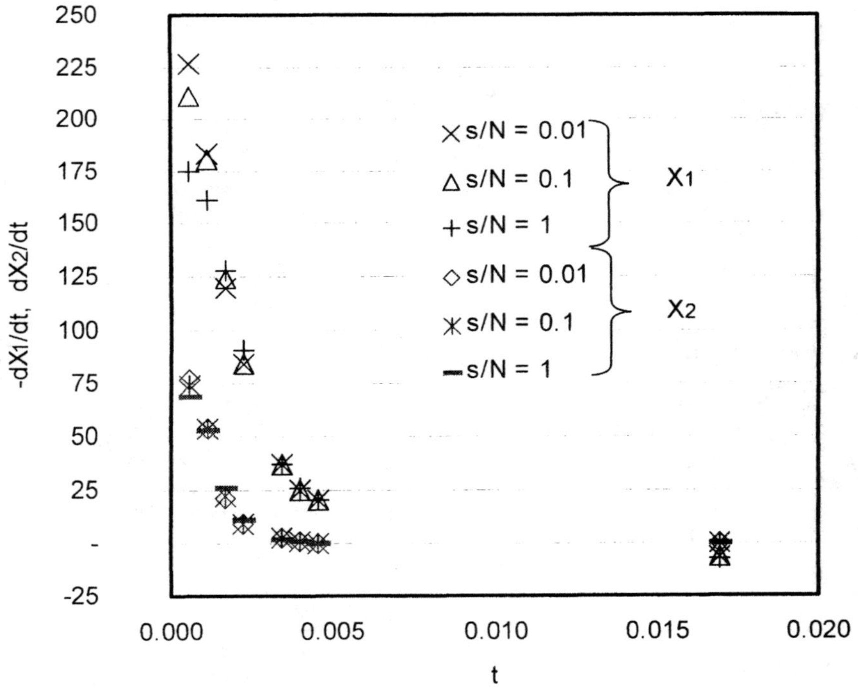

Figure 7.2: Computed time derivatives of x_1 and x_2 using smooth cubic splines for three different values of the smoothing parameter (s/N=0.01, 0.1 and 1).

8

Practical Guidelines for Algorithm Implementation

Besides the basic estimation algorithm, one should be aware of several details that should be taken care of for a successful and meaningful estimation of unknown parameters. Items such as quality of the data at hand (detection of outliers), generation of initial guesses, overstepping and potential remedies to overcome possible ill-conditioning of the problem are discussed in the following sections. Particular emphasis is given to implementation guidelines for differential equation models. Finally, we touch upon the problem of autocorrelation.

8.1 INSPECTION OF THE DATA

One of the very first things that the analyst should do prior to attempting to fit a model to the data, is to inspect the data at hand. *Visual inspection* is a very powerful tool as the eye can often pick up an inconsistent behavior. The primary goal of this data-inspection is to spot potential *outliers*.

Barnett et al. (1994) define outliers as *the observations in a sample which appear to be inconsistent with the remainder of the sample*. In engineering applications, an outlier is often the result of gross measurement error. This includes a mistake in the calculations or in data coding or even a copying error. An outlier could also be the result of inherent variability of the process although *chances are it is not!*

Besides visual inspection, there are several statistical procedures for detecting outliers (Barnett and Lewis, 1978). These tests are based on the examination of the residuals. Practically this means that if a residual is bigger than 3 or 4 standard deviations, it should be considered a potential outlier since the probability of occurrence of such a data point (due to the inherent variability of the process only) is less than 0.1%.

A potential outlier should always be examined carefully. First we check if there was a simple mistake during the recording or the early manipulation of the data. If this is not the case, we proceed to a careful examination of the experimental circumstances during the particular experiment. We should be very careful not to discard ("reject") an outlier unless we have strong *non-statistical* reasons for doing so. In the worst case, we should report both results, one with the outlier and the other without it.

Instead of a detailed presentation of the effect of extreme values and outliers on least squares estimation, the following common sense approach is recommended in the analysis of engineering data:

(i) Start with careful visual observation of the data provided in tabular and graphical form.
(ii) Identify "first" set of potential outliers.
(iii) Examine whether they should be discarded due to non-statistical reasons. Eliminate those from the data set.
(iv) Estimate the parameters and obtain the values of response variables and calculate the residuals.
(v) Plot the residuals and examine whether there any data point lie beyond 3 standard deviations around the mean response estimated by the mathematical model.
(vi) Having identified the second set of potential outliers, examine whether they should be discarded due to non-statistical reasons and eliminate them from the data set.
(vii) For all remaining "outliers" examine their effect on the model response and the estimated parameter values. This is done by performing the parameter estimation calculations twice, once with the outlier included in the data set and once without it.
(viii) Determine which of the outliers are highly informative compared to the rest data points. The remaining outliers should have little on no effect on the model parameters or the mean estimated response of the model and hence, they can be ignored.
(ix) Prepare a final report where the effect of highly informative outliers is clearly presented.
(x) Perform replicate experiments at the experimental conditions where the outliers were detected, if of course, time, money and circumstances allow it.

In summary, the possibility of detecting and correcting an outlier in the data is of paramount importance particularly if it is spotted by an early inspection.

Therefore, careful inspection of the data is highly advocated prior to using any parameter estimation software.

8.2 GENERATION OF INITIAL GUESSES

One of the most important tasks prior to employing an iterative parameter estimation method for the solution of the nonlinear least squares problem is the generation of initial guesses (starting values) for the unknown parameters. A good initial guess facilitates quick convergence of any iterative method and particularly of the Gauss-Newton method to the optimum parameter values. There are several approaches that can be followed to arrive at reasonably good starting values for the parameters and they are briefly discussed below.

8.2.1 Nature and Structure of the Model

The nature of the mathematical model that describes a physical system may dictate a range of acceptable values for the unknown parameters. Furthermore, repeated computations of the response variables for various values of the parameters and subsequent plotting of the results provides valuable experience to the analyst about the behavior of the model and its dependency on the parameters. As a result of this exercise, we often come up with fairly good initial guesses for the parameters. The only disadvantage of this approach is that it could be time consuming. This counterbalanced by the fact that one learns a lot about the structure and behavior of the model at hand.

8.2.2 Asymptotic Behavior of the Model Equations

Quite often the asymptotic behavior of the model can aid us in determining sufficiently good initial guesses. For example, let us consider the Michaelis-Menten kinetics for enzyme catalyzed reactions,

$$y_i = \frac{k_1 x_i}{k_2 + x_i} \tag{8.1}$$

When x_i tends to zero, the model equation reduces to

$$y_i \approx \frac{k_1}{k_2} x_i \tag{8.2}$$

and hence, from the very first data points at small values of x_i, we can obtain k_1/k_2 as the slope of the straight line $y_i = \beta x_i$. Furthermore, as x_i tends to infinity, the model equation reduces to

$$y_i \approx k_1 \quad (8.3)$$

and k_1 is obtained as the asymptotic value of y_i at large values of x_i.

Let us also consider the following exponential decay model, often encountered in analyzing environmental samples,

$$y_i = k_1 + k_2\, exp(-k_3 x_i) \quad (8.4)$$

At large values of x_i, (i.e., as $x_i \rightarrow \infty$) the model reduces to

$$y_i \approx k_1 \quad (8.5)$$

whereas at values of x_i, near zero (i.e., as $x_i \rightarrow 0$) the model reduces to

$$y_i = k_1 + k_2(1 - k_3 x_i) \quad (8.6a)$$

or

$$y_i = k_1 + k_2 - k_2 k_3 x_i \quad (8.6b)$$

which is of the form $y_i = \beta_0 + \beta_1 x_i$ and hence, by performing a linear regression with the values of x_i near zero we obtain estimates of $k_1 + k_2$ and $k_2 k_3$. Combined with our estimate of k_1 we obtain starting values for all the unknown parameters.

In some cases we may not be able to obtain estimates of all the parameters by examining the asymptotic behavior of the model. However, it can still be used to obtain estimates of some of the parameters or even establish an approximate relationship between some of the parameters as functions of the rest.

8.2.3 Transformation of the Model Equations

A suitable transformation of the model equations can simplify the structure of the model considerably and thus, initial guess generation becomes a trivial task. The most interesting case which is also often encountered in engineering applications, is that of *transformably linear models*. These are nonlinear models that reduce to simple linear models after a suitable transformation is performed. These models have been extensively used in engineering particularly before the wide availability of computers so that the parameters could easily be obtained with linear least squares estimation. Even today such models are also used to reveal characteristics of the behavior of the model in a graphical form.

If we have a transformably linear system, our chances are outstanding to get quickly very good initial guesses for the parameters. Let us consider a few typical examples.

The integrated form of substrate utilization in an enzyme catalyzed batch bioreactor is given by the implicit equation

$$k_1 \left(x_i - x_0\right) = y_0 y_i + k_2 \, ln\left(\frac{y_0}{y_i}\right) \tag{8.7}$$

where x_i is time and y_i is the substrate concentration. The initial conditions (x_0, y_0) are assumed to be known precisely. The above model is implicit in y_i and we should use the implicit formulation discussed in Chapter 2 to solve it by the Gauss-Newton method. Initial guesses for the two parameters can be obtained by noticing that this model is transformably linear since Equation 8.7 can be written as

$$\frac{y_0 - y_i}{x_i - x_0} = k_1 - k_2\left(\frac{1}{x_i - x_0}\right) ln\left(\frac{y_0}{y_i}\right) \tag{8.8}$$

which is of the form $Y_i = \beta_0 + \beta_1 X_i$ where

$$Y_i = \frac{y_0 - y_i}{x_i - x_0} \tag{8.9a}$$

and

$$X_i = \left(\frac{1}{x_i - x_0}\right) ln\left(\frac{y_0}{y_i}\right) \tag{8.9b}$$

Initial estimates for the parameters can be readily obtained using linear least squares estimation with the transformed model.

The famous Michaelis-Menten kinetics expression shown below

$$y_i = \frac{k_1 x_i}{k_2 + x_i} \tag{8.10}$$

can become linear by the following transformations also known as

Lineweaver-Burk plot: $$\left(\frac{1}{y_i}\right) = \frac{1}{k_1} + \frac{k_2}{k_1}\left(\frac{1}{x_i}\right) \tag{8.11}$$

Eadie-Hofstee plot:
$$y_i = k_1 - k_2\left(\frac{y_i}{x_i}\right) \tag{8.12}$$

and *Hanes* plot:
$$\left(\frac{x_i}{y_i}\right) = \frac{1}{k_1}x_i + \frac{k_2}{k_1} \tag{8.13}$$

All the above transformations can readily produce initial parameter estimates for the kinetic parameters k_1 and k_2 by performing a simple linear regression. Another class of models that are often transformably linear arise in heterogeneous catalysis. For example, the rate of dehydrogeneration of ethanol into acetaldehyde over a Cu-Co catalyst is given by the following expression assuming that the reaction on two adjacent sites is the rate controlling step (Franckaerts and Froment, 1964; Froment and Bischoff, 1990).

$$y_i = \frac{k_5\left(x_{1i} - \dfrac{x_{2i}x_{3i}}{K_{eq}}\right)}{\left(1 + k_1x_{1i} + k_2x_{2i} + k_3x_{3i} + k_4x_{4i}\right)^2} \tag{8.14}$$

where y_i is the measured overall reaction rate and x_1, x_2, x_3 and x_4 are the partial pressures of the chemical species. Good initial guesses for the unknown parameters, $k_1,\ldots,k_5$, can be obtained by linear least squares estimation of the transformed equation,

$$\sqrt{\frac{x_{1i} - \dfrac{x_{2i}x_{3i}}{K_{eq}}}{y_i}} = \frac{1}{\sqrt{k_5}} + \frac{k_1}{\sqrt{k_5}}x_{1i} + \frac{k_2}{\sqrt{k_5}}x_{2i} + \frac{k_3}{\sqrt{k_5}}x_{3i} + \frac{k_4}{\sqrt{k_5}}x_{4i} \tag{8.15}$$

which is of the form $Y = \beta_0+\beta_1x_1+\beta_2x_2+\beta_3x_3+\beta_4x_4$.

8.2.4 Conditionally Linear Systems

In engineering we often encounter conditionally linear systems. These were defined in Chapter 2 and it was indicated that special algorithms can be used which exploit their conditional linearity (see Bates and Watts, 1988). In general, we need to provide initial guesses only for the nonlinear parameters since the conditionally linear parameters can be obtained through linear least squares estimation.

For example let us consider the three-parameter model,

$$y_i = k_1 + k_2\,exp(-k_3 x_i) \tag{8.16}$$

where both k_1 and k_2 are the conditionally linear parameters. These can be readily estimated by linear least squares once the value of k_3 is known. Conditionally linear parameters arise naturally in reaction rates with an Arrhenius temperature dependence.

8.2.5 Direct Search Approach

If we have very little information about the parameters, direct search methods, like the LJ optimization technique presented in Chapter 5, present an excellent way to generate very good initial estimates for the Gauss-Newton method. Actually, for algebraic equation models, direct search methods can be used to determine the optimum parameter estimates quite efficiently. However, if estimates of the uncertainty in the parameters are required, use of the Gauss-Newton method is strongly recommended, even if it is only for a couple of iterations.

8.3 OVERSTEPPING

Quite often the direction determined by the Gauss-Newton method, or any other gradient method for that matter, is towards the optimum, however, the length of the suggested increment of the parameters could be too large. As a result, the value of the objective function at the new parameter estimates could actually be higher than its value at the previous iteration.

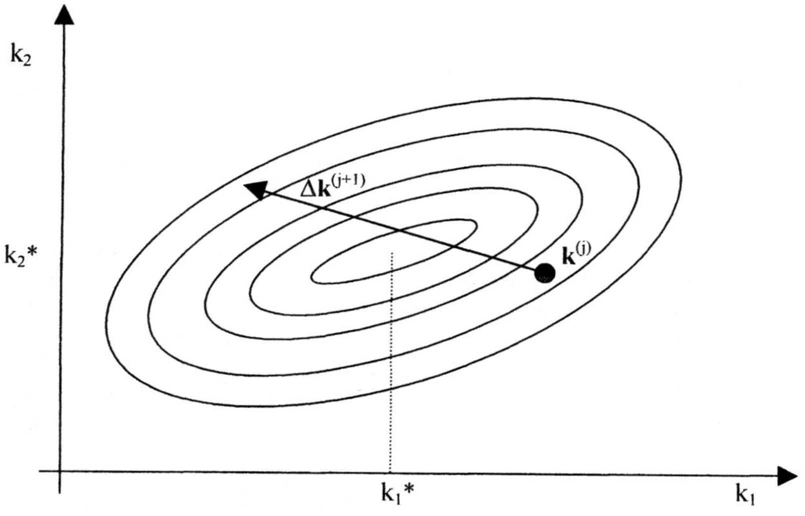

Figure 8.1 Contours of the objective function in the vicinity of the optimum. Potential problems with overstepping are shown for a two-parameter problem.

The classical solution to this problem is by limiting the step-length through the introduction of a stepping parameter, μ $(0<\mu\leq 1)$, namely

$$\mathbf{k}^{(j+1)} = \mathbf{k}^{(j)} + \mu\Delta\mathbf{k}^{(j+1)} \tag{8.17}$$

The easiest way to arrive at an acceptable value of μ, μ_a, is by employing the bisection rule as previously described. Namely, we start with $\mu=1$ and we keep on halving μ until the objective function at the new parameter values becomes less than that obtained in the previous iteration, i.e., we reduce μ until

$$S(\mathbf{k}^{(j)}+\mu_a\Delta\mathbf{k}^{(j+1)}) < S(\mathbf{k}^{(j)}). \tag{8.18}$$

Normally, we stop the step-size determination here and we proceed to perform another iteration of Gauss-Newton method. This is what has been implemented in the computer programs accompanying this book.

8.3.1 An Optimal Step-Size Policy

Once an acceptable value for the step-size has been determined, we can continue and with only one additional evaluation of the objective function, we can obtain the optimal step-size that should be used along the direction suggested by the Gauss-Newton method.

Essentially we need to perform a simple line search along the direction of $\Delta\mathbf{k}^{(j+1)}$. The simplest way to do this is by approximating the objective function by a quadratic along this direction. Namely,

$$S(\mu) \equiv S(\mathbf{k}^{(j)}+\mu\Delta\mathbf{k}^{(j+1)}) = \beta_0 + \beta_1\mu + \beta_2\mu^2 \tag{8.19}$$

The first coefficient is readily obtained at $\mu=0$ as

$$\beta_0 = S_0(\mathbf{k}^{(j)}) \tag{8.20}$$

where $S_0 = S(\mathbf{k}^{(j)})$ which has already been calculated. The other two coefficients, β_1 and β_2, can be computed from information already available at $\mu=\mu_a$ and with an additional evaluation at $\mu=\mu_a/2$. If we denote by S_1 and S_2 the values of the objective function at $\mu=\mu_a/2$ and $\mu=\mu_a$ respectively, we have

$$S_1 = S_0 + \beta_1/2\ \mu_a + \beta_2/4\ \mu_a^2 \tag{8.21a}$$

and

$$S_2 = S_0 + \beta_1\mu_a + \beta_2\mu_a^2 \tag{8.21b}$$

Solution of the above two equations yields,

$$\beta_1 = \frac{4S_1 - 3S_0 - S_2}{\mu_a} \tag{8.22a}$$

and

$$\beta_2 = 2\frac{S_0 + S_2 - 2S_1}{\mu_a^2} \tag{8.22b}$$

Having estimated β_1 and β_2, we can proceed and obtain the optimum step-size by using the stationary criterion

$$\frac{\partial S(\mathbf{k}^{(j)} + \mu\Delta\mathbf{k}^{(j+1)})}{\partial\mu} = 0 \tag{8.23}$$

which yields

$$\mu_{opt} = -\frac{\beta_1}{2\beta_2} \tag{8.24}$$

Substitution of the expressions for β_1 and β_2 yields

$$\mu_{opt} = \frac{\mu_a(4S_1 - 3S_0 - S_2)}{2(2S_1 - S_0 - S_2)} \tag{8.25}$$

The above expression for the optimal step-size is used in the calculation of the next estimate of the parameters to be used in the next iteration of the Gauss-Newton method,

$$\mathbf{k}^{(j+1)} = \mathbf{k}^{(j)} + \mu_{opt}\Delta\mathbf{k}^{(j+1)} \tag{8.26}$$

If we wish to avoid the additional objective function evaluation at $\mu=\mu_a/2$, we can use the extra information that is available at $\mu=0$. This approach is preferable for differential equation models where evaluation of the objective function requires the integration of the state equations. It is presented later in Section 8.7 where we discuss the implementation of Gauss-Newton method for ODE models.

8.4 ILL-CONDITIONING OF MATRIX A AND PARTIAL REMEDIES

If two or more of the unknown parameters are highly correlated, or one of the parameters does not have a measurable effect on the response variables, matrix **A** may become singular or near-singular. In such a case we have a so called *ill-posed* problem and matrix **A** is *ill-conditioned.*

A measure of the degree of ill-conditioning of a nonsingular square matrix is through the *condition number* which is defined as

$$cond(\mathbf{A}) = \|\mathbf{A}\| \, \|\mathbf{A}^{-1}\| \tag{8.27}$$

The condition number is always greater than one and it represents the maximum amplification of the errors in the right hand side in the solution vector. The condition number is also equal to the square root of the ratio of the largest to the smallest singular value of **A**. In parameter estimation applications, **A** is a positive definite *symmetric* matrix and hence, the *cond*(**A**) is also equal to the ratio of the largest to the smallest eigenvalue of **A**, i.e.,

$$cond(\mathbf{A}) = \frac{\lambda_{max}}{\lambda_{min}} \tag{8.28}$$

Generally speaking, for condition numbers less than 10^3 the parameter estimation problem is well-posed. For condition numbers greater than 10^{10} the problem is relatively ill-conditioned whereas for condition numbers 10^{30} or greater the problem is very ill-conditioned and we may encounter computer overflow problems.

The condition number of a matrix **A** is intimately connected with the sensitivity of the solution of the linear system of equations $\mathbf{A}\,\mathbf{x} = \mathbf{b}$. When solving this equation, the error in the solution can be magnified by an amount as large as *cond*(**A**) times the norm of the error in **A** and **b** due to the presence of the error in the data.

$$\frac{\|\Delta\mathbf{x}\|}{\|\mathbf{x}\|} \le cond(\mathbf{A}) \left(\frac{\|\Delta\mathbf{b}\|}{\|\mathbf{b}\|} + \frac{\|\Delta\mathbf{A}\|}{\|\mathbf{A}\|} \right) \tag{8.29}$$

Thus, the error in the solution vector is expected to be large for an ill-conditioned problem and small for a well-conditioned one. In parameter estimation, vector **b** is comprised of a linear combination of the response variables (measurements) which contain the error terms. Matrix **A** does not depend explicitly on the response variables, it depends only on the parameter sensitivity coefficients which depend only on the independent variables (assumed to be known precisely) and on the estimated parameter vector **k** which incorporates the uncertainty in the data. As a result, we expect most of the uncertainty in Equation 8.29 to be present in $\Delta\mathbf{b}$.

If matrix **A** is ill-conditioned at the optimum (i.e., at $\mathbf{k}=\mathbf{k}^*$), there is not much we can do. We are faced with a "truly" ill-conditioned problem and the estimated parameters will have highly questionable values with unacceptably large estimated variances. Probably, the most productive thing to do is to reexamine the structure and dependencies of the mathematical model and try to reformulate a better posed problem. Sequential experimental design techniques can also aid us in this direction to see whether additional experiments will improve the parameter estimates significantly (see Chapter 12).

If however, matrix **A** is reasonably well-conditioned at the optimum, **A** could easily be ill-conditioned when the parameters are away from their optimal values. This is quite often the case in parameter estimation and it is particularly true for highly nonlinear systems. In such cases, we would like to have the means to move the parameters estimates from the initial guess to the optimum even if the condition number of matrix **A** is excessively high for these initial iterations.

In general there are two remedies to this problem: (i) use a pseudoinverse and/or (ii) use Levenberg-Marquardt's modification.

8.4.1 Pseudoinverse

The eigenvalue decomposition of the positive definite symmetric matrix **A** yields

$$\mathbf{A} = \mathbf{V}^T \mathbf{\Lambda} \mathbf{V} \tag{8.30}$$

where **V** is orthogonal (i.e., $\mathbf{V}^{-1} = \mathbf{V}^T$) and $\mathbf{\Lambda} = diag(\lambda_1, \lambda_2, ..., \lambda_p)$. Furthermore, it is assumed that the eigenvalues are in descending order ($\lambda_1 > \lambda_2 > ... > \lambda_p$).

Having decomposed matrix **A**, we can readily compute the inverse from

$$\mathbf{A}^{-1} = \mathbf{V}^T \mathbf{\Lambda}^{-1} \mathbf{V} \tag{8.31}$$

If matrix **A** is well-conditioned, the above equation should be used. If however, **A** is ill-conditioned, we have the option without any additional computation effort, to use instead the *pseudoinverse* of **A**. Essentially, instead of $\mathbf{\Lambda}^{-1}$ in Equation 8.31, we use the pseudoinverse of $\mathbf{\Lambda}$, $\mathbf{\Lambda}^{\#}$.

The *pseudoinverse,* $\mathbf{\Lambda}^{\#}$ is obtained by inverting only the eigenvalues which are greater than a user-specified small number, δ, and setting the rest equal to zero. Namely,

$$\mathbf{\Lambda}^{\#} = diag\,(1/\lambda_1, 1/\lambda_2, \ldots, 1/\lambda_\pi, 0, \ldots, 0) \tag{8.32}$$

where $\lambda_j \geq \delta$, $j=1,2,\ldots,\pi$ and $\lambda_j < \delta$, $j=\pi+1,\pi+2,\ldots,p$. In other words, by using the pseudoinverse of **A**, we are neglecting the contribution of the eigenvalues that are very close to zero. Typically, we have used for δ a value of 10^{-20} when λ_1 is of order 1.

Finally, it should be noted that this approach yields an approximate solution to $\mathbf{A}\Delta\mathbf{k}^{(j+1)} = \mathbf{b}$ that has the additional advantage of keeping $\|\Delta\mathbf{k}^{(j+1)}\|$ as small as possible.

8.4.2 Marquardt's Modification

In order to improve the convergence characteristics and robustness of the Gauss-Newton method, Levenberg in 1944 and later Marquardt (1963) proposed to modify the normal equations by adding a small positive number, γ^2, to the diagonal elements of **A**. Namely, at each iteration the increment in the parameter vector is obtained by solving the following equation

$$(\mathbf{A} + \gamma^2\mathbf{I}\,)\,\Delta\mathbf{k}^{(j+1)} = \mathbf{b} \tag{8.33}$$

Levenberg-Marquardt's modification is often given a geometric interpretation as being a compromise between the direction of the Gauss-Newton method and that of steepest descent. The increment vector, $\Delta\mathbf{k}^{(j+1)}$, is a weighted average between that by the Gauss-Newton and the one by steepest descent.

A more interesting interpretation of Levenberg-Marquardt's modification can be obtained by examining the eigenvalues of the modified matrix $(\mathbf{A}+\gamma^2\mathbf{I})$. If we consider the eigenvalue decomposition of **A**, $\mathbf{V}^T\mathbf{\Lambda V}$ we have,

$$\mathbf{A}+\gamma^2\mathbf{I} = \mathbf{V}^T\mathbf{\Lambda V}+\gamma^2\mathbf{V}^T\mathbf{V} = \mathbf{V}^T(\mathbf{\Lambda}+\gamma^2\mathbf{I})\mathbf{V} = \mathbf{V}^T\mathbf{\Lambda}^M\mathbf{V} \tag{8.34}$$

where

$$\Lambda^M = diag\,(\lambda_1+\gamma^2, \lambda_2+\gamma^2,\ldots,\lambda_p+\gamma^2) \tag{8.35}$$

The net result is that all the eigenvalues of matrix **A** are all increased by γ^2, i.e., the eigenvalues are now $\lambda_1+\gamma^2$, $\lambda_2+\gamma^2,\ldots,\lambda_p+\gamma^2$. Obviously, the large eigenvalues will be hardly changed whereas the small ones that are much smaller than γ^2 become essentially equal to γ^2 and hence, the condition number of matrix **A** is reduced from $\lambda_{max}/\lambda_{min}$ to

$$cond(\mathbf{A}) = \frac{\lambda_{max} + \gamma^2}{\lambda_{min} + \gamma^2} \approx \frac{\lambda_{max}}{\gamma^2} \tag{8.36}$$

There are some concerns about the implementation of the method (Bates and Watts, 1988) since one must decide how to manipulate both Marquardt's γ^2 and the step-size λ. In our experience, a good implementation strategy is to decompose matrix A and examine the magnitude of the small eigenvalues. If some of them are excessively small, we use a γ^2 which is larger compared to the small eigenvalues and we compute $\|\Delta\mathbf{k}^{(j+1)}\|$. Subsequently, using the bisection rule we obtain an acceptable or even an optimal step-size along the direction of $\Delta\mathbf{k}^{(j+1)}$.

8.4.3 Scaling of Matrix A

When the parameters differ by more than one order of magnitude, matrix **A** may appear to be ill-conditioned even if the parameter estimation problem is well-posed. The best way to overcome this problem is by introducing the *reduced sensitivity coefficients,* defined as

$$G_{Rij} = \left(\frac{\partial y_i}{\partial k_j}\right) k_j \tag{8.37}$$

and hence, the reduced parameter sensitivity matrix, $\mathbf{G}_R$, is related to our usual sensitivity matrix, **G**, as follows

$$\mathbf{G}_R = \mathbf{G}\,\mathbf{K} \tag{8.38}$$

where

$$\mathbf{K} = diag(k_1, k_2, \ldots, k_p) \tag{8.39}$$

As a result of this scaling, the normal equations $\mathbf{A}\Delta\mathbf{k}^{(j+1)} = \mathbf{b}$ become

$$\mathbf{A}_R \Delta\mathbf{k}_R^{(j+1)} = \mathbf{b}_R \tag{8.40}$$

where

$$\mathbf{A}_R = \mathbf{K}\left[\sum_{i=1}^{N} \mathbf{G}_i^T \mathbf{Q}_i \mathbf{G}_i\right]\mathbf{K} = \mathbf{KAK} \tag{8.41}$$

$$\mathbf{b}_R = \mathbf{K}\left[\sum_{i=1}^{N} \mathbf{G}_i^T \mathbf{Q}_i \left[\hat{\mathbf{y}}_i - \mathbf{f}(\mathbf{x}_i, \mathbf{k}^{(j)})\right]\right] = \mathbf{Kb} \tag{8.42}$$

and

$$\Delta\mathbf{k}_R^{(j+1)} = \mathbf{K}^{-1}\Delta\mathbf{k}^{(j+1)} \tag{8.43}$$

The parameter estimates for the next iteration are now obtained (using a step-size μ, $0<\mu\leq 1$) as follows

$$\mathbf{k}^{(j+1)} = \mathbf{k}^{(j)} + \mu\,\mathbf{K}\,\Delta\mathbf{k}_R^{(j+1)} \tag{8.44}$$

or equivalently

$$k_i^{(j+1)} = k_i^{(j)} \left[1 + \mu \Delta k_{R,i}^{(j+1)}\right] \quad ; \quad i=1,\ldots,p \tag{8.45}$$

With this modification the conditioning of matrix **A** is significantly improved and *cond*($\mathbf{A}_R$) gives a more reliable measure of the ill-conditioning of the parameter estimation problem. This modification has been implemented in all computer programs provided with this book.

8.5 USE OF "PRIOR" INFORMATION

Under certain conditions we may have some prior information about the parameter values. This information is often summarized by assuming that each parameter is distributed normally with a given mean and a small or large variance depending on how trustworthy our prior estimate is. The Bayesian objective function, $S_B(\mathbf{k})$, that should be minimized for algebraic equation models is

$$S_B(\mathbf{k}) = \sum_{i=1}^{N} \left[\hat{\mathbf{y}}_i - \mathbf{f}(\mathbf{x}_i,\mathbf{k})\right]^T \mathbf{Q}_i \left[\hat{\mathbf{y}}_i - \mathbf{f}(\mathbf{x}_i,\mathbf{k})\right] + (\mathbf{k} - \mathbf{k}_B)^T \mathbf{V}_B^{-1} (\mathbf{k} - \mathbf{k}_B) \tag{8.46}$$

and for differential equation models it takes the form,

$$S_B(\mathbf{k}) = \sum_{i=1}^{N} \left[\hat{\mathbf{y}}_i - \mathbf{y}(t_i,\mathbf{k})\right]^T \mathbf{Q}_i \left[\hat{\mathbf{y}}_i - \mathbf{y}(t_i,\mathbf{k})\right] + (\mathbf{k} - \mathbf{k}_B)^T \mathbf{V}_B^{-1} (\mathbf{k} - \mathbf{k}_B) \tag{8.47}$$

We have assumed that the prior information can be described by the multivariate normal distribution, i.e., **k** is normally distributed with mean $\mathbf{k}_B$ and covariance matrix $\mathbf{V}_B$.

The required modifications to the Gauss-Newton algorithm presented in Chapter 4 are rather minimal. At each iteration, we just need to add the following terms to matrix **A** and vector **b**,

$$\mathbf{A} = \mathbf{A}_{GN} + \mathbf{V}_B^{-1} \tag{8.48}$$

and

$$\mathbf{b} = \mathbf{b}_{GN} - \mathbf{V}_B^{-1} (\mathbf{k}^{(j)} - \mathbf{k}_B) \tag{8.49}$$

where $\mathbf{A}_{GN}$ and $\mathbf{b}_{GN}$ are matrix **A** and vector **b** for the Gauss-Newton method as given in Chapter 4 for algebraic or differential equation models. The prior covariance matrix of the parameters ($\mathbf{V}_B$) is often a diagonal matrix and since in the solution of the problem only the inverse of $\mathbf{V}_B$ is used, it is preferable to use as input to the program the inverse itself.

From a computer implementation point of view, this provides some extra flexibility to handle simultaneously parameters for which we have some prior knowledge and others for which no information is available. For the latter we simply need to input *zero* as the inverse of their prior variance.

Practical experience has shown that (i) if we have a relatively large number of data points, the prior has an insignificant effect on the parameter estimates (ii) if the parameter estimation problem is ill-posed, use of "prior" information has a stabilizing effect. As seen from Equation 8.48, all the eigenvalues of matrix **A** are increased by the addition of positive terms in its diagonal. It acts almost like Marquadt's modification as far as convergence characteristics are concerned.

8.6 SELECTION OF WEIGHTING MATRIX Q IN LEAST SQUARES ESTIMATION

As we mentioned in Chapter 2, the user specified matrix $\mathbf{Q}_i$ should be equal to the inverse of $COV(\mathbf{e}_i)$. However, in many occasions we have very little information about the nature of the error in the measurements. In such cases, we have found it very useful to use $\mathbf{Q}_i$ as a normalization matrix to make the measured responses of the same order of magnitude. If the measurements do not change substantially from data point to data point, we can use a constant $\mathbf{Q}$. The simplest form of $\mathbf{Q}$ that we have found adequate is to use a diagonal matrix whose j^{th} element in the diagonal is the inverse of the squared mean response of the jth variable,

$$Q_{jj} = \frac{1}{\left(\frac{1}{N}\sum_{i=1}^{N}\hat{y}_{j,i}\right)^2} \tag{8.50}$$

This is equivalent to assuming a constant standard error in the measurement of the jth response variable, and at the same time the standard errors of different response variables are proportional to the average value of the variables. This is a "safe" assumption when no other information is available, and least squares estimation pays equal attention to the errors from different response variables (e.g., concentration, versus pressure or temperature measurements).

If however the measurements of a response variable change over several orders of magnitude, it is better to use the non-constant diagonal weighting matrix $\mathbf{Q}_i$ given below

$$Q_{i,jj} = \frac{1}{\left(\hat{y}_{j,i}\right)^2} \tag{8.51}$$

This is equivalent to assuming that the standard error in the ith measurement of the jth response variable is proportional to its value, again a rather "safe" assumption as it forces least squares to pay equal attention to all data points.

8.7 IMPLEMENTATION GUIDELINES FOR ODE MODELS

For models described by a set of ordinary differential equations there are a few modifications we may consider implementing that enhance the performance (robustness) of the Gauss-Newton method. The issues that one needs to address more carefully are (i) numerical instability during the integration of the state and sensitivity equations, (ii) ways to enlarge the region of convergence.

8.7.1 Stiff ODE Models

Stiff differential equations appear quite often in engineering. The wide range of the prevailing time constants creates additional numerical difficulties which tend to shrink even further the region of convergence. It is noted that if the state equations are stiff, so are the sensitivity differential equations since both of them have the same Jacobean. It has been pointed out by many numerical analysts that deal with the stability and efficiency of integration algorithms (e.g., Edsberg, 1976) that the scaling of the problem is very crucial for the numerical solver to behave at its best. Therefore, it is preferable to scale the state equations according to the order of magnitude of the given measurements, whenever possible. In addition, to normalize the sensitivity coefficients we introduce the reduced sensitivity coefficients,

$$G_R(i,j) = \left(\frac{\partial x_i}{\partial k_j}\right) k_j \tag{8.52}$$

and hence, the reduced parameter sensitivity matrix, $\mathbf{G}_R$, is related to our usual sensitivity matrix, $\mathbf{G}$, as follows

$$\mathbf{G}_R(t) = \mathbf{G}(t)\,\mathbf{K} \tag{8.53}$$

where $\mathbf{K} = diag(k_1, k_2, \ldots, k_p)$. By this transformation the sensitivity differential equations shown below

$$\frac{d\mathbf{G}}{dt} = \left(\frac{\partial \mathbf{f}^T}{\partial \mathbf{x}}\right)^T \mathbf{G}(t) + \left(\frac{\partial \mathbf{f}^T}{\partial \mathbf{k}}\right)^T \; ; \; \mathbf{G}(t_0) = 0 \tag{8.54}$$

become

$$\frac{d\mathbf{G}_R}{dt} = \left(\frac{\partial \mathbf{f}^T}{\partial \mathbf{x}}\right)^T \mathbf{G}_R(t) + \left(\frac{\partial \mathbf{f}^T}{\partial \mathbf{k}}\right)^T \mathbf{K} \;;\; \mathbf{G}_R(t_0) = 0 \tag{8.55}$$

With this transformation, the normal equations now become,

$$\mathbf{A}_R \Delta\mathbf{k}_R^{(j+1)} = \mathbf{b}_R \tag{8.56}$$

where

$$\mathbf{A}_R = \left[\sum_{i=1}^{N} \mathbf{G}_R(t_i)^T \mathbf{C}^T \mathbf{Q}_i \mathbf{C} \mathbf{G}_R(t_i)\right] \tag{8.57}$$

$$\mathbf{b}_R = \left[\sum_{i=1}^{N} \mathbf{G}_R(t_i)^T \mathbf{C}^T \mathbf{Q}_i \left[\hat{\mathbf{y}}(t_i) - \mathbf{C}\mathbf{x}(t_i, \mathbf{k}^{(j)})\right]\right] \tag{8.58}$$

and

$$\Delta\mathbf{k}_R^{(j+1)} = \mathbf{K}^{-1} \Delta\mathbf{k}^{(j+1)} \tag{8.59}$$

The parameter estimates for the next iteration are now obtained as

$$\mathbf{k}^{(j+1)} = \mathbf{k}^{(j)} + \mu \mathbf{K} \Delta\mathbf{k}_R^{(j+1)} \tag{8.60}$$

or

$$k_i^{(j+1)} = k_i^{(j)} \left[1 + \mu \Delta k_{R,i}^{(j+1)}\right] \;;\quad i=1,\ldots,p \tag{8.61}$$

As we have already pointed out in this chapter for systems described by algebraic equations, the introduction of the reduced sensitivity coefficients results in a reduction of *cond*(**A**). Therefore, the use of the reduced sensitivity coefficients should also be beneficial to non-stiff systems.

We strongly suggest the use of the reduced sensitivity whenever we are dealing with differential equation models. Even if the system of differential equations is non-stiff at the optimum (when **k**=**k***), when the parameters are far from their optimal values, the equations may become stiff temporarily for a few iterations of the Gauss-Newton method. Furthermore, since this transformation also results in better conditioning of the normal equations, we propose its use at all times. This transformation has been implemented in the program for ODE systems provided with this book.

8.7.2 Increasing the Region of Convergence

A well known problem of the Gauss-Newton method is its relatively small region of convergence. Unless the initial guess of the unknown parameters is in the vicinity of the optimum, divergence may occur. This problem has received much attention in the literature. For example, Ramaker et al. (1970) suggested the incorporation of Marquadt's modification to expand the region of convergence. Donnely and Quon (1970) proposed a procedure whereby the measurements are perturbed if divergence occurs and a series of problems are solved until the model trajectories "match" the original data. Nieman and Fisher (1972) incorporated linear programming into the parameter estimation procedure and suggested the solution of a series of constrained parameter estimation problems where the search is restricted in a small parameter space around the chosen initial guess. Wang and Luus (1980) showed that the use of a shorter data-length can enlarge substantially the region of convergence of Newton's method. Similar results were obtained by Kalogerakis and Luus (1980) using the quasilinearization method.

In this section we first present an efficient step-size policy for differential equation systems and we present two approaches to increase the region of convergence of the Gauss-Newton method. One through the use of the Information Index and the other by using a two-step procedure that involves direct search optimization.

8.7.2.1 An Optimal Step-Size Policy

The proposed step-size policy for differential equation systems is fairly similar to our approach for algebraic equation models. First we start with the bisection rule. We start with $\mu=1$ and we keep on halving it until an acceptable value, μ_a, has been found, i.e., we reduce μ until

$$S(\mathbf{k}^{(j)}+\mu_a\Delta\mathbf{k}^{(j+1)}) < S(\mathbf{k}^{(j)}). \tag{8.18}$$

It should be emphasized here that it is unnecessary to integrate the state equations for the entire data length for each value of μ. Once the objective function becomes greater than $S(\mathbf{k}^{(j)})$, a smaller value for μ can be chosen. By this procedure, besides the savings in computation time, numerical instability is also avoided since the objective function often becomes large very quickly and integration is stopped well before computer overflow is threatened.

Having an acceptable value for μ, we can either stop here and proceed to perform another iteration of Gauss-Newton method or we can attempt to locate the optimal step-size along the direction suggested by the Gauss-Newton method.

The main difference with differential equation systems is that every evaluation of the objective function requires the integration of the state equations. In this section we present an optimal step size policy proposed by Kalogerakis and Luus (1983b) which uses information only at $\mu=0$ (i.e., at $\mathbf{k}^{(j)}$) and at $\mu=\mu_a$ (i.e., at

$\mathbf{k}^{(j)}+\mu_a\Delta\mathbf{k}^{(j+1)}$). Let us consider a Taylor series expansion of the state vector with respect to μ, retaining up to second order terms,

$$\mathbf{x}(t, \mathbf{k}^{(j)} + \mu\Delta\mathbf{k}^{(j+1)}) = \mathbf{x}(t, \mathbf{k}^{(j)}) + \mu\mathbf{G}(t)\Delta\mathbf{k}^{(j+1)} + \mu^2\mathbf{r}(t) \tag{8.62}$$

where the *n-dimensional* residual vector $\mathbf{r}(t)$ is not a function of μ. Vector $\mathbf{r}(t)$ is easily computed by the already performed integration at μ_a and using the state and sensitivity coefficients already computed at $\mathbf{k}^{(j)}$; namely,

$$\mathbf{r}(t) = \frac{1}{\mu_a^2}\left(\mathbf{x}(t, \mathbf{k}^{(j)} + \mu_a\Delta\mathbf{k}^{(j+1)}) - \mathbf{x}(t, \mathbf{k}^{(j)}) - \mu_a\mathbf{G}(t)\Delta\mathbf{k}^{(j+1)}\right) \tag{8.63}$$

Thus, having $\mathbf{r}(t_i)$, i=1,2,...,N, the objective function S(μ) becomes

$$S(\mu) = \sum_{i=1}^{N}\left[\hat{\mathbf{y}}(t_i) - \mathbf{C}\mathbf{x}(t_i,\mathbf{k}^{(j)}) - \mu\mathbf{C}\mathbf{G}(t_i)\Delta\mathbf{k}^{(j+1)} - \mu^2\mathbf{C}\mathbf{r}(t_i)\right]^T\mathbf{Q}_i$$
$$\left[\hat{\mathbf{y}}(t_i) - \mathbf{C}\mathbf{x}(t_i,\mathbf{k}^{(j)}) - \mu\mathbf{C}\mathbf{G}(t_i)\Delta\mathbf{k}^{(j+1)} - \mu^2\mathbf{C}\mathbf{r}(t_i)\right] \tag{8.64}$$

Subsequent use of the stationary criterion ∂S/∂μ=0, yields the following 3rd order equation for μ,

$$\beta_3\mu^3 + \beta_2\mu^2 + \beta_1\mu + \beta_0 = 0 \tag{8.65}$$

where

$$\beta_3 = 2\sum_{i=1}^{N}\mathbf{r}(t_i)^T\mathbf{C}^T\mathbf{Q}_i\mathbf{C}\,\mathbf{r}(t_i) \tag{8.66a}$$

$$\beta_2 = 3\sum_{i=1}^{N}\mathbf{r}(t_i)^T\mathbf{C}^T\mathbf{Q}_i\mathbf{C}\,\mathbf{G}(t_i)\Delta\mathbf{k}^{(j+1)} \tag{8.66b}$$

$$\beta_1 = \Delta\mathbf{k}^{(j+1)^T}\sum_{i=1}^{N}\mathbf{G}(t_i)^T\mathbf{C}^T\mathbf{Q}_i\mathbf{C}\,\mathbf{G}(t_i)\Delta\mathbf{k}^{(j+1)} -$$
$$2\sum_{i=1}^{N}\mathbf{r}(t_i)^T\mathbf{C}^T\mathbf{Q}_i\left[\hat{\mathbf{y}}(t_i) - \mathbf{C}\mathbf{x}(t_i,\mathbf{k}^{(j)})\right] \tag{8.66c}$$

$$\beta_0 = -\Delta\mathbf{k}^{(j+1)^T}\sum_{i=1}^{N}\mathbf{G}(t_i)^T\mathbf{C}^T\mathbf{Q}_i\left[\hat{\mathbf{y}}(t_i)-\mathbf{C}\mathbf{x}(t_i,\mathbf{k}^{(j)})\right] \tag{8.66d}$$

Solution of Equation 8.65 yields the optimum value for the step-size. The solution can be readily obtained by Newton's method within 3 or 4 iterations using μ_a as a starting value. This optimal step-size policy was found to yield very good results. The only problem that it has is that one needs to store the values of state and sensitivity equations at each iteration. For high dimensional systems this is not advisable.

Finally it is noted that in the above equations we can substitute $\mathbf{G}(t)$ with $\mathbf{G}_R(t)$ and $\Delta\mathbf{k}^{(j+1)}$ with $\Delta\mathbf{k}_R^{(j+1)}$ in case we wish to use the reduced sensitivity coefficient formulation.

8.7.2.2 Use of the Information Index

The remedies to increase the region of convergence include the use of a pseudoinverse or Marquardt's modification that overcome the problem of ill-conditioning of matrix **A**. However, if the basic sensitivity information is not there, the estimated direction $\Delta\mathbf{k}^{(j+1)}$ cannot be obtained reliably.

A careful examination of matrix **A** or $\mathbf{A}_R$, shows that **A** depends not only on the sensitivity coefficients but also on their values *at the times when the given measurements have been taken.* When the parameters are far from their optimum values, some of the sensitivity coefficients may be excited and become large in magnitude at times which fall outside the range of the given data points. This means that the output vector will be insensitive to these parameters at the given measurement times, resulting in loss of sensitivity information captured in matrix **A**. The normal equation, $\mathbf{A}\Delta\mathbf{k}^{(j+1)}=\mathbf{b}$, becomes ill-conditioned and any attempt to change the equation is rather self-defeating since the parameter estimates will be unreliable.

Therefore, instead of modifying the normal equations, we propose a direct approach whereby the conditioning of matrix **A** can be significantly improved by using an appropriate section of the data so that most of the available sensitivity information is captured for the current parameter values. To be able to determine the proper section of the data where sensitivity information is available, Kalogerakis and Luus (1983b) introduced the *Information Index* for each parameter, defined as

$$I_j(t) = k_j\left(\frac{\partial\mathbf{y}^T}{\partial k_j}\right)\mathbf{Q}\left(\frac{\partial\mathbf{y}}{\partial k_j}\right)k_j \quad ; \; j=1,\ldots,p \tag{8.67}$$

Or equivalently, using the sensitivity coefficient matrix,

$$I_j(t) = k_j \boldsymbol{\delta}_j^T \mathbf{G}^T(t)\mathbf{C}^T\mathbf{Q}\mathbf{C}\mathbf{G}(t)\boldsymbol{\delta}_j k_j \quad ; \; j=1,\dots,p \tag{8.68}$$

where $\boldsymbol{\delta}_j$ is a *p-dimensional* vector with 1 in the jth element and zeros elsewhere. The scalar $I_j(t)$ should be viewed as an index measuring the overall sensitivity of the output vector to parameter k_j at time t.

Thus given an initial guess for the parameters, we can integrate the state and sensitivity equations and compute the Information Indices, $I_j(t)$, $j=1,\dots,p$ as functions of time. Subsequently by plotting $I_j(t)$ versus time, preferably on a semi-log scale, we can spot immediately where they become excited and large in magnitude. If observations are not available within this time interval, artificial data can be generated by data smoothing and interpolation to provide the missing sensitivity information. In general, the best section of the data could be determined at each iteration, as during the course of the iterations different sections of the data may become the most appropriate. However, it is also expected that during the iterations the most appropriate sections of the data will lie somewhere in between the range of the given measurements and the section determined by the initial parameter estimates. It is therefore suggested, that the section selected based on the initial parameter estimates be combined with the section of the given measurements and the entire data length is used for all iterations. At the last iteration all artificial data are dropped and only the given data are used.

The use of the Information Index and the optimal step size policy are illustrated in Chapter 16 where we present parameter estimation examples for ODE models. For illustration purposes we present here the information indices for the two-parameter model describing the pyrolytic dehydrogenation of benzene to diphenyl and triphenyl (introduced first in Section 6.5.2). In Figure 8.2 the Information Indices are presented in graphical form with parameter values (k_1=355,400 and k_2=403,300) three orders of magnitude away from the optimum. With this initial guess the Gauss-Newton method fails to converge as all the available sensitivity information falls outside the range of the given measurements. By generating artificial data by interpolation, subsequent use of the Gauss-Newton method brings the parameters to the optimum (k_1=355.4 and k_2=403.3) in nine iterations. For the sake of comparison the Information Indices are also presented when the parameters have their optimal values. As seen in Figure 8.3, in terms of experimental design the given measurements have been taken at proper times, although some extra information could have been gained by having a few extra data points in the interval $[10^{-4}, 5.83\times10^{-4}]$.

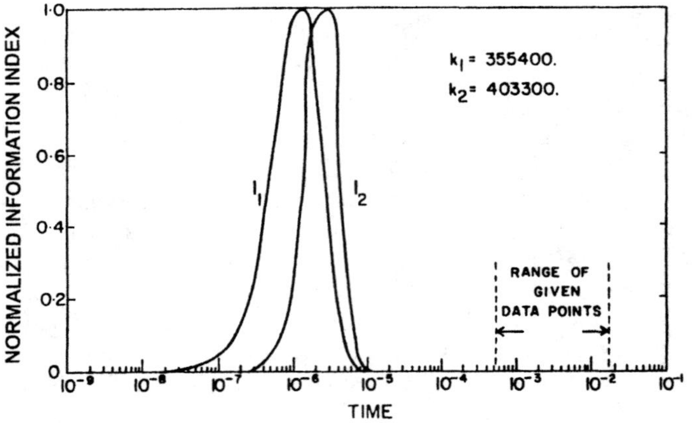

Figure 8.2 *Normalized Information Index of k_1 and k_2 versus time to determine the best section of data to be used by the Gauss-Newton method (I_{1max}=0.0884, I_{2max}=0.0123) [reprinted from Industrial Engineering Chemistry Fundamentals with permission from the American Chemical Society].*

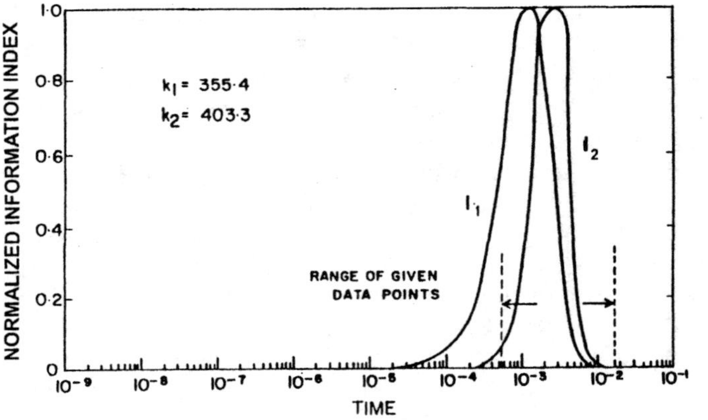

Figure 8.3 *Normalized Information Index of k_1 and k_2 versus time to determine whether the measurements have been collected at proper times (I_{1max}=0.0885, I_{2max}=0.0123) [reprinted from Industrial Engineering Chemistry Fundamentals with permission from the American Chemical Society].*

When the dynamic system is described by a set of stiff ODEs and observations during the fast transients are not available, generation of artificial data by interpolation at times close to the origin may be very risky. If however, we observe all state variables (i.e., **C**=**I**), we can overcome this difficulty by redefining the initial state (Kalogerakis and Luus, 1983b). Instead of restricting the use of data close to the initial state, the data can be shifted to allow any of the early data points to constitute the initial state where the fast transients have died out. At the shifted origin, generation of artificial data is considerably easier. Illustrative examples of this technique are provided by Kalogerakis and Luus (1983b).

8.7.2.3 Use of Direct Search Methods

A simple procedure to overcome the problem of the small region of convergence is to use a two-step procedure whereby direct search optimization is used to initially to bring the parameters in the vicinity of the optimum, followed by the Gauss-Newton method to obtain the best parameter values and estimates of the uncertainty in the parameters (Kalogerakis and Luus, 1982).

For example let us consider the estimation of the two kinetic parameters in the Bodenstein-Linder model for the homogeneous gas phase reaction of NO with O_2 (first presented in Section 6.5.1). In Figure 8.4 we see that the use of direct search (LJ optimization) can increase the overall size of the region of convergence by at least two orders of magnitude.

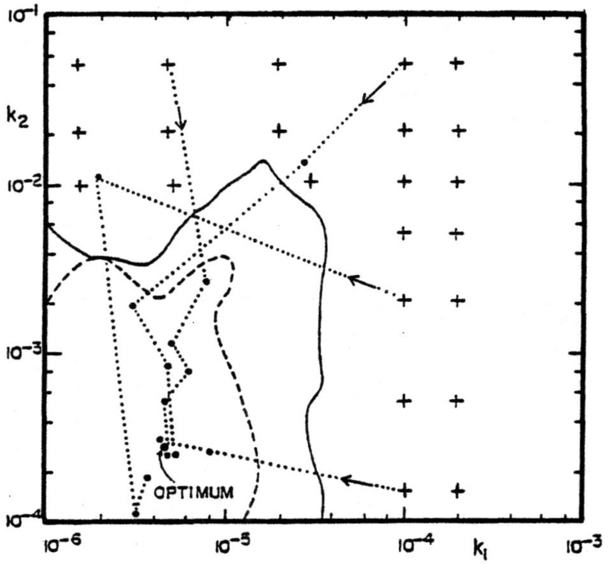

Figure 8.4 Use of the LJ optimization procedure to bring the first parameter estimates inside the region of convergence of the Gauss-Newton method (denoted by the solid line). All test points are denoted by +. Actual path of some typical runs is shown by the dotted line.

8.8 AUTOCORRELATION IN DYNAMIC SYSTEMS

When experimental data are collected over time or distance there is always a chance of having autocorrelated residuals. Box et al. (1994) provide an extensive treatment of correlated disturbances in discrete time models. The structure of the disturbance term is often *moving average* or *autoregressive* models. Detection of autocorrelation in the residuals can be established either from a *time series plot* of the residuals versus time (or experiment number) or from a *lag plot*. If we can see a pattern in the residuals over time, it probably means that there is correlation between the disturbances.

An autoregressive model (AR) of order 1 for the residuals has the form

$$e_i = \rho e_{i-1} + \xi_i \tag{8.69}$$

where e_i is the residual of the ith measurement, ρ is an unknown constant and ξ_i are independent random disturbances with zero mean and constant variance (also known as *white noise*). In engineering, when we are dealing with parameter estimation of mechanistic models, we do not encounter highly correlated residuals. This is not the case however when we use the *black box* approach (Box et al., 1994).

If we have data collected at a constant sampling interval, any autocorrelation in the residuals can be readily detected by plotting the residual autocorrelation function versus lag number. The latter is defined as,

$$r_k = \sum_{i=k+1} \frac{e_i e_{i-k}}{N\hat{\sigma}_\varepsilon^2} \qquad \text{at lag } k=1,2,... \tag{8.70}$$

The 95% confidence intervals of the autocorrelation function beyond lag 2 is simply given by $\pm 2/\sqrt{N}$. If the estimated autocorrelation function has values at greater than $2/\sqrt{N}$, we should correct the data for autocorrelation.

If we assume that the residuals satisfy Equation 8.69, we must transform the data so that the new residuals are not correlated. Substituting the definition of the residual (for the single-response case) into Equation 8.69 we obtain upon rearrangement,

$$\hat{y}_i = \rho\,\hat{y}_{i-1} + f(\mathbf{x}_i, \mathbf{k}) - \rho\, f(\mathbf{x}_{i-1}, \mathbf{k}) + \xi_i \tag{8.71}$$

The above model equation now satisfies all the criteria for least squares estimation. As an initial guess for **k** we can use our estimated parameter values when it was assumed that no correlation was present. Of course, in the second step we have to include ρ in the list of the parameters to be determined.

Finally it is noted that the above equations can be readily extended to the multi-response case especially if we assume that there is no cross-correlation between different response variables.

9

Constrained Parameter Estimation

In some cases besides the governing algebraic or differential equations, the mathematical model that describes the physical system under investigation is accompanied with a set of constraints. These are either *equality* or *inequality constraints* that must be satisfied when the parameters converge to their best values. The constraints may be simply on the parameter values, e.g., a reaction rate constant must be positive, or on the response variables. The latter are often encountered in thermodynamic problems where the parameters should be such that the calculated thermophysical properties satisfy all constraints imposed by thermodynamic laws. We shall first consider equality constraints and subsequently inequality constraints.

9.1 EQUALITY CONSTRAINTS

Equality constraints are rather seldom in parameter estimation. If there is an equality constraint among the parameters, one should first attempt to eliminate one of the unknown parameters simply by solving explicitly for one of the parameters and then substituting that relationship in the model equations. Such an action reduces the dimensionality of the parameter estimation problem which aids significantly in achieving convergence.

If the equality constraint involves independent variables and parameters in an algebraic model, i.e., it is of the form, $\varphi(\mathbf{x}, \mathbf{k}) = 0$, and if we can solve explicitly for one of the unknown parameters, simple substitution of the expression into the model equations reduces the number of unknown parameters by one.

If the equality constraint involves the response variables, then we have the option to either substitute the experimental measurement of the response variables into the constraint or to use the error-in-variables method to obtain besides the parameters the noise-free value of the response variables. The use of the error-in-variables method is discussed in Chapter 14 (parameter estimation with equations of state). The general approach to handle any equality constraint is through the use of Lagrange multipliers discussed next.

9.1.1 Lagrange Multipliers

Let us consider constrained least squares estimation of unknown parameters in algebraic equation models first. The problem can be formulated as follows:

Given a set of data points $\{(\mathbf{x}_i,\mathbf{y}_i),\ i=1,\ldots,N\}$ and a mathematical model of the form, $\mathbf{y} = \mathbf{f}(\mathbf{x},\mathbf{k})$, the objective is to determine the unknown parameter vector $\mathbf{k}$ by minimizing the least squares objective function subject to the equality constraint, namely

$$\textit{minimize} \qquad S_{LS}(\mathbf{k}) = \sum_{i=1}^{N}[\hat{\mathbf{y}}_i - \mathbf{f}(\mathbf{x}_i,\mathbf{k})]^T \mathbf{Q}\,[\hat{\mathbf{y}}_i - \mathbf{f}(\mathbf{x}_i,\mathbf{k})] \tag{9.1}$$

$$\textit{subject to} \qquad \varphi(\mathbf{x}_0,\mathbf{y}_0,\mathbf{k}) = 0 \tag{9.2}$$

The point where the constraint is satisfied, $(\mathbf{x}_0,\mathbf{y}_0)$, may or may not belong to the data set $\{(\mathbf{x}_i,\hat{\mathbf{y}}_i),\ i=1,\ldots,N\}$. The above constrained minimization problem can be transformed into an unconstrained one by introducing the Lagrange multiplier, ω and augmenting the least squares objective function to form the Lagrangian,

$$S_{LG}(\mathbf{k},\omega) = S_{LS}(\mathbf{k}) + \omega\varphi(\mathbf{x}_0,\mathbf{y}_0,\mathbf{k}) \tag{9.3}$$

The above unconstrained estimation problem can be solved by a small modification of the Gauss-Newton method. Let us assume that we have an estimate $\mathbf{k}^{(j)}$ of the parameters at the jth iteration. Linearization of the model equation and the constraint around $\mathbf{k}^{(j)}$ yields,

$$\mathbf{f}(\mathbf{x}_i,\mathbf{k}^{(j+1)}) = \mathbf{f}(\mathbf{x}_i,\mathbf{k}^{(j)}) + \left(\frac{\partial \mathbf{f}^T}{\partial \mathbf{k}}\right)^T \Delta\mathbf{k}^{(j+1)} \tag{9.4}$$

and

$$\varphi(\mathbf{x}_0,\mathbf{y}_0,\mathbf{k}^{(j+1)}) = \varphi(\mathbf{x}_0,\mathbf{y}_0,\mathbf{k}^{(j)}) + \left(\frac{\partial \varphi}{\partial \mathbf{k}}\right)^T \Delta\mathbf{k}^{(j+1)} \tag{9.5}$$

Substitution of the Equations 9.4 and 9.5 into the Lagrangian given by Equation 9.3 and use of the stationary conditions

$$\frac{\partial S_{LG}(\mathbf{k}^{(j+1)},\omega)}{\partial \mathbf{k}^{(j+1)}} = \mathbf{0} \tag{9.6}$$

and

$$\frac{\partial S_{LG}(\mathbf{k}^{(j+1)},\omega)}{\partial \omega} = 0 \tag{9.7}$$

yield the following system of linear equations

$$\mathbf{A}\Delta\mathbf{k}^{(j+1)} = \mathbf{b} \tag{9.8a}$$

and

$$\mathbf{c}^{T}\Delta\mathbf{k}^{(j+1)} = -\varphi_0 \tag{9.8b}$$

where

$$\mathbf{A} = \sum_{i=1}^{N}\mathbf{G}_i^{T}\mathbf{Q}\mathbf{G}_i \tag{9.9a}$$

$$\mathbf{b} = \mathbf{b}_{GN} - \frac{\omega}{2}\mathbf{c} \tag{9.9b}$$

$$\mathbf{b}_{GN} = \sum_{i=1}^{N}\mathbf{G}_i^{T}\mathbf{Q}[\mathbf{y}_i - \mathbf{f}(\mathbf{x}_i,\mathbf{k}^{(j)})] \tag{9.9c}$$

$$\varphi_0 = -\varphi(\mathbf{x}_0,\mathbf{y}_0,\mathbf{k}^{(j)}) \tag{9.9d}$$

$$\mathbf{c} = \left(\frac{\partial \varphi}{\partial \mathbf{k}}\right) \tag{9.9e}$$

Equation 9.8a is solved with respect to $\Delta\mathbf{k}^{(j+1)}$ to yield

$$\Delta\mathbf{k}^{(j+1)} = \mathbf{A}^{-1}[\mathbf{b}_{GN} - \frac{\omega}{2}\mathbf{c}] \tag{9.10}$$

Subsequent substitution into Equation 9.8b yields

$$\mathbf{c}^{T}\mathbf{A}^{-1}\mathbf{b}_{GN} - \frac{\omega}{2}\mathbf{c}^{T}\mathbf{A}^{-1}\mathbf{c} = -\varphi_0 \tag{9.11}$$

which upon rearrangement results in

$$\omega = 2\left(\frac{\varphi_0 + \mathbf{c}^T\mathbf{A}^{-1}\mathbf{b}_{GN}}{\mathbf{c}^T\mathbf{A}^{-1}\mathbf{c}}\right) \tag{9.12}$$

substituting the above expression for the Lagrange multiplier into Equation 9.8a we arrive at the following linear equation for $\Delta\mathbf{k}^{(j+1)}$,

$$\mathbf{A}\Delta\mathbf{k}^{(j+1)} = \mathbf{b}_{GN} - \left(\frac{\varphi_0 + \mathbf{c}^T\mathbf{A}^{-1}\mathbf{b}_{GN}}{\mathbf{c}^T\mathbf{A}^{-1}\mathbf{c}}\right)\mathbf{c} \tag{9.13}$$

The above equation leads to the following steps of the modified Gauss-Newton method.

Modified Gauss-Newton Algorithm for Constrained Least Squares

The implementation of the modified Gauss-Newton method is accomplished by following the steps given below,

Step 1. Generate/assume an initial guess for the parameter vector **k**.

Step 2. Given the current estimate of the parameters, $\mathbf{k}^{(j)}$, compute the parameter sensitivity matrix, $\mathbf{G}_i$, the response variables $\mathbf{f}(\mathbf{x}_i,\mathbf{k}^{(j)})$, and the constraint, φ_0.

Step 3. Set up matrix **A**, vector $\mathbf{b}_{GN}$ and compute vector **c**.

Step 4. Perform an eigenvalue decomposition of $\mathbf{A}=\mathbf{V}^T\mathbf{\Lambda}\mathbf{V}$ and compute $\mathbf{A}^{-1}=\mathbf{V}^T\mathbf{\Lambda}^{-1}\mathbf{V}$.

Step 5. Compute the right hand side of Equation 9.13 and then solve Equation 9.13 with respect to $\Delta\mathbf{k}^{(j+1)}$.

Step 6. Use the bisection rule to determine an acceptable step-size and then update the parameter estimates.

Step 7. Check for convergence. If converged, estimate *COV*(**k***) and stop; else go back to Step 2.

The above constrained parameter estimation problem becomes much more challenging if the location where the constraint must be satisfied, $(\mathbf{x}_0,\mathbf{y}_0)$, is not known *a priori*. This situation arises naturally in the estimation of binary interaction parameters in cubic equations of state (see Chapter 14). Furthermore, the above development can be readily extended to several constraints by introducing an equal number of Lagrange multipliers.

If the mathematical model is described by a set of differential equations, constrained parameter estimation becomes a fairly complicated problem. Normally we may have two different types of constraints on the state variables: (i) isoperimetric constraints, expressed as an integral (or summation) constraint on the state variables, and/or (ii) simple constraints on the state variables. In the first case, we generate the Langrangian by introducing the Lagrange multiplier ω which is obtained in a similar fashion as the described previously. If however, we have a simple constraint on the state variables, the Lagrange multiplier ω, becomes an unknown functional, $\omega(t)$, that must be determined. This problem can be tackled using elements from the calculus of variations but it is beyond the scope of this book and it is not considered here.

9.2 INEQUALITY CONSTRAINTS

There are two types of inequality constraints. Those that involve only the parameters (e.g., the parameters must be positive and less than one) and those that involve not only the parameters but also the dependent variables (e.g., the predicted concentrations of all species must always be positive or zero and the unknown reaction rate constants must all be positive).

We shall examine each case independently.

9.2.1 Optimum Is Internal Point

Most of the constrained parameter estimation problems belong to this case. Based on scientific considerations, we arrive quite often at constraints that the parameters of the mathematical model should satisfy. Most of the time these are of the form,

$$k_{min,i} < k_i < k_{max,i} \quad ; \ i=1,\ldots,p \tag{9.14}$$

and our objective is to ensure that the optimal parameter estimates satisfy the above constraints. If our initial guesses are very poor, during the early iterations of the Gauss-Newton method the parameters may reach beyond these boundaries where the parameter values may have no physical meaning or the mathematical model breaks down which in turn may lead to severe convergence problems. If the optimum is indeed an internal point, we can readily solve the problem using one of the following three approaches.

9.2.1.1 Reparameterization

The simplest way to deal with constraints on the parameters is to ignore them! We use the unconstrained algorithm of our choice and if the converged parameter estimates satisfy the constraints no further action is required. Unfortunately we cannot always use this approach. A smart way of imposing simple constraints on the parameters is through *reparameterization*. For example, if we have the simple constraint (often encountered in engineering problems),

$$k_i > 0 \tag{9.15}$$

the following reparameterization always enforces the above constraint (Bates and Watts, 1988),

$$k_i = exp(\kappa_i) \tag{9.16}$$

By conducting our search over κ_i regardless of its value, $exp(\kappa_i)$ and hence k_i is always positive. For the more general case of the interval constraint on the parameters given by Equation 9.14, we can perform the following transformation,

$$k_i = k_{min,i} + \frac{k_{max,i} - k_{min,i}}{1 + exp(\kappa_i)} \tag{9.17}$$

Using the above transformation, we are able to perform an unconstrained search over κ_i. For any value of κ_i, the original parameter k_i remains within its limits. When κ_i approaches very large values (tends to infinity), k_i approaches its lower limit, $k_{min,i}$ whereas when κ_i approaches very large negative values (tends to minus infinity), k_i approaches its upper limit, $k_{max,i}$. Obviously, the above transformation increases the complexity of the mathematical model; however, there are no constraints on the parameters.

9.2.1.2 Penalty Function

In this case instead of reparameterizing the problem, we augment the objective function by adding extra terms that tend to explode when the parameters approach near the boundary and become negligible when the parameters are far. One can easily construct such functions.

One of the simplest and yet very effective penalty function that keeps the parameters in the interval ($k_{min,i}$, $k_{max,i}$) is

$$\Theta_i(k_i) = \frac{k_{max,i} - k_{min,i}}{k_i - k_{min,i}} + \frac{k_{max,i} - k_{min,i}}{k_{max,i} - k_i} \quad ; \ i = 1,\dots,p \tag{9.18}$$

The functions essentially place an equally weighted penalty for small or large-valued parameters on the overall objective function. If penalty functions for constraints on all unknown parameters are added to the objective function, we obtain,

$$S_p(\mathbf{k},\zeta) = \sum_{i=1}^{N} \mathbf{e}_i^T \mathbf{Q}_i \mathbf{e}_i + 2\zeta \sum_{i=1}^{p} \Theta_i(k_i) \tag{9.19}$$

or

$$S_p(\mathbf{k},\zeta) = \sum_{i=1}^{N} \mathbf{e}_i^T \mathbf{Q}_i \mathbf{e}_i + 2\zeta \sum_{i=1}^{p} \left[\frac{k_{max,i} - k_{min,i}}{k_i - k_{min,i}} + \frac{k_{max,i} - k_{min,i}}{k_{max,i} - k_i} \right] \tag{9.20}$$

The user supplied weighting constant, ζ (≥ 0), should have a large value during the early iterations of the Gauss-Newton method when the parameters are away from their optimal values. As the parameters approach the optimum, ζ should be reduced so that the contribution of the penalty function is essentially negligible (so that no bias is introduced in the parameter estimates).

With a few minor modifications, the Gauss-Newton method presented in Chapter 4 can be used to obtain the unknown parameters. If we consider Taylor series expansion of the penalty function around the current estimate of the parameter we have,

$$\Theta_i(k_i^{(j+1)}) = \Theta_i(k_i^{(j)}) + \left(\frac{\partial \Theta_i}{\partial k_i}\right)^{(j)} \Delta k_i^{(j+1)} + \\ + \frac{1}{2} \Delta k_i^{(j+1)} \left(\frac{\partial^2 \Theta_i}{\partial k_i^2}\right)^{(j)} \Delta k_i^{(j+1)} \tag{9.21}$$

where

$$\left(\frac{\partial \Theta_i}{\partial k_i}\right)^{(j)} = -\frac{k_{max,i} - k_{min,i}}{\left(k_i^{(j)} - k_{min,i}\right)^2} + \frac{k_{max,i} - k_{min,i}}{\left(k_{max,i} - k_i^{(j)}\right)^2} \tag{9.22}$$

and

$$\left(\frac{\partial^2 \Theta_i}{\partial k_i^2}\right)^{(j)} = 2\left(\frac{k_{max,i} - k_{min,i}}{\left(k_i^{(j)} - k_{min,i}\right)^3} - \frac{k_{max,i} - k_{min,i}}{\left(k_{max,i} - k_i^{(j)}\right)^3}\right) \tag{9.23}$$

Subsequent use of the stationary condition $(\partial S_p/\partial \mathbf{k}^{(j+1)}) = \mathbf{0}$, yields the normal equations

$$\mathbf{A}\Delta\mathbf{k}^{(j+1)} = \mathbf{b} \tag{9.24}$$

The *diagonal* elements of matrix **A** are given by

$$\mathbf{A}(i,i) = \mathbf{A}_{GN}(i,i) + \left(\frac{\partial^2 \Theta_i}{\partial k_i^{\,2}}\right)^{(j)} \tag{9.25}$$

and elements of vector **b** are given by

$$\mathbf{b}(i) = \mathbf{b}_{GN}(i) - \left(\frac{\partial \Theta_i}{\partial k_i}\right)^{(j)} \tag{9.26}$$

where matrix $\mathbf{A}_{GN}$ and vector $\mathbf{b}_{GN}$ are those given in Chapter 4 for the Gauss-Newton method. The above equations apply equally well to differential equation models. In this case $\mathbf{A}_{GN}$ and $\mathbf{b}_{GN}$ are those given in Chapter 6.

9.2.1.3 Bisection Rule

If we are certain that the optimum parameter estimates lie well within the constraint boundaries, the simplest way to ensure that the parameters stay within the boundaries is through the use of the bisection rule. Namely, during each iteration of the Gauss-Newton method, if anyone of the new parameter estimates lie beyond its boundaries, then vector $\Delta\mathbf{k}^{(j+1)}$ is halved, until all the parameter constraints are satisfied. Once the constraints are satisfied, we proceed with the determination of the step-size that will yield a reduction in the objective function as already discussed in Chapters 4 and 6.

Our experience with algebraic and differential equation models has shown that this is indeed the easiest and most effective approach to use. It has been implemented in the computer programs provided with this book.

9.2.2 The Kuhn-Tucker Conditions

The most general case is covered by the well-known *Kuhn-Tucher conditions* for optimality. Let us assume the most general case where we seek the unknown parameter vector **k**, that will

Minimize

$$S(\mathbf{k}) = \sum_{i=1}^{N} \mathbf{e}_i^T \mathbf{Q}_i \mathbf{e}_i \tag{9.27a}$$

Subject to $\varphi_i(\mathbf{k}) = 0 \quad ; \quad i=1,2,\ldots,n_\varphi$ (9.27b)

and $\psi_i(\mathbf{k}) \geq 0 \; ; \; i= n_\varphi+1,\, n_\varphi+2,\ldots,\, n_\varphi+n_\psi$ (9.27c)

This constrained minimization problem is again solved by introducing n_φ Lagrange multipliers for the equality constraints, n_ψ Lagrange multipliers for the inequality constraints and by forming the augmented objective function (Langrangian)

$$S_{LG}(\mathbf{k},\boldsymbol{\omega}) = \sum_{i=1}^{N} \mathbf{e}_i^T \mathbf{Q}_i \mathbf{e}_i + \sum_{i=1}^{n_\varphi} \omega_i \varphi_i - \sum_{i=1}^{n_\psi} \omega_{n_\varphi+i} \psi_{n_\varphi+i} \tag{9.28}$$

The necessary conditions for $\mathbf{k}^*$ to be the optimal parameter values corresponding to a minimum of the augmented objective function $S_{LG}(\mathbf{k},\boldsymbol{\omega})$ are given by Edgar and Himmelblau (1988) and Gill et al. (1981) and are briefly presented here.

The Langrangian function must be at a stationary point, i.e.,

$$\frac{\partial S_{LG}(\mathbf{k},\boldsymbol{\omega})}{\partial \mathbf{k}} = \mathbf{0} \quad \text{at } \mathbf{k}^*, \boldsymbol{\omega}^* \tag{9.29}$$

The constraints are satisfied at $\mathbf{k}^*$, i.e.,

$\varphi_i(\mathbf{k}^*) = 0 \quad ; \quad i=1,2,\ldots,n_\varphi$ (9.27b)

and

$\psi_i(\mathbf{k}^*) \geq 0 \quad ; \quad i= n_\varphi+1,\, n_\varphi+2,\ldots,\, n_\varphi+n_\psi$ (9.27c)

The Lagrange multipliers corresponding to the inequality constraints (ω_i, $i=n_\varphi+1, \ldots, n_\varphi+n_\psi$) are non-negative, in particular,

$\omega_i > 0$; for all active inequality constraints (when $\psi_i(\mathbf{k}^*)=0$) (9.30a)

and

$\omega_i = 0$; for all inactive inequality constraints (when $\psi_i(\mathbf{k}^*)>0$) (9.30b)

Based on the above, we can develop an "adaptive" Gauss-Newton method for parameter estimation with equality constraints whereby the set of active constraints (which are all equalities) is updated at each iteration. An example is provided in Chapter 14 where we examine the estimation of binary interactions parameters in cubic equations of state subject to predicting the correct phase behavior (i.e., avoiding erroneous two-phase split predictions under certain conditions).

10

Gauss-Newton Method for Partial Differential Equation (PDE) Models

In this chapter we concentrate on dynamic, distributed systems described by partial differential equations. Under certain conditions, some of these systems, particularly those described by linear PDEs, have analytical solutions. If such a solution does exist and the unknown parameters appear in the solution expression, the estimation problem can often be reduced to that for systems described by algebraic equations. However, most of the time, an analytical solution cannot be found and the PDEs have to be solved numerically. This case is of interest here. Our general approach is to convert the partial differential equations (PDEs) to a set of ordinary differential equations (ODEs) and then employ the techniques presented in Chapter 6 taking into consideration the high dimensionality of the problem.

10.1 FORMULATION OF THE PROBLEM

When dealing with systems described by PDEs two additional complications arise. First, the unknown parameters may appear in the PDE or in the boundary conditions or both. Second, the measurements are in general a function of time and space. However, we could also have measurements which are integrals (average values) over space or time or both.

Let us consider the general class of systems described by a system of *n* non-linear parabolic or hyperbolic partial differential equations. For simplicity we assume that we have only one spatial independent variable, z.

$$\frac{\partial w_j}{\partial t} = f_j(t, z, \mathbf{w}(t,z), \frac{\partial \mathbf{w}}{\partial z}, \frac{\partial^2 \mathbf{w}}{\partial z^2}; \mathbf{k}) \quad ; \quad j=1,\ldots,n \tag{10.1}$$

with the given initial conditions,

$$w_j(t_0, z) = w_{j0}(z) \quad ; \quad j=1,\ldots,n \tag{10.2}$$

where **k** is the *p-dimensional* vector of unknown parameters. Furthermore, the appropriate boundary conditions are also given, i.e.,

$$\psi_j(t, z, \mathbf{w}(t,z), \frac{\partial \mathbf{w}}{\partial z}; \mathbf{k}) = 0 \quad ; \quad z \in \Omega \ , \quad j=1,\ldots,n \tag{10.3}$$

The distributed state variables $w_j(t,z)$, $j=1,\ldots,n$, are generally not all measured. Furthermore the measurements could be taken at certain points in space and time or they could be averages over space or time. If we define as **y** *the m-dimensional* output vector, each measured variable, $y_j(t)$, $j=1,\ldots,m$, is related to the state vector $\mathbf{w}(t,z)$ by any of the following relationships (Seinfeld and Lapidus, 1974):

(a) Measurements taken at a particular point in space, z_j, at the sampling times, $t_1, t_2,\ldots,t_N$, i.e.,

$$y_j(t_i) = h_j(t_i, z_j, \mathbf{w}(t_i, z_j)) \quad ; \quad j=1,\ldots,m_1 \tag{10.4}$$

(b) Measurements taken as an average over particular subspaces, Ω_j, at the sampling times, $t_1, t_2,\ldots,t_N$, i.e.,

$$y_j(t_i) = \int_{\Omega_j} h_j(t_i, z, w(t_i, z))dz \quad ; \quad j= m_1+1,\ldots,m_2 \tag{10.5}$$

(c) Measurements taken at a particular point in space, z_j, as a time average over successive sampling times, $t_1, t_2,\ldots,t_N$, i.e.,

$$y_j(t_i) = \int_{t_{i-1}}^{t_i} h_j(t, z_j, w(t, z_j))dt \quad ; \quad j= m_2+1,\ldots,m_3 \tag{10.6}$$

(d) Measurements taken as an average over particular subspaces, Ω_j, and as a time average over successive sampling times, $t_1, t_2, \ldots, t_N$, i.e.,

$$y_j(t_i) = \int_{\Omega_j} \int_{t_{i-1}}^{t_i} h_j(t, z, w(t, z)) dt dz \quad ; \; j = m_3+1, \ldots, m \tag{10.7}$$

As usual, it is assumed that the actual measurements of the output vector are related to the model calculated values by

$$\hat{y}_j(t_i) = y_j(t_i) + \varepsilon_{j,i} \quad ; \; i=1,\ldots,N, \;\; j=1,\ldots,m \tag{10.8a}$$

or in vector form

$$\hat{\mathbf{y}}(t_i) = \mathbf{y}(t_i) + \boldsymbol{\varepsilon}_i \quad ; \; i=1,\ldots,N \tag{10.8b}$$

The unknown parameter vector $\mathbf{k}$ is obtained by minimizing the corresponding least squares objective function where the weighting matrix $\mathbf{Q}_i$ is chosen based on the statistical characteristics of the error term $\boldsymbol{\varepsilon}_i$ as already discussed in Chapter 2.

10.2 THE GAUSS-NEWTON METHOD FOR PDE MODELS

Following the same approach as in Chapter 6 for ODE models, we linearize the output vector around the current estimate of the parameter vector $\mathbf{k}^{(j)}$ to yield

$$\mathbf{y}(t, z, \mathbf{k}^{(j+1)}) = \mathbf{y}(t, z, \mathbf{k}^{(j)}) + \left(\frac{\partial \mathbf{y}^T}{\partial \mathbf{k}}\right)^T \Delta\mathbf{k}^{(j+1)} \tag{10.9}$$

The output sensitivity matrix $(\partial \mathbf{y}^T/\partial \mathbf{k})^T$ is related to the sensitivity coefficient matrix defined as

$$\mathbf{G}(t, z) = \left(\frac{\partial \mathbf{w}^T}{\partial \mathbf{k}}\right)^T \tag{10.10a}$$

or equivalently

$$G_{ji}(t, z) = \left(\frac{\partial w_j}{\partial k_i}\right) \quad ; \; j=1,\ldots,n \;\; i=1,\ldots,p \tag{10.10b}$$

The relationship between $(\partial y_j/\partial k_i)$ and G_{ji} is obtained by implicit differentiation of Equations 10.4 - 10.7 depending on the type of measurements we have. Namely,

(a) if the output relationship is given by Equation 10.4, then

$$\frac{\partial y_j}{\partial k_i} = \sum_{r=1}^{n} \frac{\partial h_j}{\partial w_r}\frac{\partial w_r}{\partial k_i} \quad ; \; j=1,\ldots,m_1, \; i=1,\ldots,p \tag{10.11a}$$

or

$$\frac{\partial y_j}{\partial k_i} = \sum_{r=1}^{n} \frac{\partial h_j}{\partial w_r} G_{ri} \quad ; \; j=1,\ldots,m_1, \; i=1,\ldots,p \tag{10.11b}$$

(b) if the output relationship is given by Equation 10.5, then

$$\frac{\partial y_j}{\partial k_i} = \int_{\Omega_j} \left(\sum_{r=1}^{n} \frac{\partial h_j}{\partial w_r}\frac{\partial w_r}{\partial k_i} \right) dz \quad ; \; j=m_1+1,\ldots,m_2, \; i=1,\ldots,p \tag{10.12a}$$

or

$$\frac{\partial y_j}{\partial k_i} = \int_{\Omega_j} \left(\sum_{r=1}^{n} \frac{\partial h_j}{\partial w_r} G_{ri} \right) dz \quad ; \; j=m_1+1,\ldots,m_2, \; i=1,\ldots,p \tag{10.12b}$$

(c) if the output relationship is given by Equation 10.6, then

$$\frac{\partial y_j}{\partial k_i} = \int_{t_{i-1}}^{t_i} \left(\sum_{r=1}^{n} \frac{\partial h_j}{\partial w_r}\frac{\partial w_r}{\partial k_i} \right) dt \quad j=m_2+1,\ldots,m_3, \; i=1,\ldots,p \tag{10.13a}$$

or

$$\frac{\partial y_j}{\partial k_i} = \int_{t_{i-1}}^{t_i} \left(\sum_{r=1}^{n} \frac{\partial h_j}{\partial w_r} G_{ri} \right) dt \quad j=m_2+1,\ldots,m_3, \; i=1,\ldots,p \tag{10.13b}$$

(d) if the output relationship is given by Equation 10.7, then

$$\frac{\partial y_j}{\partial k_i} = \int_{\Omega_j} \int_{t_{i-1}}^{t_i} \left(\sum_{r=1}^{n} \frac{\partial h_j}{\partial w_r}\frac{\partial w_r}{\partial k_i} \right) dtdz \quad ; j=m_3+1,\ldots,m, \; i=1,\ldots,p \tag{10.14a}$$

or

$$\frac{\partial y_j}{\partial k_i} = \int_{\Omega_j}\int_{t_{i-1}}^{t_i}\left(\sum_{r=1}^{n}\frac{\partial h_j}{\partial w_r}G_{ri}\right)dtdz \quad ;j=m_3+1,\ldots,m,\ i=1,\ldots,p \qquad (10.14b)$$

Obviously in order to implement the Gauss-Newton method we have to compute the sensitivity coefficients $G_{ji}(t,z)$. In this case however, the sensitivity coefficients are obtained by solving a set of PDEs rather than ODEs.

The governing partial differential equation for $\mathbf{G}(t,z)$ is obtained by differentiating both sides of Equation 10.1 with respect to $\mathbf{k}$ and reversing the order of differentiation. The resulting PDE for $G_{ji}(t,z)$ is given by (Seinfeld and Lapidus, 1974),

$$\frac{\partial G_{ji}}{\partial t} = \sum_{r=1}^{n}\left[\left(\frac{\partial f_j}{\partial w_r}\right)G_{ri} + \frac{\partial f_j}{\partial\left(\frac{\partial w_r}{\partial z}\right)}\left(\frac{\partial G_{ri}}{\partial z}\right) + \frac{\partial f_j}{\partial\left(\frac{\partial^2 w_r}{\partial z^2}\right)}\left(\frac{\partial^2 G_{ri}}{\partial z^2}\right)\right] + \left(\frac{\partial f_j}{\partial k_i}\right) \quad ;\ j=1,\ldots,n,\quad i=1,\ldots,p \qquad (10.15)$$

The initial condition for $G_{ji}(t,z)$ is obtained by differentiating both sides of Equation 10.2 with respect to k_i. Equivalently we can simply argue that since the initial condition for $w_j(t,z)$ is known, it does not depend on the parameter values and hence,

$$G_{ji}(t_0,x) = 0 \quad ;\ j=1,\ldots,n,\quad i=1,\ldots,p \qquad (10.16)$$

The boundary condition for $G_{ji}(t,z)$ is obtained by differentiating both sides of Equation 10.3 with respect to k_i to yield,

$$\sum_{r=1}^{n}\left[\left(\frac{\partial \psi_j}{\partial w_r}\right)G_{ri} + \frac{\partial \psi_j}{\partial\left(\frac{\partial w_r}{\partial z}\right)}\left(\frac{\partial G_{ri}}{\partial z}\right)\right] + \frac{\partial \psi_j}{\partial k_i} = 0 \quad ;\ j=1,\ldots,n \qquad (10.17)$$

Equations 10.15 to 10.17 define a set of $(n\times p)$ partial differential equations for the sensitivity coefficients that need to be solved at each iteration of the Gauss-Newton method together with the n PDEs for the state variables.

Having computed $(\partial\mathbf{y}^T/\partial\mathbf{k})^T$ we can proceed and obtain a linear equation for $\Delta\mathbf{k}^{(j+1)}$ by substituting Equation 10.9 into the least squares objective function and using the stationary criterion $(\partial S/\partial\mathbf{k}^{(j+1)}) = \mathbf{0}$. The resulting equation is of the form

$$\mathbf{A}\Delta\mathbf{k}^{(j+1)} = \mathbf{b} \tag{10.18}$$

where

$$\mathbf{A} = \sum_{i=1}^{N} \left(\frac{\partial \mathbf{y}^T}{\partial \mathbf{k}}\right) \mathbf{Q}_i \left(\frac{\partial \mathbf{y}^T}{\partial \mathbf{k}}\right)^T \tag{10.19}$$

and

$$\mathbf{b} = \sum_{i=1}^{N} \left(\frac{\partial \mathbf{y}^T}{\partial \mathbf{k}}\right) \mathbf{Q}_i \left(\hat{\mathbf{y}}(t_i) - \mathbf{y}(t_i)\right)^T \tag{10.20}$$

At this point we can summarize the steps required to implement the Gauss-Newton method for PDE models. At each iteration, given the current estimate of the parameters, $\mathbf{k}^{(j)}$, we obtain $\mathbf{w}(t,z)$ and $\mathbf{G}(t,z)$ by solving numerically the state and sensitivity partial differential equations. Using these values we compute the model output, $\mathbf{y}(t_i,\mathbf{k}^{(j)})$, and the output sensitivity matrix, $(\partial\mathbf{y}^T/\partial\mathbf{k})^T$ for each data point i=1,...,N. Subsequently, these are used to set up matrix **A** and vector **b**. Solution of the linear equation yields $\Delta\mathbf{k}^{(j+1)}$ and hence $\mathbf{k}^{(j+1)}$ is obtained. The bisection rule to yield an acceptable step-size at each iteration of the Gauss-Newton method should also be used.

This approach is useful when dealing with relatively simple partial differential equation models. Seinfeld and Lapidus (1974) have provided a couple of numerical examples for the estimation of a single parameter by the steepest descent algorithm for systems described by one or two simultaneous PDEs with simple boundary conditions.

In our opinion the above formulation does not provide any computational advantage over the approach described next since the PDEs (state and sensitivity equations) need to be solved numerically.

10.3 THE GAUSS-NEWTON METHOD FOR DISCRETIZED PDE MODELS

Rather than discretizing the PDEs for the state variables and the sensitivity coefficients in order to solve them numerically at each iteration of the Gauss-Newton method, it is preferable to discretize first the governing PDEs and then estimate the unknown parameters. Essentially, instead of a PDE model, we are now dealing with an ODE model of high dimensionality.

If for example we discretize the region over which the PDE is to be solved into *M* grid blocks, use of finite differences (or any other discretization scheme) to approximate the spatial derivatives in Equation 10.1 yields the following system of ODEs:

$$\frac{d\mathbf{x}}{dt} = \boldsymbol{\varphi}(t, \mathbf{x}; \mathbf{k}) \tag{10.21}$$

where the new *nM-dimensional* state vector **x** is defined as

$$\mathbf{x}(t) = \begin{bmatrix} \mathbf{w}(t, z_1) \\ \mathbf{w}(t, z_2) \\ \mathbf{w}(t, z_3) \\ \vdots \\ \mathbf{w}(t, z_M) \end{bmatrix} \tag{10.22}$$

where $\mathbf{w}(t,z_i)$ is the *n-dimensional* vector of the original state variables at the ith grid block. The solution of this parameter estimation problem can readily be obtained by the Gauss-Newton method for ODE models presented in Chapter 6 as long as the high dimensionality of the problem is taken into consideration.

10.3.1 Efficient Computation of the Sensitivity Coefficients

The computation of the sensitivity coefficients in PDE models is a highly demanding operation. For example if we consider a typical three-dimensional three-phase reservoir simulation model, we could easily have 1,000 to 10,000 grid blocks with 20 unknown parameters representing unknown average porosities and permeabilties in a ten zone reservoir structure. This means that the state vector (Equation 10.22) will be comprised of 3,000 to 30,000 state variables since for each grid block we have three state variables (pressure and two saturations). The corresponding number of sensitivity coefficients will be 60,000 to 600,000! Hence, if one implements the Gauss-Newton method the way it was suggested in Chapter 6, 63,000 to 630,000 ODEs would have to be solved simultaneously. A rather formidable task.

Therefore, efficient computation schemes of the state and sensitivity equations are of paramount importance. One such scheme can be developed based on the *sequential integration of the sensitivity coefficients*. The idea of decoupling the direct calculation of the sensitivity coefficients from the solution of the model equations was first introduced by Dunker (1984) for stiff chemical mechanisms such as the oxidation of hydrocarbons in the atmosphere, the pyrolysis of ethane, etc. Leis and Kramer (1988) presented implementation guidelines and an error control strategy that ensures the independent satisfaction of local error criteria by both numerical solutions of the model and the sensitivity equations. The procedure has been adapted and successfully used for automatic history matching (i.e., parameter estimation) in reservoir engineering (Tan and Kalogerakis, 1991; Tan, 1991).

Due to the high dimensionality of 3-D models, Euler's method is often used for the integration of the spatially discretized model equations. There are several implementations of Euler's method (e.g., semi-implicit, fully implicit or adaptive implicit). Fully implicit methods require much more extensive coding and significant computational effort in matrix operations with large storage requirements. The work required during each time-step is considerably more than other solution methods. However, these disadvantages are fully compensated by the stability of the solution method which allows the use of much larger time-steps in situations that exhibit large pore volume throughputs, well coning, gas percolation or high transmissibility variation (Tan, 1991). Obviously, in parameter estimation fully implicit formulations are preferable since stable integration of the state and sensitivity ODEs for a wide range of parameter values is highly desirable.

As we have already pointed out in Chapter 6, the Jacobean matrix of the state and sensitivity equations is the same. As a result, we can safely assume that the maximum allowable time-step that has been determined during the implicit integration of state equations should also be acceptable for the integration of the sensitivity equations. If the state and sensitivity equations were simultaneously integrated, for each reduction in the time-step any work performed in the integration of the sensitivity equations would have been in vain. Therefore, it is proposed to integrate the sensitivity equations only after the integration of the model equations has converged for each time-step. This is shown schematically in Figure 10.1

The integration of the state equations (Equation 10.21) by the fully implicit Euler's method is based on the iterative determination of $\mathbf{x}(t_{i+1})$. Thus, having $\mathbf{x}(t_i)$ we solve the following difference equation for $\mathbf{x}(t_{i+1})$.

$$\frac{\mathbf{x}(t_{i+1})-\mathbf{x}(t_i)}{\Delta t} = \boldsymbol{\varphi}(t_{i+1},\mathbf{x}(t_{i+1})) \tag{10.23}$$

where Δt is $t_{i+1} - t_i$.

If we denote by $\mathbf{x}^{(j)}(t_{i+1})$, $\mathbf{x}^{(j+1)}(t_{i+1})$, $\mathbf{x}^{(j+2)}(t_{i+1})$... the iterates for the determination of $\mathbf{x}(t_{i+1})$, linearization of the right hand side of Equation 10.23 around $\mathbf{x}^{(j)}(t_{i+1})$ yields

$$\frac{\mathbf{x}^{(j+1)}(t_{i+1})-\mathbf{x}(t_i)}{\Delta t} = \boldsymbol{\varphi}(t_{i+1},\mathbf{x}^{(j)}(t_{i+1})) + \left(\frac{\partial \boldsymbol{\varphi}^T}{\partial \mathbf{x}}\right)^T \Delta\mathbf{x}^{(j+1)} \tag{10.24}$$

which upon rearrangement yields

$$\left[\left(\frac{\partial \boldsymbol{\varphi}^T}{\partial \mathbf{x}}\right)^T - \frac{1}{\Delta t}\mathbf{I}\right]\Delta\mathbf{x}^{(j+1)} = \frac{\mathbf{x}^{(j)}(t_{i+1}) - \mathbf{x}(t_i)}{\Delta t} - \boldsymbol{\varphi}(t_{i+1}, \mathbf{x}^{(j)}(t_{i+1})) \quad (10.25)$$

where $\Delta\mathbf{x}^{(j+1)} = \mathbf{x}^{(j+1)}(t_{i+1}) - \mathbf{x}^{(j)}(t_{i+1})$ and the Jacobean matrix $(\partial\boldsymbol{\varphi}^T/\partial\mathbf{x})^T$ is evaluated at $\mathbf{x}^{(j)}(t_{i+1})$.

Equation 10.25 is of the form $\mathbf{A}\Delta\mathbf{x}=\mathbf{b}$ which can be solved for $\Delta\mathbf{x}^{(j+1)}$ and thus $\mathbf{x}^{(j+1)}(t_{i+1})$ is obtained. Normally, we converge to $\mathbf{x}(t_{i+1})$ in very few iterations. If however, convergence is not achieved or the integration error tolerances are not satisfied, the time-step is reduced and the computations are repeated.

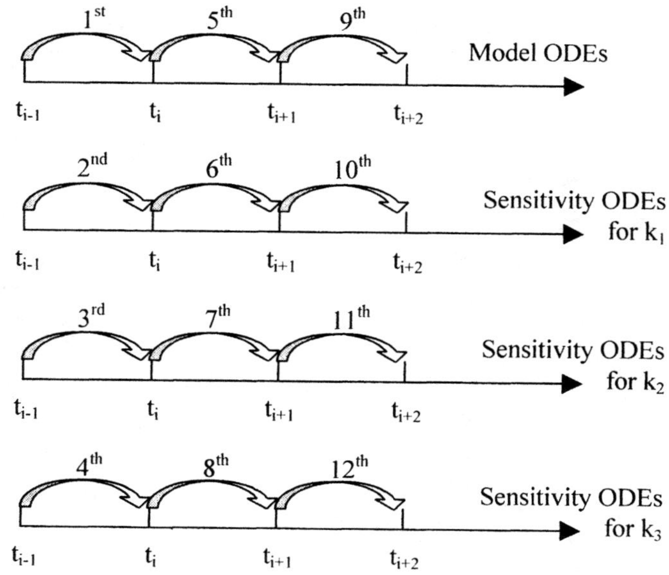

Figure 10.1 Schematic diagram of the sequential solution of model and sensitivity equations. The order is shown for a three parameter problem. Steps 1, 5 and 9 involve iterative solution that requires a matrix inversion at each iteration of the fully implicit Euler's method. All other steps (i.e., the integration of the sensitivity equations) involve only one matrix multiplication each.

For the solution of Equation 10.25 the inverse of matrix **A** is computed by iterative techniques as opposed to direct methods often employed for matrices of low order. Since matrix **A** is normally very large, its inverse is more economically found by an iterative method. Many iterative methods have been published such as successive over-relaxation (SOR) and its variants, the strongly implicit procedure (SIP) and its variants, Orthomin and its variants (Stone, 1968), nested factorization (Appleyard and Chesire, 1983) and iterative D4 with minimization (Tan and Letkeman, 1982) to name a few.

Once $\mathbf{x}(t_{i+1})$ has been computed, we can proceed for the computation of the sensitivity coefficients at t_{i+1}. Namely, if we denote as $\mathbf{g}_r(t_{i+1})$ the rth column of sensitivity matrix $\mathbf{G}(t_{i+1})$, $\mathbf{g}_r(t_{i+1})$ is to be obtained by solving the linear ODE

$$\frac{d\mathbf{g}_r}{dt} = \left(\frac{\partial \boldsymbol{\varphi}^T}{\partial \mathbf{x}}\right)^T \mathbf{g}_r(t) + \left(\frac{\partial \boldsymbol{\varphi}}{\partial k_r}\right) \tag{10.26}$$

which for the fully implicit Euler's method yields the following difference equation

$$\frac{\mathbf{g}_r(t_{i+1}) - \mathbf{g}_r(t_i)}{\Delta t} = \left(\frac{\partial \boldsymbol{\varphi}^T}{\partial \mathbf{x}}\right)^T \mathbf{g}_r(t_{i+1}) + \left(\frac{\partial \boldsymbol{\varphi}}{\partial k_r}\right) \tag{10.27}$$

where the Jacobean matrix $(\partial\boldsymbol{\varphi}^T/\partial\mathbf{x})^T$ and vector $(\partial\boldsymbol{\varphi}/\partial k_r)$ are evaluated at $\mathbf{x}(t_{i+1})$. Upon rearrangement we have,

$$\left[\left(\frac{\partial \boldsymbol{\varphi}^T}{\partial \mathbf{x}}\right)^T - \frac{1}{\Delta t}\mathbf{I}\right] \mathbf{g}_r(t_{i+1}) = -\frac{\mathbf{g}_r(t_i)}{\Delta t} - \left(\frac{\partial \boldsymbol{\varphi}}{\partial k_r}\right) \tag{10.28}$$

The solution of Equation 10.28 is obtained *in one step by performing a simple matrix multiplication* since the inverse of the matrix on the left hand side of Equation 10.28 is already available from the integration of the state equations. Equation 10.28 is solved for r=1,...,p and thus the whole sensitivity matrix $\mathbf{G}(t_{i+1})$ is obtained as $[\mathbf{g}_1(t_{i+1}), \mathbf{g}_2(t_{i+1}),..., \mathbf{g}_p(t_{i+1})]$. The computational savings that are realized by the above procedure are substantial, especially when the number of unknown parameters is large (Tan and Kalogerakis, 1991). With this modification the computational requirements of the Gauss-Newton method for PDE models become reasonable and hence, the estimation method becomes implementable.

A numerical example for the estimation of unknown parameters in PDE models is provided in Chapter 18 where we discuss automatic history matching of reservoir simulation models.

11

Statistical Inferences

Once we have estimated the unknown parameters that appear in an algebraic or ODE model, it is quite important to perform a few additional calculations to establish estimates of the standard error in the parameters and in the expected response variables. These additional computational steps are very valuable as they provide us with a quantitative measure of the quality of the overall fit and inform us how trustworthy the parameter estimates are.

11.1 INFERENCES ON THE PARAMETERS

When the Gauss-Newton method is used to estimate the unknown parameters, we linearize the model equations and at each iteration we solve the corresponding linear least squares problem. As a result, the estimated parameter values have linear least squares properties. Namely, the parameter estimates are normally distributed, unbiased (i.e., $E(\mathbf{k}^*)=\mathbf{k}$) and their covariance matrix is given by

$$COV(\mathbf{k}^*) = \sigma_\varepsilon^2 [\mathbf{A}^*]^{-1} \tag{11.1}$$

where $\mathbf{A}^*$ is matrix $\mathbf{A}$ evaluated at $\mathbf{k}^*$. It should be noted that for linear least squares matrix $\mathbf{A}$ is independent of the parameters while this is clearly not the case for nonlinear least squares problems. The required estimate of the variance σ_ε^2 is obtained from

$$\hat{\sigma}_\varepsilon^2 = \frac{S(\mathbf{k}^*)}{(\text{d.f.})} = \frac{S(\mathbf{k}^*)}{\text{N}\,m-p} \tag{11.2}$$

where (d.f.)= Nm–p are the degrees of freedom, namely the total number of measurements minus the number of unknown parameters.

The above expressions for the $COV(\mathbf{k}^*)$ and $\hat{\sigma}_\varepsilon^2$ are valid, if the statistically correct choice of the weighting matrix $\mathbf{Q}_i$, (i=1,...,N) is used in the formulation of the problem. Namely, if the errors in the response variables (ε_i, i=1,...,N) are normally distributed with zero mean and covariance matrix,

$$COV(\varepsilon_i) = \sigma_\varepsilon^2 \mathbf{M}_i \tag{11.3}$$

we should use $[\mathbf{M}_i]^{-1}$ as the weighting matrix $\mathbf{Q}_i$ where the matrices $\mathbf{M}_i$, i=1,...,N are *known* whereas the scaling factor, σ_ε^2, could be unknown. Based on the structure of $\mathbf{M}_i$ we arrive at the various cases of least squares estimation (Simple LS, Weighted LS or Generalized LS) as described in detail in Chapter 2.

Although the computation of $COV(\mathbf{k}^*)$ is a simple extra step after convergence of the Gauss-Newton method, we are not obliged to use the Gauss-Newton method for the search of the best parameter values. Once **k*** has been obtained using any search method, one can proceed and compute the sensitivity coefficients by setting up matrix **A** and thus quantify the uncertainty in the estimated parameter values by estimating $COV(\mathbf{k}^*)$.

Approximate inference regions for nonlinear models are defined by analogy to the linear models. In particular, *the (1–α)100% joint confidence region* for the parameter vector **k** is described by the ellipsoid,

$$[\mathbf{k}-\mathbf{k}^*]^T [\mathbf{A}^*]^{-1}[\mathbf{k}-\mathbf{k}^*] = p\hat{\sigma}_\varepsilon^2 F^\alpha_{p,Nm-p} \tag{11.4a}$$

or

$$[\mathbf{k}-\mathbf{k}^*]^T [\mathbf{A}^*]^{-1}[\mathbf{k}-\mathbf{k}^*] = \frac{p\,S(\mathbf{k}^*)}{N\,m-p} F^\alpha_{p,Nm-p} \tag{11.4b}$$

where α is the selected probability level in Fisher's F-distribution and $F^\alpha_{p,Nm-p}$ is obtained from the F-distribution tables with ν_1=p and ν_2=(Nm–p) degrees of freedom. The corresponding *(1–α)100% marginal confidence interval* for each parameter, k_i, i=1,2,...,p, is given by

$$k_i^* - t^\nu_{\alpha/2}\hat{\sigma}_{k_i} \le k_i \le k_i^* + t^\nu_{\alpha/2}\hat{\sigma}_{k_i} \tag{11.5}$$

where $t^\nu_{\alpha/2}$ is obtained from the tables of Student's T-distribution with ν=(Nm–p) degrees of freedom. The standard error of parameter k_i, $\hat{\sigma}_{k_i}$, is obtained as the square root of the corresponding diagonal element of the inverse of matrix **A*** multiplied by $\hat{\sigma}_\varepsilon$, i.e.,

$$\hat{\sigma}_{k_i} = \hat{\sigma}_\varepsilon \sqrt{\left\{[\mathbf{A}^*]^{-1}\right\}_{ii}} \tag{11.6}$$

It is reminded that for v≥30 the approximation $t^{v}_{\alpha/2} \approx z_{\alpha/2}$ can be used where $z_{\alpha/2}$ is obtained from the standard normal distribution tables. Simply put, when we have many data points, we can simply take as the 95% confidence interval twice the standard error (recall that $z_{0.025}$=1.96 whereas $t^{30}_{0.025}$ =2.042).

The linear approximation employed by the Gauss-Newton method in solving the nonlinear least squares problem enables us to obtain inference regions for the parameters very easily. However, *these regions are only approximate* and practice has shown that in many cases they can be very misleading (Watts and Bates, 1988). Nonetheless, even if the regions are not exact, we obtain a very good idea of the correlation among the parameters.

We can compute the *exact (1−α)100% joint parameter likelihood region* using the equation given below

$$S(\mathbf{k}) = S(\mathbf{k}^*)\left[1+\frac{p}{Nm-p}F^{\alpha}_{p,Nm-p}\right] \tag{11.7}$$

The computation of the above surface in the parameter space is not trivial. For the two-parameter case (*p*=2), the joint confidence region on the k_1-k_2 plane can be determined by using any contouring method. The contour line is approximated from many function evaluations of S(**k**) over a dense grid of (k_1, k_2) values.

11.2 INFERENCES ON THE EXPECTED RESPONSE VARIABLES

Having determined the uncertainty in the parameter estimates, we can proceed and obtain confidence intervals for the expected mean response. Let us first consider models described by a set of nonlinear algebraic equations, **y**=**f**(**x**,**k**). The *100(1−α)% confidence interval of the expected mean response* of the variable y_j at $\mathbf{x}_0$ is given by

$$f_j(\mathbf{x}_0,\mathbf{k}^*) - t^{v}_{\alpha/2}\hat{\sigma}_{y_{j0}} \le \mu_{y_{j0}} \le f_j(\mathbf{x}_0,\mathbf{k}^*) + t^{v}_{\alpha/2}\hat{\sigma}_{y_{j0}} \tag{11.8}$$

where $t^{v}_{\alpha/2}$ is obtained from the tables of Student's T-distribution with v=(N*m*−*p*) degrees of freedom. Based on the linear approximation of the model equations, we have

$$\mathbf{y}_0(\mathbf{k}) = \mathbf{f}(\mathbf{x}_0,\mathbf{k}^*) + \left(\frac{\partial \mathbf{f}^T}{\partial \mathbf{k}}\right)^T [\mathbf{k} - \mathbf{k}^*] \tag{11.9}$$

with the partial derivative $(\partial\mathbf{f}^T/\partial\mathbf{k})^T$ evaluated at $\mathbf{x}_0$ and $\mathbf{k}^*$. Taking variances from both sides we have

$$COV(\mathbf{y}_0) = \left(\frac{\partial \mathbf{f}^T}{\partial \mathbf{k}}\right)^T COV(\mathbf{k}^*) \left(\frac{\partial \mathbf{f}^T}{\partial \mathbf{k}}\right) \tag{11.10a}$$

Substitution of the expression for $COV(\mathbf{k}^*)$ yields,

$$COV(\mathbf{y}_0) = \hat{\sigma}^2_{\varepsilon} \left(\frac{\partial \mathbf{f}^T}{\partial \mathbf{k}}\right)^T [\mathbf{A}^*]^{-1} \left(\frac{\partial \mathbf{f}^T}{\partial \mathbf{k}}\right) \tag{11.10b}$$

The standard prediction error of y_{j0}, $\hat{\sigma}_{y_{j0}}$, is the square root of the jth diagonal element of $COV(\mathbf{y}_0)$, namely,

$$\hat{\sigma}_{y_{j0}} = \hat{\sigma}_{\varepsilon} \sqrt{\left(\frac{\partial f_j}{\partial \mathbf{k}}\right)^T [\mathbf{A}^*]^{-1} \left(\frac{\partial f_j}{\partial \mathbf{k}}\right)} \tag{11.11}$$

Equation 11.8 represents the confidence interval for the *mean* expected response rather than a *future observation (future measurement)* of the response variable, $\hat{\mathbf{y}}_0$. In this case, besides the uncertainty in the estimated parameters, we must include the uncertainty due to the measurement error (ε_0). The *(1−α)100% confidence interval* of $\hat{y}_{j0}$ is

$$f_j(\mathbf{x}_0,\mathbf{k}^*) - t^{v}_{\alpha/2}\hat{\sigma}_{\hat{y}_{j0}} \le \hat{y}_{j0} \le f_j(\mathbf{x}_0,\mathbf{k}^*) + t^{v}_{\alpha/2}\hat{\sigma}_{\hat{y}_{j0}} \tag{11.12}$$

where the standard prediction error of $\hat{y}_{j0}$ is given by

$$\hat{\sigma}_{y_{j0}} = \hat{\sigma}_{\varepsilon} \sqrt{1 + \left(\frac{\partial f_j}{\partial \mathbf{k}}\right)^T [\mathbf{A}^*]^{-1} \left(\frac{\partial f_j}{\partial \mathbf{k}}\right)} \tag{11.13}$$

Next let us turn our attention to models described by a set of ordinary differential equations. We are interested in establishing confidence intervals for each of the response variables y_j , $j=1,\ldots,m$ at any time $t=t_o$. The linear approximation of the output vector at time t_0,

$$\mathbf{y}(t_0,\mathbf{k}) = \mathbf{Cx}(t_0,\mathbf{k}^*) + \mathbf{CG}(t_0)[\mathbf{k}-\mathbf{k}^*] \tag{11.14}$$

yields the expression for $COV(\mathbf{y}(t_o))$,

$$COV(\mathbf{y}_0) = \mathbf{CG}(t_0)COV(\mathbf{k}^*)\mathbf{C}^T\mathbf{G}^T(t_0) \tag{11.15}$$

which can be rewritten as

$$COV(\mathbf{y}_0) = \hat{\sigma}_{\varepsilon}^2\,\mathbf{CG}(t_0)[\mathbf{A}^*]^{-1}\mathbf{C}^T\mathbf{G}^T(t_0) \tag{11.16}$$

with the sensitivity coefficients matrix $\mathbf{G}(t_0)$ evaluated at $\mathbf{k}^*$. The estimated standard prediction error of $y_j(t_0)$ is obtained as the square root of the jth diagonal element of $COV(\mathbf{y}(t_o))$.

$$\hat{\sigma}_{y_{jo}} = \hat{\sigma}_{\varepsilon}\sqrt{\left\{\mathbf{CG}(t_0)[\mathbf{A}^*]^{-1}\mathbf{C}^T\mathbf{G}^T(t_0)\right\}_{jj}} \tag{11.17}$$

Based on the latter, we can compute the *(1−α)100% confidence interval of the expected mean response* of y_j at $t=t_o$,

$$y_j(t_o,\mathbf{k}^*) - t^{\nu}_{\alpha/2}\hat{\sigma}_{y_{jo}} \le \mu_{y_{jo}} \le y_j(t_o,\mathbf{k}^*) + t^{\nu}_{\alpha/2}\hat{\sigma}_{y_{j0}} \tag{11.18}$$

If on the other hand we wish to compute the *(1−α)100% confidence interval of the response* of y_j at $t=t_o$, we must include the error term (ε_o) in the calculation of the standard error, namely we have

$$y_j(t_o,\mathbf{k}^*) - t^{\nu}_{\alpha/2}\hat{\sigma}_{\hat{y}_{jo}} \le \hat{y}_j(t_o) \le y_j(t_o,\mathbf{k}^*) + t^{\nu}_{\alpha/2}\hat{\sigma}_{\hat{y}_{j0}} \tag{11.19}$$

with

$$\hat{\sigma}_{\hat{y}_{jo}} = \hat{\sigma}_{\varepsilon}\sqrt{1 + \left\{\mathbf{C}\mathbf{G}(t_o)[\mathbf{A}^*]^{-1}\mathbf{C}^T\mathbf{G}^T(t_o)\right\}_{jj}} \tag{11.20}$$

11.3 MODEL ADEQUACY TESTS

There is a plethora of model adequacy tests that the user can employ to decide whether the assumed mathematical model is indeed adequate. Generally speaking these tests are based on the comparison of the experimental error variance estimated by the model to that obtained experimentally or through other means.

11.3.1 Single Response Models

Let us first consider models that have only one measured variable (m=1). We shall consider two cases. One where we know *precisely* the value of the experimental error variance and the other when we have an estimate of it. Namely, there is quantifiable uncertainty in our estimate of the experimental error variance.

<u>CASE 1:</u> σ_ε^2 is known precisely:

In this case we assume that we know precisely the value of the standard experimental error in the measurements (σ_ε). Using Equation 11.2 we obtain an estimate of the experimental error variance under the assumption that the model is adequate. Therefore, to test whether the model is adequate we simply need to test the hypothesis

$$H_0: \quad \sigma^2_{model} = \sigma^2_\varepsilon$$

$$H_1: \quad \sigma^2_{model} > \sigma^2_\varepsilon$$

at any desirable level of significance, e.g., α=0.05. Here with σ^2_{model} we denote the error variance estimated by the model equations (Equation 11.2); namely, $\hat{\sigma}^2_\varepsilon$ is an estimate of σ^2_{model} .

Since σ^2_ε is known exactly (i.e., there is no uncertainty in its value, it is a given number) the above hypothesis test is done through a χ^2-test. Namely,

$$\text{If} \quad \chi^2_{data} > \chi^2_{v=(Nm-p),1-\alpha} \quad => \quad \text{Reject } H_0$$

where

$$\chi^2_{data} = (Nm - p)\frac{\hat{\sigma}^2_\varepsilon}{\sigma^2_\varepsilon} = \frac{S(\mathbf{k}^*)}{\sigma^2_\varepsilon} \tag{11.21}$$

and $\chi^2_{v=(Nm-p),1-\alpha}$ is obtained from the tables of the χ^2-distribution with degrees of freedom $v=(Nm-p)$.

<u>CASE 2:</u> σ^2_ε is known approximately:

Let us assume that σ^2_ε is not known exactly, however, we have performed n *repeated* measurements of the response variable. From this small sample of multiple measurements we can determine the sample mean and sample variance. If s^2_ε is the sample estimate of σ^2_ε, estimated from the n repeated measurements it is given by

$$s^2_\varepsilon = \frac{1}{n-1}\sum_{i=1}^{n}(y_i - \bar{y})^2 \tag{11.22}$$

where the sample mean is obtained from

$$\bar{y} = \frac{1}{n}\sum_{i=1}^{n} y_i \tag{11.23}$$

Again, we test the hypothesis at any desirable level of significance, for example $\alpha=0.05$

$$H_0: \quad \sigma^2_{model} = \sigma^2_\varepsilon$$
$$H_1: \quad \sigma^2_{model} > \sigma^2_\varepsilon$$

In this case, since σ^2_ε is known only approximately, the above hypothesis is tested using an F-test, i.e.,

$$\text{If} \quad F_{data} > F^{v_1=(Nm-p),v_2=n-1}_{1-\alpha} \quad \Rightarrow \quad \text{Reject } H_0$$

where

$$F_{data} = \frac{\hat{\sigma}_{\varepsilon}^{2}}{s_{\varepsilon}^{2}} \tag{11.24}$$

and $F_{1-\alpha}^{\nu_1=d.f.,\nu_2=n-1}$ is obtained from the tables of the F-distribution.

11.3.2 Multivariate Models

Let us now consider models that have only more than one measured variable (m>1). The previously described model adequacy tests have multivariate extensions that can be found in several advanced statistics textbooks. For example, the book *Introduction to Applied Multivariate Statistics* by Srivastava and Carter (1983) presents several tests on covariance matrices.

In many engineering applications, however, we can easily reduce the problem to the univariate tests presented in the previous section by assuming that the covariance matrix of the errors can be written as

$$COV(\varepsilon_i) = \sigma_{\varepsilon}^{2}\mathbf{M}_i \quad ; \quad i=1,\ldots,N \tag{11.25}$$

where $\mathbf{M}_i$ are known matrices. Actually quite often we can further assume that the matrices $\mathbf{M}_i$, i=1,…,N are the same and equal to matrix $\mathbf{M}$.

An independent estimate of $COV(\varepsilon)$, $\hat{\boldsymbol{\Sigma}}$, that is required for the adequacy tests can be obtained by performing N_R repeated experiments as

$$\hat{\boldsymbol{\Sigma}} = \frac{1}{N_R}\sum_{l=1}^{N_R}[\hat{\mathbf{y}}_l - \bar{\mathbf{y}}][\hat{\mathbf{y}}_l - \bar{\mathbf{y}}]^T \tag{11.26}$$

or for the case of univariate tests, s_{ε}^{2}, the sample estimate of σ_{ε}^{2} in Equation 11.25, can be obtained from

$$s_{\varepsilon}^{2} = \frac{1}{(N_R - m)}\sum_{l=1}^{N_R}[\hat{\mathbf{y}}_l - \bar{\mathbf{y}}]^T M^{-1}[\hat{\mathbf{y}}_l - \bar{\mathbf{y}}] \tag{11.27}$$

12

Design of Experiments

It is quite obvious by now that the quality of the parameter estimates that have been obtained with any of the previously described techniques ultimately depends on "how good" the data at hand is. It is thus very important, when we do have the option, to design our experiments in such a way so that the information content of the data is the highest possible. Generally speaking, there are two approaches to experimental design: (1) factorial design and (2) sequential design. In sequential experimental design we attempt to satisfy one of the following two objectives: (i) estimate the parameters as accurately as possible or (ii) discriminate among several rival models and select the best one. In this chapter we briefly discuss the design of preliminary experiments (factorial designs), and then we focus our attention on the sequential experimental design where we examine algebraic and ODE models separately.

12.1 PRELIMINARY EXPERIMENTAL DESIGN

There are many books that address experimental design and present *factorial experimental design* in detail (for example *Design and Analysis of Experiments* by Montgomery (1997) or *Design of Experiments* by Anderson and McLean (1974)). As engineers, we are faced quite often with the need to design a set of preliminary experiments for a process that very little or essentially no information is available.

Factorial design, and in particular 2^k designs, represent a generally sound strategy. The independent variables (also called *factors* in experimental design) are assigned two values (a high and a low value) and experiments are conducted in all possible combinations of the independent variables (also called *treatments*). These experiments are easy to design and if the levels are chosen appropriately very valuable information about the model can be gathered.

For example, if we have two independent variables (x_1 and x_2) the following four (2^2) experiments can be readily designed:

Run	*Independent Variables*
1	(x_1-Low, x_2-Low)
2	(x_1-Low, x_2-High)
3	(x_1-High, x_2-Low)
4	(x_1-High, x_2-High)

If we have three independent variables to vary, the complete 2^3 factorial design corresponds to the following 8 experiments:

Run	*Independent Variables*
1	(x_1-Low, x_2-Low, x_3-Low)
2	(x_1-Low, x_2-Low, x_3-High)
3	(x_1-Low, x_2-High, x_3-Low)
4	(x_1-High, x_2-Low, x_3-Low)
5	(x_1-Low, x_2-High, x_3-High)
6	(x_1-High, x_2-Low, x_3-High)
7	(x_1-High, x_2-High, x_3-Low)
8	(x_1-High, x_2-High, x_3-High)

These designs are extremely powerful (from a statistical point of view) if we do not have a mathematical model of the system under investigation and we simply wish to establish the effect of each of these three independent variables (or their interaction) on the measured response variables.

This is rarely the case in engineering. Most of the time we do have some form of a mathematical model (simple or complex) that has several unknown parameters that we wish to estimate. In these cases the above designs are very straightforward to implement; however, the information may be inadequate if the mathematical model is nonlinear and comprised of several unknown parameters. In such cases, *multilevel factorial designs* (for example, 3^k or 4^k designs) may be more appropriate.

A typical 3^2 design for the case of two independent variables (x_1 and x_2) that each can assume three values (Low, Medium and High) takes the form:

Run	*Independent Variables*
1	(x_1-Low, x_2-Low)
2	(x_1-Low, x_2-Medium)
3	(x_1-Low, x_2-High)
4	(x_1-Medium, x_2-Low)
5	(x_1-Medium, x_2-Medium)
6	(x_1-Medium, x_2-High)
7	(x_1-High, x_2-Low)
8	(x_1-High, x_2-Medium)
9	(x_1-High, x_2-High)

The above experimental design constitute an excellent set of "preliminary experiments" for nonlinear models with several unknown parameters. Based on the analysis of these experiments we obtain estimates of the unknown parameters that we can use to design subsequent experiments in a rational manner taking advantage of all information gathered up to that point.

12.2 SEQUENTIAL EXPERIMENTAL DESIGN FOR PRECISE PARAMETER ESTIMATION

Let us assume that N experiments have been conducted up to now and given an estimate of the parameter vector based on the experiments performed up to now, we wish to design the next experiment so that we maximize the information that shall be obtained. In other words, the problem we are attempting to solve is:

What are the best conditions to perform the next experiment so that the variance of the estimated parameters is minimized?

Let us consider the case of an algebraic equation model (i.e., $\mathbf{y} = \mathbf{f}(\mathbf{x},\mathbf{k})$). The problem can be restated as "find the best experimental conditions (i.e., $\mathbf{x}_{N+1}$) where the next experiment should be performed so that the variance of the parameters is minimized."

It was shown earlier that if N experiments have been performed, the covariance matrix of the parameters is estimated by

$$COV(\mathbf{k}) = \hat{\sigma}_\varepsilon^2 \mathbf{A}^{-1} \tag{12.1}$$

where

$$\mathbf{A} = \sum_{i=1}^{N} \mathbf{G}_i^T \mathbf{Q}_i \mathbf{G}_i \tag{12.2}$$

and where $\mathbf{G}_i$ is the sensitivity coefficients matrix, $(\partial\mathbf{f}/\partial\mathbf{k})_i$, for the ith experiment and $\mathbf{Q}_i$ is a suitably selected weighting matrix based on the distribution of the errors in the measurement of the response variables. Obviously, the sensitivity matrix $\mathbf{G}_i$ is *only* a function of $\mathbf{x}_i$ and the current estimate of $\mathbf{k}$.

If for a moment we assume that the next experiment at conditions $\mathbf{x}_{N+1}$ has been performed, the *new* matrix $\mathbf{A}$ would be:

$$\mathbf{A}^{new} = \sum_{i=1}^{N} \mathbf{G}_i^T \mathbf{Q}_i \mathbf{G}_i + \mathbf{G}_{N+1}^T \mathbf{Q}_{N+1} \mathbf{G}_{N+1} = \mathbf{A}^{old} + \mathbf{G}_{N+1}^T \mathbf{Q}_{N+1} \mathbf{G}_{N+1} \tag{12.3}$$

and the resulting parameter covariance matrix would be

$$COV(\mathbf{k}) = \hat{\sigma}_\varepsilon^2 \left[\mathbf{A}^{new}\right]^{-1} \tag{12.4}$$

The implication here is that if the parameter values do not change significantly from their current estimates when the additional measurements are included in the estimation, we can *quantify* the effect of each additional experiment *before* it has been carried out! Hence, we can search all over the *operability region* (i.e., over all the potential values of $\mathbf{x}_{N+1}$) to find the best conditions for the next experiment. The operability region is defined as the set of the feasible experimental conditions and can usually be adequately represented by a small number of grid points. The size and form of the operability region is dictated by the feasibility of attaining these conditions. For example, the thermodynamic equilibrium surface at a given experimental temperature limits the potential values of the partial pressure of butene, butadiene and hydrogen (the three independent variables) in a butene dehydrogenation reactor (Dumez and Froment, 1976).

Next, we shall discuss the actual optimality criteria that can be used in determining the conditions for the next experiment.

12.2.1 The Volume Design Criterion

If we do not have any particular preference for a specific parameter or a particular subset of the parameter vector, we can minimize the variance of all parameters simultaneously by *minimizing the volume of the joint 95% confidence region*. Obviously a small joint confidence region is highly desirable.

Minimization of the volume of the ellipsoid

$$\left[\mathbf{k}-\mathbf{k}^*\right]^T \mathbf{A}^{new}\left[\mathbf{k}-\mathbf{k}^*\right] = p\hat{\sigma}_\varepsilon^2 F^{\alpha}_{p,(N+1)m-p} \qquad (12.5)$$

is equivalent to *maximization* of $det(\mathbf{A}^{new})$ which in turn is equivalent to *maximization* of the product

$$\Pi = \lambda_1\lambda_2\ldots\lambda_p \qquad (12.6)$$

where λ_i, i=1,...,p are the eigenvalues of matrix $\mathbf{A}^{new}$. Using any eigenvalue decomposition routine for real symmetric matrices, we can calculate the eigenvalues and hence, the determinant of $\mathbf{A}^{new}$ for each potential value of $\mathbf{x}_{N+1}$ in the operability region. The conditions that yield the maximum value for $det(\mathbf{A}^{new})$ should be used to conduct the next experiment.

12.2.2 The Shape Design Criterion

In certain occasions the volume criterion is not appropriate. In particular when we have an ill-conditioned problem, use of the volume criterion results in an elongated ellipsoid (like a cucumber) for the joint confidence region that has a small volume; however, the variance of the individual parameters can be very high. We can determine the shape of the joint confidence region by examining the *cond*(**A**) which is equal to $\lambda_{max}/\lambda_{min}$ and represents *the ratio of the principal axes* of the ellipsoid.

In this case, it is best to choose the experimental conditions which will yield the *minimum length for the largest principal axis of the ellipsoid.* This is equivalent to

$$maximization \text{ of } \lambda_{min}$$

Again, we can determine the *condition number* and λ_{min} of matrix $\mathbf{A}^{new}$ using any eigenvalue decomposition routine that computes the eigenvalues of a real symmetric matrix and use the conditions ($\mathbf{x}_{N+1}$) that correspond to a maximum of λ_{min}.

When the parameters differ by several orders of magnitude between them, the joint confidence region will have a long and narrow shape even if the parameter estimation problem is well-posed. To avoid unnecessary use of the shape criterion, instead of investigating the properties of matrix **A** given by Equation 12.2, it is better to use the normalized form of matrix **A** given below (Kalogerakis and Luus, 1984) as $\mathbf{A}_R$.

$$\mathbf{A}_R = \mathbf{K}\left[\sum_{i=1}^{N} \mathbf{G}_i^T \mathbf{Q}_i \mathbf{G}_i\right]\mathbf{K} \qquad (12.7)$$

where $\mathbf{K}=diag(k_1,k_2,\ldots,k_p)$. Therefore, $\mathbf{A}^{new}$ should be determined from

$$\mathbf{A}^{new} = \mathbf{K}\left[\sum_{i=1}^{N}\mathbf{G}_i^T\mathbf{Q}_i\mathbf{G}_i\right]\mathbf{K} + \mathbf{K}\mathbf{G}_{N+1}^T\mathbf{Q}_{N+1}\mathbf{G}_{N+1}\mathbf{K} \tag{12.8}$$

or

$$\mathbf{A}^{new} = \mathbf{A}^{old} + \mathbf{K}\mathbf{G}_{N+1}^T\mathbf{Q}_{N+1}\mathbf{G}_{N+1}\mathbf{K} \tag{12.9}$$

Essentially this is equivalent to using $(\partial f_i/\partial k_j)k_j$ instead of $(\partial f_i/\partial k_j)$ for the sensitivity coefficients. By this transformation the sensitivity coefficients are normalized with respect to the parameters and hence, the covariance matrix calculated using Equation 12.4 yields the standard deviation of each parameter as a percentage of its current value.

12.2.3 Implementation Steps

Based on the material presented up to now the steps that need to be followed to design the next experiment for the precise estimation of the model parameters is given below:

Step 1. Perform a series of initial experiments (based on a factorial design) to obtain initial estimates for the parameters and their covariance matrix.
Step 2. For each grid point of the operability region, compute the sensitivity coefficients and generate $\mathbf{A}^{new}$ given by Equation 12.9.
Step 3. Perform an eigenvalue decomposition of matrix $\mathbf{A}^{new}$ to determine its *condition number, determinant* and λ_{min}.
Step 4. Select the experimental conditions that correspond to a maximum $det(\mathbf{A}^{new})$ or maximum λ_{min} when the volume or the shape criterion is used respectively. The computed condition number indicates whether the volume or the shape criterion is the most appropriate to use.
Step 5. Perform the experiment at the selected experimental conditions.
Step 6. Based on the additional measurement of the response variables, estimate the parameter vector and its covariance matrix.
Step 7. If the obtained accuracy is satisfactory, stop; else go back to Step 2 and select the conditions for an additional experiment.

In general, the search for the optimal $\mathbf{x}_{N+1}$ is made all over the operability region. Experience has shown however, that the best conditions are always found on the boundary of the operability region (Froment and Bischoff, 1990). This simplification can significantly reduce the computational effort required to determine the best experimental conditions for the next experiment.

12.3 SEQUENTIAL EXPERIMENTAL DESIGN FOR MODEL DISCRIMINATION

Based on alternative assumptions about the mechanism of the process under investigation, one often comes up with a set of alternative mathematical models that could potentially describe the behavior of the system. Of course, it is expected that only one of them is the correct model and the rest should prove to be inadequate under certain operating conditions. Let us assume that we have conducted a set of preliminary experiments and we have fitted several rival models that could potentially describe the system. The problem we are attempting to solve is:

What are the best conditions to perform the next experiment so that with the additional information we will maximize our ability to discriminate among the rival models?

Let us assume that we have r rival models

$$\left.\begin{aligned} \mathbf{y}^{(1)} &= \mathbf{f}^{(1)}(\mathbf{x}, \mathbf{k}^{(1)}) \\ \mathbf{y}^{(2)} &= \mathbf{f}^{(2)}(\mathbf{x}, \mathbf{k}^{(2)}) \\ &\vdots \\ \mathbf{y}^{(r)} &= \mathbf{f}^{(r)}(\mathbf{x}, \mathbf{k}^{(r)}) \end{aligned}\right\} \tag{12.10}$$

that could potentially describe the behavior of the system. Practically this means that if we perform a model adequacy test (as described in Chapter 11) based on the experimental data at hand, none of the these models gets rejected. Therefore, we must perform additional experiments in an attempt to determine which one of the rival models is the correct one *by rejecting all the rest.*

Obviously, it is very important that the next experiment has *maximum discriminating power*. Let us illustrate this point with a very simple example where simple common sense arguments can lead us to a satisfactory design. Assume that we have the following two rival single-response models, each with two parameters and one independent variable:

Model 1: $$y^{(1)} = k_1^{(1)} x + k_2^{(1)} \tag{12.11a}$$

Model 2: $$y^{(2)} = k_2^{(2)} e^{k_1^{(2)} x} \tag{12.11b}$$

Design of the next experiment simply means selection of the best value of x for the next run. Let us assume that based on information up to now, we have estimated the parameters in each of the two rival models and computed their predicted response shown in Figure 12.1. It is apparent that if the current parameter estimates will not change significantly, both models will be able to fit satisfactorily data taken in the vicinity of x_2 or x_4 where the predicted response of the two models is about the same. On the other hand, if we conduct the next experiment near x_5, only one of the two models will most likely be able to fit it satisfactorily. The vicinity of x_5 as shown in Figure 12.1, corresponds to the area of the operability region (defined as the interval $[x_{min}, x_{max}]$) where the divergence between the two models is maximized.

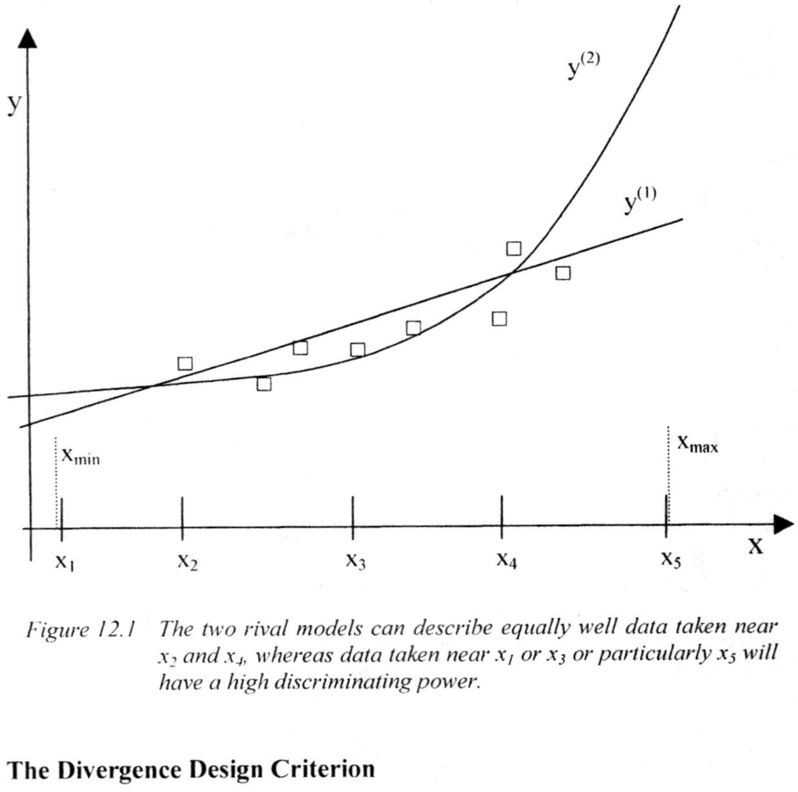

Figure 12.1 The two rival models can describe equally well data taken near x_2 and x_4, whereas data taken near x_1 or x_3 or particularly x_5 will have a high discriminating power.

12.3.1 The Divergence Design Criterion

If the structure of the models is more complex and we have more than one independent variable or we have more than two rival models, selection of the best experimental conditions may not be as obvious as in the above example. A straightforward design to obtain the best experimental conditions is based on the divergence criterion.

Hunter and Reimer (1965) proposed the *simple Divergence* criterion that can readily be extended to multi-response situations. In general, if we have performed N experiments; the experimental conditions $\mathbf{x}_{N+1}$ for the next one are obtained by maximizing the *weighted Divergence* between the rival models, defined as

$$D(\mathbf{x}_{N+1}) = \sum_{i=1}^{r}\sum_{j=i+1}^{r}\left[\mathbf{y}^{(i)} - \mathbf{y}^{(j)}\right]^T \mathbf{Q}\left[\mathbf{y}^{(i)} - \mathbf{y}^{(j)}\right] \tag{12.12}$$

where r is the number of the rival models and $\mathbf{Q}$ is a user-supplied weighting matrix. The role of $\mathbf{Q}$ is to normalize the contribution of the response variables that may be of a different order of magnitude (e.g., the elements of the output vector, $\mathbf{y}$, could be comprised of the process temperature in degrees K, pressure in Pa and concentration as mole fraction).

Box and Hill (1967) proposed a criterion that incorporates the uncertainties associated with model predictions. For two rival single-response models the proposed divergence expression takes the form,

$$D(\mathbf{x}_{N+1}) = P_{0,1}P_{0,2}\left[\frac{(\hat{\sigma}^2_{\varepsilon,1} - \hat{\sigma}^2_{\varepsilon,2})^2}{(\sigma^2_\varepsilon + \hat{\sigma}^2_{\varepsilon,1})(\sigma^2_\varepsilon + \hat{\sigma}^2_{\varepsilon,2})} + \left(y^{(1)} - y^{(2)}\right)^2\left(\frac{1}{\sigma^2_\varepsilon + \hat{\sigma}^2_{\varepsilon,1}} + \frac{1}{\sigma^2_\varepsilon + \hat{\sigma}^2_{\varepsilon,2}}\right)\right] \tag{12.13}$$

where σ^2_ε is the variance of the experimental error in the measurement (obtained through replicate experiments) and $\sigma^2_{\varepsilon,1}, \sigma^2_{\varepsilon,2}$ are the variance of y as calculated based on models 1 and 2 respectively. $P_{0,1}$ is the prior probability (based on the N experiments already performed) of model 1 being the correct model. Similarly, $P_{0,2}$ is the prior probability of model 2. Box and Hill (1967) expressed the adequacy in terms of posterior probabilities. The latter serve as prior probabilities for the design of the next experiment.

Buzzi-Ferraris et al., (1983, 1984) proposed the use of a more powerful statistic for the discrimination among rival models; however, computationally it is more intensive as it requires the calculation of the sensitivity coefficients at each grid point of the operability region. Thus, the simple divergence criterion of Hunter and Reimer (1965) appears to be the most attractive.

12.3.2 Model Adequacy Tests for Model Discrimination

Once the new experiment has been conducted, all models are tested for adequacy. Depending on our knowledge of σ^2_ε we may perform a χ^2-test, an F-test or Bartlett's χ^2-test (when σ^2_ε is completely unknown). The first two tests have been discussed previously in Chapter 11 as Cases I and II. Bartlett's χ^2-test is presented next.

Case III: Bartlett's χ^2-test

Let us assume that $\sigma^2_{\varepsilon,i}$ is the estimate of σ^2_ε from the ith model. The associated degrees of freedom for the ith model are denoted with $(d.f)_i$ and are equal to the total number of measurements minus the number of parameters that appear in the ith model. The idea is that the minimum sum of squares of residuals divided by the appropriate degrees of freedom is an unbiased estimate of the experimental error variance *only* for the correct mathematical model. All other estimates based on the other models are biased due to lack of fit. Hence, one simply needs to examine the homogeneity of the estimated variances by the rival models. Namely, Bartlett's χ^2-test is based on testing the hypothesis

$$H_0: \quad \sigma^2_{\varepsilon,1} = \sigma^2_{\varepsilon,2} = \sigma^2_{\varepsilon,3} = \ldots = \sigma^2_{\varepsilon,r}$$

$$H_1: \quad \text{not all of the } \sigma^2_{\varepsilon,i} \ (i=1,\ldots,r) \text{ are equal.}$$

with a user-selected level of significance (usually α=0.01 or α=0.05 is employed).

The above hypothesis is tested using an χ^2-test. Namely,

$$\text{if } \chi^2_{data} > \chi^2_{v=r-1,1-\alpha} \quad \Rightarrow \quad \text{Reject } H_0$$

where $\chi^2_{v=r-1,1-\alpha}$ is obtained from the tables and χ^2_{data} is computed from the data as follows:

First, we generate the *pooled variance*, $\hat{\sigma}^2_p$,

$$\hat{\sigma}^2_p = \frac{\sum_{i=1}^{r} [(d.f.)_i - 1]\hat{\sigma}^2_{\varepsilon,i}}{\sum_{i=1}^{r} (d.f.)_i} \tag{12.14}$$

which contains lack of fit, and then we compute χ^2_{data} from

$$\chi^2_{data} = \frac{ln[\hat{\sigma}_p^2]\sum_{i=1}^{r}(d.f.)_i - \sum_{i=1}^{r}(d.f.)_i\, ln[\hat{\sigma}^2_{\varepsilon,i}]}{1 + \frac{1}{3(r-1)}\left(\sum_{i=1}^{r}\frac{1}{(d.f.)_i} - \frac{1}{\sum_{i=1}^{r}(d.f.)_i}\right)} \tag{12.15}$$

When the hypothesis H_o is rejected, we drop the model with the highest $\hat{\sigma}^2_{\varepsilon,i}$ and we repeat the test with one less model. We keep on removing models until H_o cannot be rejected any more. These models are now used in the determination of the overall divergence for the determination of the experimental conditions for the next experiment.

Although all the underlying assumptions (local linearity, statistical independence, etc.) are rarely satisfied, Bartlett's χ^2-test procedure has been found adequate in both simulated and experimental applications (Dumez et al., 1977; Froment, 1975). However, it should be emphasized that only the χ^2-test and the F-test are true model adequacy tests. Consequently, they may eliminate all rival models if none of them is truly adequate. On the other hand, Bartlett's χ^2-test does not guarantee that the retained model is truly adequate. It simply suggests that it is the best one among a set of inadequate models!

12.3.3 Implementation Steps for Model Discrimination

Based on the material presented above, the implementation steps to design the next experiment for the discrimination among *r* rival model are:

Step 1. Perform a series of initial experiments (based on a factorial design) to obtain initial estimates for the parameters and their covariance matrix for each of the *r* rival models.

Step 2. For each grid point of the operability region, compute the Weighted Divergence, *D*, given by Equation 12.12. In the computations we consider only the models which are adequate at the present time (not all the rival models).

Step 3. Select the experimental conditions that correspond to a maximum *D*.

Step 4. Perform the experiment at the selected experimental conditions.

Step 5. Based on the new measurement of the response variables, estimate the parameter vector and its covariance matrix for *all* rival models.

Step 6. Perform the appropriate model adequacy test (χ^2-test, F-test or Bartlett's χ^2-test) for *all* rival models (just in case one of the models was rejected prematurely). If more than one model remains adequate, go back to Step 2 to select the conditions for another experiment.

The idea behind the computation of the Divergence *only* among adequate models in Step 2, is to base the decision for the next experiment on the models that are still competing. Models that have been found inadequate should not be included in the divergence computations. However, once the new data point becomes available it is good practice to update the parameter estimates for all models (adequate or inadequate ones). Practice has shown that under conditions of high experimental errors, as additional information becomes available some models may become adequate again!

12.4 SEQUENTIAL EXPERIMENTAL DESIGN FOR ODE SYSTEMS

Let us now turn our attention to systems described by ordinary differential equations (ODEs). Namely, the mathematical model is of the form,

$$\frac{d\mathbf{x}(t)}{dt} = \mathbf{f}(\mathbf{x}(t), \mathbf{u}, \mathbf{k}) \quad ; \quad \mathbf{x}(t_0) = \mathbf{x}_0 \tag{12.16}$$

$$\mathbf{y}(t) = \mathbf{C}\mathbf{x}(t) \tag{12.17}$$

where $\mathbf{k}=[k_1,k_2,\ldots,k_p]^T$ is a *p-dimensional* vector of unknown parameters; $\mathbf{x}=[x_1,x_2,\ldots,x_n]^T$ is an *n-dimensional* vector of state variables; $\mathbf{x}_0=[x_{10},x_{20},\ldots,x_{n0}]^T$ is an *n-dimensional* vector of initial conditions for the state variables and $\mathbf{u}=[u_1,u_2,\ldots,u_r]^T$ is an *r-dimensional* vector of manipulated variables which are normally set by the operator/experimentalist.

Design of the next experiment in this case requires the following:

(i) Selection of the initial conditions, $\mathbf{x}_0$.
(ii) Selection of the sampling rate and the final time.
(iii) Selection of the manipulated variables, $\mathbf{u}$, which could be kept either constant or even be optimally manipulated over time.

The sequential experimental design can be made either for precise parameter estimation or for model discrimination purposes.

12.4.1 Selection of Optimal Sampling Interval and Initial State for Precise Parameter Estimation

The optimality criteria based on which the conditions for the next experiment are determined are the same for dynamic and algebraic systems. However, for a dynamic system we determine the conditions not just of the next measurement but rather the conditions under which a complete experimental run will be carried out. For precise parameter estimation we examine the determinant and the eigenvalues of matrix $\mathbf{A}^{new}$,

$$\mathbf{A}^{new} = \mathbf{A}^{old} + \mathbf{K}\left[\sum_{i=1}^{N_P} \mathbf{G}^T(t_i)\mathbf{C}^T\mathbf{QCG}(t_i)\right]\mathbf{K} \tag{12.18}$$

where N_P is the number of data points to be collected, $\mathbf{K}=diag(k_1,k_2,\ldots,k_p)$, $\mathbf{G}(t)$ is the parameter sensitivity matrix, $(\partial\mathbf{x}^T/\partial\mathbf{k})^T$ and matrix $\mathbf{A}^{old}$ is our usual matrix $\mathbf{A}$ of the normal equations based on the experimental runs performed up to now. If the number of runs is N_R and the number of data points gathered in each run is N_P, $\mathbf{A}^{old}$ is of the form,

$$\mathbf{A}^{old} = \mathbf{K}\sum_{j=1}^{N_R}\left[\sum_{i=1}^{N_P} \mathbf{G}^T(t_i)\mathbf{C}^T\mathbf{QCG}(t_i)\right]_j \mathbf{K} \tag{12.19}$$

Our selection of the initial state, $\mathbf{x}_0$, and the value of the manipulated variables vector, $\mathbf{u}(t)$ determine a particular experiment. Here we shall assume that the input variables $\mathbf{u}(t)$ are kept constant throughout an experimental run. Therefore, the operability region is defined as a closed region in the $[x_{0,1},x_{0,2},\ldots,x_{0,n}, u_1,u_2,\ldots,u_r]^T$ –space. Due to physical constraints these independent variables are limited to a very narrow range, and hence, the operability region can usually be described with a small number of grid points.

In addition to the selection of the experimental conditions for the next experiment (i.e., selection of the best grid point in the operability region according to the volume or the shape criterion), the sampling rate and the final time must also be specified. In general, it is expected that for each particular experiment there is a corresponding optimal time interval over which the measurements should be obtained. Furthermore, given the total number of data points (N_p) that will be gathered (i.e., for a given experimental effort) there is a corresponding optimal sampling rate which should be employed.

I. Time Interval Determination

The main consideration for the choice of the most appropriate time interval is the availability of parameter sensitivity information. Kalogerakis and Luus (1984) proposed to choose the time interval over which the output vector (measured variables) is most sensitive to the parameters. In order to obtain a measure of the available sensitivity information with respect to time, they proposed the use of the Information Index introduced earlier by Kalogerakis and Luus (1983b). The information index for parameter k_j is defined as

$$I_j(t) = k_j \left(\frac{\partial \mathbf{y}^T}{\partial k_j} \right) \mathbf{Q} \left(\frac{\partial \mathbf{y}}{\partial k_j} \right) k_j \quad ; \; j=1,\ldots,p \tag{12.20}$$

where $\mathbf{Q}$ is a suitably chosen weighting matrix. As indicated by Equation 12.20, the scalar $I_j(t)$ is simply a weighted sum of squares of the sensitivity coefficients of the output vector with respect to parameter k_j at time t. Hence, $I_j(t)$ can be viewed as an index measuring the overall sensitivity of the output vector with respect to parameter k_j at time t. Using Equation 12.17 and the definition of the parameter sensitivity matrix, the above equation becomes

$$I_j(t) = k_j \boldsymbol{\delta}_j^T \mathbf{G}^T(t) \mathbf{C}^T \mathbf{Q} \mathbf{C} \mathbf{G}(t) \boldsymbol{\delta}_j k_j \quad ; \; j=1,\ldots,p \tag{12.21}$$

where $\boldsymbol{\delta}_j$ is a *p-dimensional* vector with 1 in the jth element and zeros elsewhere.

The procedure for the selection of the most appropriate time interval requires the integration of the state and sensitivity equations and the computation of $I_j(t)$, j=1,...,p at each grid point of the operability region. Next by plotting $I_j(t)$, j=1,...,p versus time (preferably on a log scale) the time interval $[t_1, t_{Np}]$ where the information indices are excited and become large in magnitude is determined. This is the time period over which measurements of the output vector should be obtained.

II. Sampling Rate Determination

Once the best time interval has been obtained, the sampling times within this interval should be obtained. Let us assume that a total number of N_P data points are to be obtained during the run. In general, the selected sampling rate should be small compared to the governing time constants of the system. However, for highly nonlinear systems where the governing time constants may change significantly over time, it is difficult to determine an appropriate constant sampling rate. The same difficulty arises when the system is governed by a set of stiff differential equations. By considering several numerical examples, Kalogerakis and Luus (1984) found that if N_P is relatively large, in order to cover the entire time interval of interest it is practical to choose $t_2, t_3, \ldots, t_{Np-1}$ by a log-linear interpolation between t_1 ($t_1>0$) and t_{Np}, namely,

$$t_{i+1} = t_i \left(\frac{t_{Np}}{t_1} \right)^{\frac{1}{Np-1}} \quad ; \; i=1,2,\ldots,N_P-2 \tag{12.22}$$

When t_1 and t_{Np} do not differ significantly (i.e., more than one order of magnitude), the proposed measurement scheme reduces practically to choosing a constant sampling rate. If, on the other hand, t_1 and t_{Np} differ by several orders of magnitude, use of the log-linear interpolation formula ensures that sensitivity information is gathered from widely different parts of the transient response covering a wide range of time scales.

The steps that should be followed to determine the best grid point in the operability region for precise parameter estimation of a dynamic system are given below:

Step 1. Compute the Information Indices, $I_j(t)$, j=1,...,p at each grid point of the operability region..

Step 2. By plotting $I_j(t)$, j=1,...,p versus time determine the optimal time interval $[t_1,t_{Np}]$ for each grid point.

Step 3. Determine the sampling times t_2, t_3,... using Equation 12.22 for each grid point.

Step 4. Integrate the state and sensitivity equations and compute matrix $\mathbf{A}^{new}$.

Step 5. Using the desired optimality criterion (volume or shape), determine the best experimental conditions for the next experiment.

III. Simplification of the Procedure

The above procedure can be modified somewhat to minimize the overall computational effort. For example, during the computation of the Information Indices in Step 1, matrix $\mathbf{A}^{new}$ can also be computed and thus an early estimate of the best grid point can be established. In addition, as is the case with algebraic systems, the optimum conditions are expected to lie on the boundary of the operability region. Thus, in Steps 2, 3 and 4 the investigation can be restricted only to the grid points which lie on the boundary surface indicated by the preliminary estimate. As a result the computation effort can be kept fairly small.

Another significant simplification of the above procedure is the use of constant sampling rate when t_1, t_{Np} are within one or two orders of magnitude and use of this sampling rate uniformly for all grid points of the operability region. Essentially, we use the information indices to see whether the chosen sampling rate is adequate for all grid points of the operability region.

If a constant sampling rate is not appropriate because of widely different time constants, a convenient simplification is the use of two sampling rates: a fast one for the beginning of the experiment followed by the slow sampling rate when the fast dynamics have died out. The selection of the fast sampling rate is more difficult and should be guided by the Information Indices and satisfy the constraints imposed by the physical limitations of the employed measuring devices. Kalogerakis and Luus (1983b) have shown that if the Information Indices suggest that there is available sensitivity information for all the parameters during the slow dynamics, we may not need to employ a fast sampling rate.

12.4.2 Selection of Optimal Sampling Interval and Initial State for Model Discrimination

If instead of precise parameter estimation, we are designing experiments for model discrimination, the best grid point of the operability region is chosen by maximizing the overall divergence, defined for dynamic systems as

$$D(\mathbf{u},\mathbf{x}_0) = \sum_{l=j+1}^{r} \sum_{j=1}^{r} \sum_{i=1}^{N_P} \left[\mathbf{y}^{(j)}(t_i) - \mathbf{y}^{(l)}(t_i)\right]^T \mathbf{Q} \left[\mathbf{y}^{(j)}(t_i) - \mathbf{y}^{(l)}(t_i)\right] \quad (12.23)$$

Again we consider that **u** is kept constant throughout an experimental run. The design procedure is the same as for algebraic systems. Of course, the time interval and sampling times must also be specified. In this case, our selection should be based on what is appropriate for all competing models, since the information gathered from these experiments will also be used to estimate more precisely the parameters in each model. Again the information index for each parameter and for each model can be used to guide us to select an overall suitable sampling rate.

12.4.3 Determination of Optimal Inputs for Precise Parameter Estimation and Model Discrimination

As mentioned previously, the independent variables which determine a particular experiment and are set by the experimentalist are the initial state, $\mathbf{x}_0$ and the vector of the manipulated variables (also known as *control input*), **u**. In the previous section we considered the case where **u** is kept constant throughout an experimental run. The case where **u** is allowed to vary as a function of time constitutes the problem of *optimal inputs*. In general, it is expected that by allowing some or all of the manipulated variables to change over time, a greater amount of information will be gathered from a run. The design of optimal inputs has been studied only for the one-parameter model case. This is fairly straightforward since maximization of the determinant of the (1×1) matrix, $\mathbf{A}^{new}$, is trivial and the optimal control problem reduces to maximization of the integral (or better summation) of the sensitivity coefficient squared. In a similar fashion, researchers have also looked at maximizing the sensitivity of *one* parameter out of all the unknown parameters present in a dynamic model.

Murray and Reiff (1984) showed that the use of an optimally selected square wave for the input variables can offer considerable improvement in parameter estimation. Of course, it should be noted that the use of constant inputs is often more attractive to the experimentalist, mainly because of the ease of experimentation. In addition, the mathematical model representing the physical system may remain simpler if the inputs are kept constant. For example, when experiments are designed for the estimation of kinetic parameters, the incorporation of the energy equation can be avoided if the experiments are carried out isothermally.

For precise parameter estimation, we need to solve several *optimal control problems* each corresponding to a grid point of the operability region. The operability region is defined now by the potential values of the initial state ($\mathbf{x}_0$). A particular optimal control problem can be formulated as follows,

> *Given the governing differential equations (the model) and the initial conditions,* $\mathbf{x}_0$*, determine the optimal inputs,* $\mathbf{u}(t)$*, so that the determinant (for the volume criterion) or the smallest eigenvalue (for the shape criterion) of matrix* $\mathbf{A}^{new}$ *(given by Equation 12.18) is maximized.*

Obviously this optimal control problem is not a typical "textbook problem". It is not even clear whether the optimal solution can be readily computed. As a result, one should consider suboptimal solutions.

One such design was proposed by Murray and Reiff (1984) where they used an optimally selected square wave for the input variables. Instead of computing an optimal profile for $\mathbf{u}(t)$, they simply investigated the magnitude and frequency of the square wave, for precise parameter estimation purposes. It should be also kept in mind that the physical constraints often limit the values of the manipulated variables to a very narrow range which suggests that the optimal solution will most likely be bang-bang.

The use of time stages of varying lengths in *iterative dynamic programming* (Luus, 2000) may indeed provide a computationally acceptable solution. Actually, such an approach may prove to be feasible particularly for model discrimination purposes. In model discrimination we seek the optimal inputs, $\mathbf{u}(t)$, that will maximize the overall divergence among r rival models given by Equation 12.23.

The original optimal control problem can also be simplified (by reducing its dimensionality) by partitioning the manipulated variables $\mathbf{u}(t)$ into two groups $\mathbf{u}_1$ and $\mathbf{u}_2$. One group $\mathbf{u}_1$ could be kept constant throughout the experiment and hence, the optimal inputs for subgroup $\mathbf{u}_2(t)$ are only determined.

Kalogerakis (1984) following the approach of Murray and Reiff (1984) suggested the use of a square wave for $\mathbf{u}(t)$ whose frequency is optimally chosen for model discrimination purposes. The rationale behind this suggestion is based on the fact that the optimal control policy is expected to be bang-bang. Thus, instead of attempting to determine the optimal switching times, one simply assumes a square wave and optimally selects its period by maximizing the divergence among the r rival models.

12.5 EXAMPLES

12.5.1 Consecutive Chemical Reactions

As an example for precise parameter estimation of dynamic systems we consider the simple consecutive chemical reactions in a batch reactor used by Hosten and Emig (1975) and Kalogerakis and Luus (1984) for the evaluation of sequential experimental design procedures of dynamic systems. The reactions are

$$A \rightarrow B \rightarrow C$$

where both steps are irreversible and kinetically of first order. The governing differential equations are

$$\frac{dx_1}{dt} = -k_1\, exp\left(-\frac{k_2}{T}\right)x_1 \quad ; \; x_1(0) = x_{0,1} \tag{12.24a}$$

$$\frac{dx_2}{dt} = k_1\, exp\left(-\frac{k_2}{T}\right)x_1 - k_3\, exp\left(-\frac{k_4}{T}\right)x_2 \quad ; \; x_2(0) = 0 \tag{12.24b}$$

where x_1 and x_2 are the concentrations of A and B (g/L) and T is the reaction temperature (K). For simplicity it is assumed that the reaction is carried out isothermally, so that there is no need to add the energy equation. Both state variables x_1 and x_2 are assumed to be measured and the standard error in the measurement (σ_ε) in both variables is equal to 0.02 (g/L). Consequently the statistically correct choice for the weighting matrix **Q** is the identity matrix.

The experimental conditions which can be set by the operator are the initial concentrations of A and B and the reaction temperature T. For simplicity the initial concentration of B is assumed to be zero, i.e., we always start with pure A. The operability region is shown in Figure 12.2 on the $x_{0,1}$-T plane where $x_{0,1}$ takes values from 1 to 4 g/L and T from 370 to 430 K.

Let us now assume that two preliminary experiments were performed at the grid points (1, 370) and (4, 430) yielding the following estimates for the parameters: $k_1 = 0.61175\times10^{11}$, $k_2 = 10000$, $k_3 = 0.62155\times10^{8}$ and $k_4 = 7500$ from 20 measurements taken with a constant sampling rate in the interval 0 to 10 h. With these preliminary parameter values the objective is to determine the best experimental conditions for the next run.

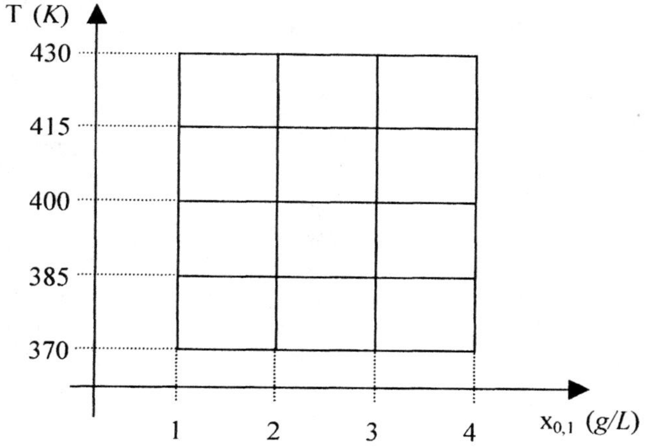

Figure 12.2 The operability region for the consecutive chemical reactions example [reprinted from the Canadian Journal of Chemical Engineering with permission].

For comparison purposes, we first determine the best grid point in the operability region using the same final time and constant sampling rate for all feasible experiments. To show the importance of the final time (or better the time interval) several values in the range 1 to 40 h were used and the results are shown in Table 12.1. A total of 20 data points were used and the best grid point was selected based on the volume criterion. As seen, the results depend heavily on the chosen final time. Furthermore, cases 4 and 5 or 6 and 7 show that for the same grid point the chosen final time has a strong effect on $det(\mathbf{A}^{new})$ and hence, there must be indeed an optimal choice for the final time for each grid point in the operability region.

This observed dependence can be readily explained by examining the behavior of the Information Indices. For example it is seen in Figure 12.3 that most of the available sensitivity information is available in the time interval 0.03 to 2 h for the grid point (4, 430). Therefore, when the final time was increased other grid point gave better results (i.e., a higher value for $det(\mathbf{A}^{new})$). On the other hand, as seen in Figure 12.4, most of the sensitivity information is available in the interval 2 to 40 h for the grid point (4, 370). Hence, only when the final time was increased to 12 h or more was this grid point selected as the best.

The above results indicate the need to assign to each grid point an appropriate time interval over which the measurements will be made. The information index provides the means to locate this time interval so that the maximum amount of information is gathered from each experiment.

Next by employing the volume criterion, the best grid point in the operability region was determined using the time intervals indicated by the information indices and the log-linear formula for the selection of the sampling times. A total of 20, 40 and 80 data points were used and the results are shown in Table 12.2. As seen, the grid point (4, 370) was consistently selected as best.

At this point it is worthwhile making a couple of comments about the Information Index. One can readily see from Figures 12.3 and 12.4 that the steady state values of the Information Indices are zero for this example. This simply means that the steady state values of the state variables do not depend on the parameters. This is expected as both reactions are irreversible and hence, regardless of their rate, at steady state all of component *A* and *B* must be converted to *C*.

From the same figures it is also seen that the normalized information indices of k_1 and k_2, as well as k_3 and k_4, overlap completely. This behavior is also not surprising since the experiments have been conducted isothermally and hence the product $k_1 exp(-k_2/T)$ behaves as one parameter which implies that $(\partial x_i/\partial k_1)$ and $(\partial x_i/\partial k_2)$ and consequently $I_1(t)$ and $I_2(t)$ differ only by a constant factor.

Finally, it is pointed out that the time interval determined by the Information Indices is subject to experimental constraints. For example, it may not be possible to collect data as early as suggested by the Information Indices. If k_1 in this example was 100 times larger, i.e., $k_1 = 0.61175\times10^{10}$, the experimentalist could be unable to start collecting data from 0.004 h (i.e., 1.5 sec) as the Information Indices would indicate for grid point (4, 430) shown in Figure 12.5.

Table 12.1 Consecutive Chemical Reactions: Effect of Final Time on the Selection of the Best Grid Point Using the Volume Criterion

Case	Final Time (h)	Best Grid Point	$det(\mathbf{A}^{new})$
1	1	(4, 430)	3.801×10^4
2	2	(4, 415)	4.235×10^4
3	4	(4, 400)	7.178×10^4
4	6	(4, 385)	7.642×10^4
5	8	(4, 385)	1.055×10^5
6	12	(4, 370)	1.083×10^5
7	40	(4, 370)	1.788×10^5

Source: Kalogerakis and Luus (1984).

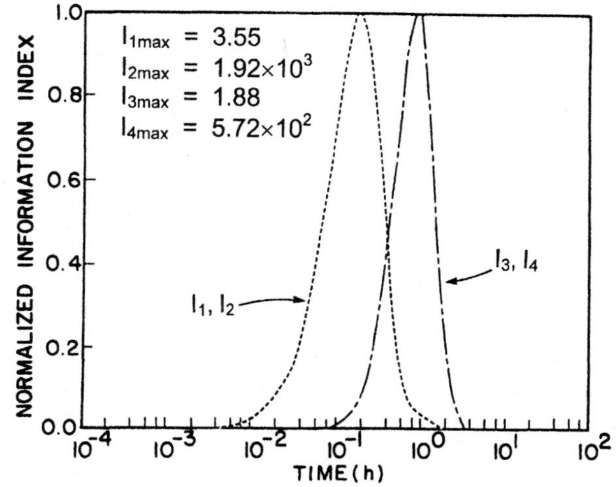

Figure 12.3 *Information indices versus time for the grid point (4, 430) [reprinted from the Canadian Journal of Chemical Engineering with permission].*

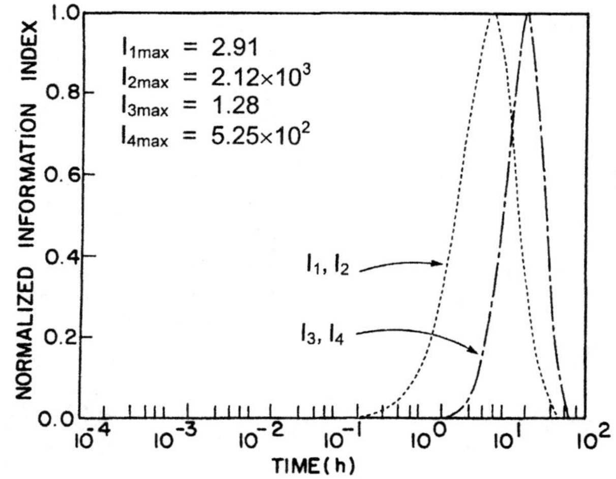

Figure 12.4 *Information indices versus time for the grid point (4, 370) [reprinted from the Canadian Journal of Chemical Engineering with permission].*

Table 12.2 Consecutive Chemical Reactions: Selection of the Best Grid Point Based on the Volume Criterion and Through the Use of the Information Index

No. of data points	λ_{min}	$det(\mathbf{A}^{new})$	Standard deviation (%) k_1	k_2	k_3	k_4
20	0.0171	1.476×10^5	15.14	0.562	8.044	0.414
40	0.0173	5.831×10^5	15.06	0.558	7.513	0.380
80	0.0174	2.316×10^6	15.02	0.556	7.236	0.362

Best grid point for all cases: (4, 370).
Source: Kalogerakis and Luus (1984).

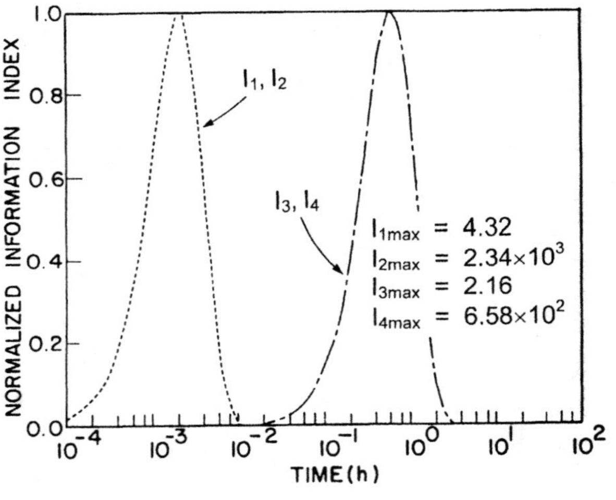

Figure 12.5 Information indices versus time for the grid point (4, 430) with Parameter k_1 100 Times Larger [reprinted from the Canadian Journal of Chemical Engineering with permission].

Finally when the selected experiment has been performed, the parameter estimates will be updated based on the new information and the predicted parameter variances will probably be somewhat different since our estimate of the parameter values will most likely be different. If the variance of the estimated parameters is acceptable we stop, or else we go back and design another experiment.

12.5.2 Fed-batch Bioreactor

As a second example let us consider the fed-batch bioreactor used by Kalogerakis and Luus (1984) to illustrate sequential experimental design methods for dynamic systems. The governing differential equations are (Lim et al., 1977):

$$\frac{dx_1}{dt} = \left(\frac{k_1 x_2}{k_1 + x_2} - D\right)x_1 - k_4 x_1 \quad ; \; x_1(0) = x_{0,1} \tag{12.25a}$$

$$\frac{dx_2}{dt} = -\frac{k_1 x_2}{(k_1 + x_2)k_3} x_1 + D(c_F - x_2) \quad ; \; x_2(0) = 0.01 \tag{12.25b}$$

where x_1 and x_2 are the biomass and limiting substrate (glucose) concentrations (*g/L*) in the bioreactor, c_F is the substrate concentration in the feed stream (*g/L*) and D is the dilution factor (h^{-1}) defined as the feed flowrate over the volume of the liquid phase in the bioreactor. The dilution factor is kept constant with respect to time to allow x_1 and x_2 to reach steady state values while the volume in the bioreactor increases exponentially (Lim et al., 1977).

It is assumed that both state variables x_1 and x_2 are measured with respect to time and that the standard experimental error (σ_ε) is 0.1 (*g/L*) for both variables. The independent variables that determine a particular experiment are (i) the inoculation density (initial biomass concentration in the bioreactor), $x_{0,1}$, with range 1 to 10 *g/L*, (ii) the dilution factor, D, with range 0.05 to 0.20 h^{-1} and (iii) the substrate concentration in the feed, c_F, with range 5 to 35 *g/L*.

In this case the operability region can be visualized as a rectangular prism in the 3-dimensional $x_{0,1}$-D-c_F - space as seen in Figure 12.6. A grid of four points in each variable has been used.

Let us now assume that from a preliminary experiment performed at $x_{0,1}$ = 7 *g/L*, D = 0.10 h^{-1} and c_F = 25 *g/L* it was found that k_1 = 0.31, k_2 = 0.18, k_3 = 0.55 and k_4 = 0.05 from 20 measurements taken at a constant sampling rate in the interval 0 to 20 *h*. Using Equation 12.1, the standard deviation (%), also known as coefficient of variation, was computed and found to be 49.86, 111.4, 8.526 and 27.75 for k_1, k_2, k_3 and k_4 respectively. With these preliminary parameter estimates, the conditions of the next experiment will be determined so that the uncertainty in the parameter estimates is minimized.

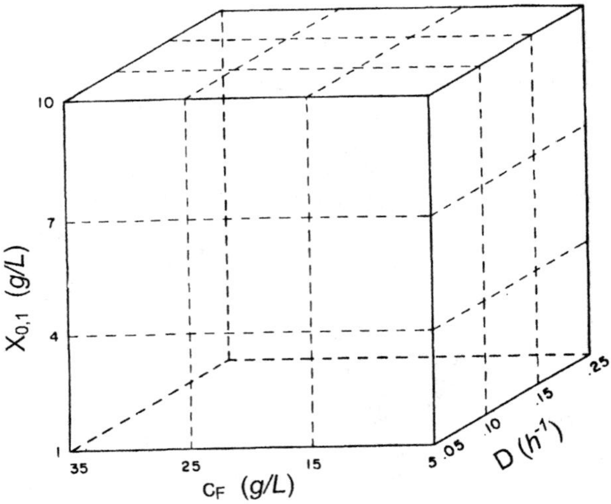

Figure 12.6 Fed-batch Bioreactor: Operability Region [reprinted from the Canadian Journal of Chemical Engineering with permission].

Table 12.3 Fed-batch Bioreactor: Effect of Final Time on the Selection of the Best Grid Point Using the Volume Criterion

Case	Final Time (h)	Best Grid Point	$det(\mathbf{A}^{new})$
1	20	(7, 0.20, 35)	1.06×10^8
2	40	(4, 0.15, 35)	1.46×10^9
3	80	(1, 0.20, 35)	2.87×10^9

Source: Kalogerakis and Luus (1984).

For comparison purposes we first determine the best grid point in the operability region using the same final time and constant sampling rate for all feasible experiments. In Table 1 the best experimental conditions are shown chosen by the volume criterion when as final time 20, 40 or 80 h is used. A total of 20 data points is used in all experiments. As seen, the selected conditions depend on our choice of the final time.

Again, this observed dependence can be readily explained by examining the behavior of the information indices. For example in Figure 12.7 it is shown that most of the available sensitivity information for grid point (1, 0.20, 35) is in the interval 25 to 75 *h*. Consequently when as final time 20 or 40 *h* is used, other grid points in the operability region gave better results. On the other hand, it is seen in Figure 12.8 that 20 *h* as final time is sufficient to capture most of the available sensitivity information (although 35 *h* would be the best choice) for the grid point (7, 0.20, 35). By allowing the final time to be 80 *h* no significant improvement in the results is expected.

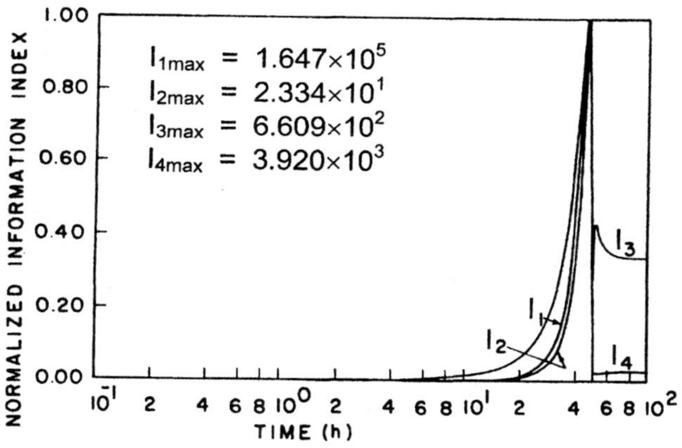

Figure 12.7 *Fed-batch Bioreactor: Information indices versus time for the grid point (1, 0.20, 35) [reprinted from the Canadian Journal of Chemical Engineering with permission].*

To illustrate the usefulness of the Information Index in determining the best time interval, let us consider the grid point (1, 0.20, 35). From Figure 12.7 we deduce that the best time interval is 25 to 75 h. In Table 12.4 the standard deviation of each parameter is shown for 7 different time intervals. From cases 1 to 4 it is seen that that measurements taken before 25 h do not contribute significantly in the reduction of the uncertainty in the parameter estimates. From case 4 to 7 it is seen that it is preferable to obtain data points within [25, 75] rather than after the steady state has been reached and the Information Indices have leveled off. Measurements taken after 75 *h* provide information only about the steady state behavior of the system.

The above results indicate again the need to assign to each grid point a different time interval over which the measurements of the output vector should be made. The Information Index can provide the means to determine this time interval. It is particularly useful when we have completely different time scales as shown by the Information Indices (shown in Figure 12.9) for the grid point (4,

0.25, 35). The log-linear interpolation for the determination of the sampling times is very useful in such cases.

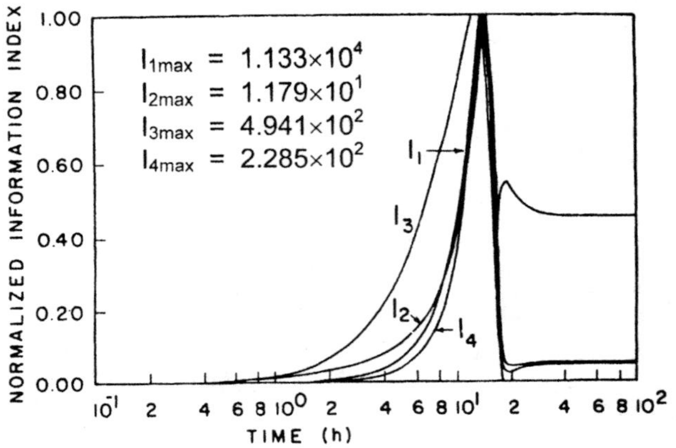

Figure 12.8 *Fed-batch Bioreactor: Information indices versus time for the grid point (7, 0.20, 35) [reprinted from the Canadian Journal of Chemical Engineering with permission].*

Table 12.4 Fed-batch Bioreactor: Standard Deviation of Parameter Estimates Versus Time Interval Used for the Grid Point (1, 0.20, 35)

Case	Time Interval	Number of Data points	Standard Deviation (%) k_1	k_2	k_3	k_4
1	[0.01, 75]	155	1.702	4.972	1.917	1.051
2	[1, 75]	75	1.704	4.972	1.919	1.052
3	[15, 75]	30	1.737	5.065	1.967	1.072
4	[25, 75]	20	1.836	5.065	2.092	1.130
5	[25, 150]	20	1.956	4.626	2.316	1.202
6	[50, 150]	20	3.316	5.262	4.057	2.053
7	[60, 150]	20	278.3	279.5	332.4	1662.

Source: Kalogerakis and Luus (1984).

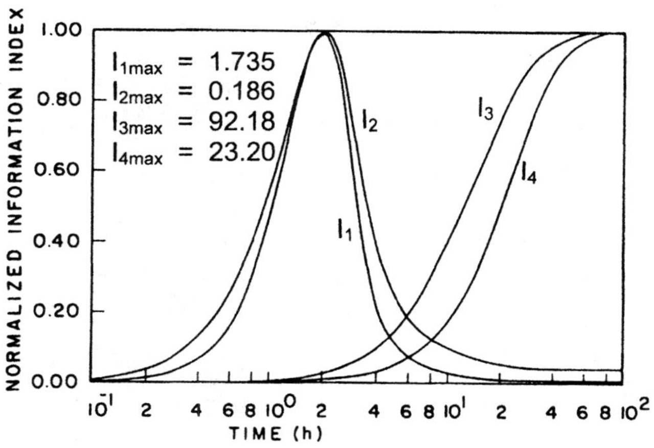

Figure 12.9 *Fed-batch Bioreactor: Information indices versus time for the grid point (4, 0.25, 35) [reprinted from the Canadian Journal of Chemical Engineering with permission].*

Table 12.5 Fed-batch Bioreactor: Selection of the Best Grid Point Based on the Volume Criterion and Use of the Information Index

No. of data points	λ_{min}	$det(\mathbf{A}^{new})$	Standard deviation (%)			
			k_1	k_2	k_3	k_4
20	0.709	6.61×10^9	1.836	5.065	2.092	11.30
40	1.405	1.07×10^{11}	1.302	3.647	1.481	8.020
80	2.816	1.74×10^{12}	0.920	2.574	1.046	5.667

Best grid point for all cases: (1, 0.20, 35).
Source: Kalogerakis and Luus (1984).

Employing the volume criterion, the best grid point in the operability region was determined using the Information Indices. A total of 20, 40 and 80 data points were used and the results are shown in Table 12.5

The condition number of matrix $\mathbf{A}^{new}$ can be used to indicate which of the optimization criteria (volume or shape) is more appropriate. In this example the condition number is rather large (~10^6) and hence, the shape criterion in expected to yield better results. Indeed, by using the shape criterion the condition number was reduced approximately by two orders of magnitude and the overall results are improved as shown in Table 12.6. As seen, the uncertainty in k_1, k_3 and k_4 was decreased while the uncertainty in k_2 was increased somewhat. The Information Indices for the selected grid point (4, 0.15, 35) are shown in Figure 12.10.

Table 12.6 Fed-batch bioreactor: Selection of the Best Grid Point Based on the Shape Criterion and Use of the Information Index

No. of data points	λ_{min}	$det(\mathbf{A}^{new})$	Standard Deviation (%)			
			k_1	k_2	k_3	k_4
20	1.408	4.71×10^8	0.954	8.357	1.175	5.248
40	2.782	7.56×10^9	0.674	5.943	0.831	3.717
80	5.604	1.22×10^{11}	0.477	4.189	0.587	2.628

Best grid point for all cases: (4, 0.15, 35).
Source: Kalogerakis and Luus (1984).

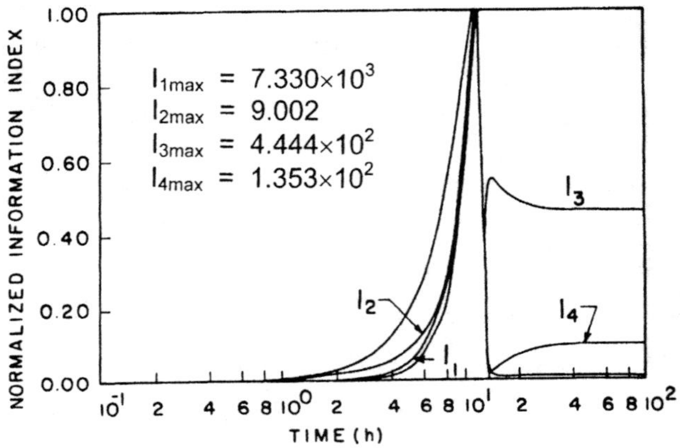

Figure 12.10 Fed-batch Bioreactor: Information indices versus time for the grid point (4, 0.15, 35) [reprinted from the Canadian Journal of Chemical Engineering with permission].

It is interesting to note that by using both design criteria, the selected best grid points lie on the boundary of the operability region. This was also observed in the previous example. The same has also been observed for systems described by algebraic equations (Rippin et al., 1980).

12.5.3 Chemostat Growth Kinetics

As a third example let us consider the growth kinetics in a chemostat used by Kalogerakis (1984) to evaluate sequential design procedures for model discrimination in dynamic systems. We consider the following four kinetic models for biomass growth and substrate utilization in the continuous baker's yeast fermentation.

Model 1 *(Monod kinetics with constant specific death rate)*

$$\frac{dx_1}{dt} = \left(\frac{k_1 x_2}{k_2 + x_2} - D\right)x_1 - k_4 x_1 \tag{12.26a}$$

$$\frac{dx_2}{dt} = -\frac{k_1 x_2}{(k_2 + x_2)k_3} x_1 + D(c_F - x_2) \tag{12.26b}$$

Model 2 *(Contois kinetics with constant specific death rate)*

$$\frac{dx_1}{dt} = \left(\frac{k_1 x_2}{k_2 x_1 + x_2} - D\right)x_1 - k_4 x_1 \tag{12.27a}$$

$$\frac{dx_2}{dt} = -\frac{k_1 x_2}{(k_2 x_1 + x_2)k_3} x_1 + D(c_F - x_2) \tag{12.27b}$$

Model 3 *(Linear specific growth rate)*

$$\frac{dx_1}{dt} = (k_1 x_2 - D)x_1 - k_3 x_1 \tag{12.28a}$$

$$\frac{dx_2}{dt} = -\frac{k_1 x_2}{k_2} x_1 + D(c_F - x_2) \tag{12.28b}$$

Model 4 *(Monod kinetics with constant maintenance energy)*

$$\frac{dx_1}{dt} = \left(\frac{k_1 x_2}{k_2 + x_2} - D\right) x_1 \tag{12.29a}$$

$$\frac{dx_2}{dt} = -\left(\frac{k_1 x_2}{(k_2 + x_2)k_3} + k_4\right) x_1 + D(c_F - x_2) \tag{12.29b}$$

In the above ODEs, x_1 and x_2 represent the biomass and substrate concentration in the chemostat, c_F is the substrate concentration in the feed stream (*g/L*) and D is the dilution factor (h^{-1}) defined as the feed flowrate over the volume of the liquid phase in the chemostat. It is assumed that both state variables, x_1 and x_2 are observed.

The experimental conditions that can be set by the operator are the initial conditions $x_1(0)$ and $x_2(0)$, and the manipulated variables (control inputs) $u_1(t)=D$ and $u_2(t)=c_F$. The range of the manipulated variables is $0.05 \le D \le 0.20$ (h^{-1}) and $5 \le c_F \le 35$ (*g/L*). For simplicity, it is assumed that the initial substrate concentration in the chemostat, $x_2(0)$, is always 0.01 *g/L*. The initial biomass concentration (inoculation) is assumed to take values in the range 1 to 10 *g/L*, i.e., $1 \le x_1(0) \le 10$ (*g/L*).

We have several options in terms of designing the next experiment that will have the maximum discriminating power among all the above four rival models:

(i) Consider both inputs as constants (i.e., D and c_F are kept constant throughout the run). In this case we have a 3-dimensional operability region ($x_1(0)$, D, c_F) that can be visualized as a rectangular prism. The best grid point is selected by maximizing the weighted divergence given by Equation 12.23.

(ii) Consider one input constant and the other one able to vary with respect to time. For example, we can allow the feed concentration to vary optimally with respect to time. The operability region is now a rectangular region on the ($x_1(0)$-D)-plane. For each grid point of the operability region we can solve the optimal control problem and determine the optimal input sequence ($c_F(t)$) that will maximize the divergence given by Equation 12.23. We can greatly simplify the problem by searching for the optimal period of a square wave sequence for c_F.

(iii) Consider both inputs variables (D and c_F) able to vary with respect to time. In this case the operability region is the one dimensional segment on the $x_1(0)$ - axis. Namely for each grid point (for example, $x_1(0)$ could take the values 1, 4, 7 & 10 *g/L*) we solve the optimal control problem whereby we determine the optimal time sequence for the dilution factor

(D) and the feed concentration (c_F) by maximizing the weighted divergence given by Equation 12.23. Again, the optimal control problem can be greatly simplified by considering square waves for both input variables with optimally selected periods.

Let us illustrate the methodology by consider cases (i) and (ii). We assume that from a preliminary experiment or from some other prior information we have the following parameter estimates for each model:

Model 1:	$\mathbf{k}^{(1)} = [0.30, 0.25, 0.56, 0.02]^T$
Model 2:	$\mathbf{k}^{(2)} = [0.30, 0.03, 0.55, 0.03]^T$
Model 3:	$\mathbf{k}^{(3)} = [0.12, 0.56, 0.02]^T$
Model 4:	$\mathbf{k}^{(4)} = [0.30, 0.25, 0.56, 0.02]^T$

Using as final time 72 *h* and a constant sampling rate of 0.75 *h*, the best grid point in the operability region was determined assuming constant inputs. The grid point (1, 0.20, 35) was found as best and the corresponding value of the weighted divergence was D=71759.

On the other hand, if we assume that c_F can change with time as a square wave, while dilution factor is kept constant throughout the run, we find as best the grid point ($x_1(0)$, D) = (1, 0.20) and an optimal period of 27 *h* when c_F is allowed to change as a square wave. In this case a maximum weighted divergence of D=84080 was computed. The dependence of the weighted divergence on the switching period of the input variable (c_F) is shown for the grid points (1, 0.15) and (1, 0.20) in Figures 12.11 and 12.12 respectively.

The ability of the sequential design to discriminate among the rival models should be examined as a function of the standard error in the measurements (σ_ε). For this reason, artificial data were generated by integrating the governing ODEs for Model 1 with "true" parameter values k_1=0.31, k_2=0.18, k_3=0.55 and k_4=0.03 and by adding noise to the noise free data. The error terms are taken from independent normal distributions with zero mean and constant standard deviation (σ_ε).

In Tables 12.7 and 12.8 the results of the χ^2-adequacy test (here σ_ε is assumed to be known) and Bartlett's χ^2-adequacy test (here σ_ε is assumed to be unknown) are shown for both designed experiments (constant inputs and square wave for $c_F(t)$).

As seen in Tables 12.7 and 12.8 when σ_ε is in the range 0.001 to 0.2 *g/L* the experiment with the square wave input has a higher discriminating power. As expected the χ^2-test is overall more powerful since σ_ε is assumed to be known.

For the design of subsequent experiments, we proceed in a similar fashion; however, in the weighted divergence only the models that still remain adequate are included.

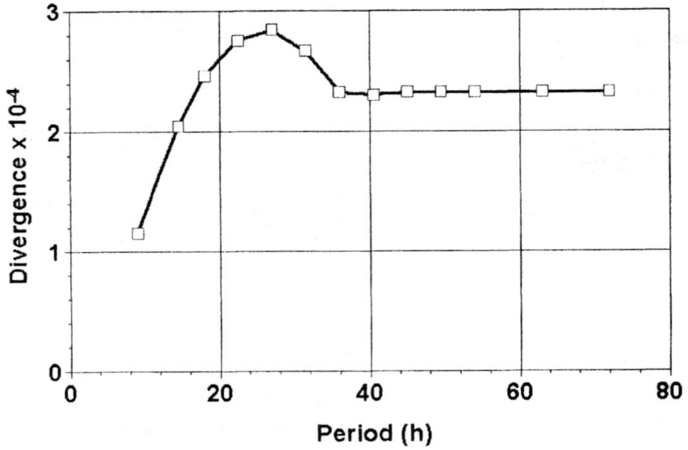

Figure 12.11 *Chemostat kinetics: Weighted divergence as a function of square wave frequency for $c_{I'}(t)$ at the grid point (1, 0.15).*

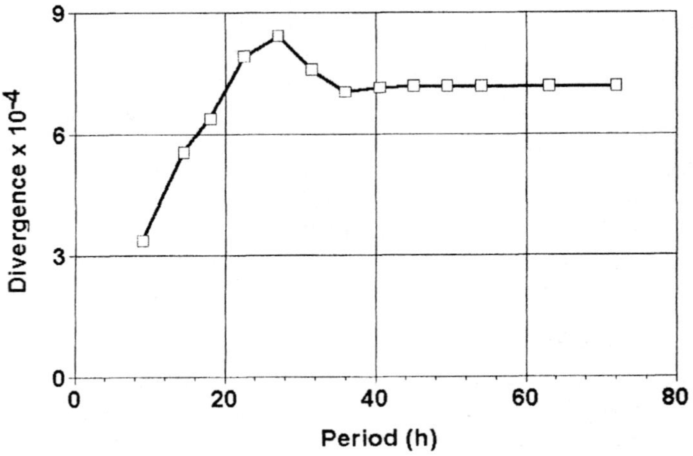

Figure 12.12 *Chemostat kinetics: Weighted divergence as a function of square wave frequency for $c_{I'}(t)$ at the grid point (1, 0.20).*

Table 12.7 Chemostat Kinetics: Results from Model Adequacy Tests Assuming σ_ε is Known (χ^2-test) Performed at α=0.01 Level of Significance

σ_ε	Constant Inputs				Square Wave Input			
	Model 1	Model 2	Model 3	Model 4	Model 1	Model 2	Model 3	Model 4
0.5	A	A	R	A	A	A	R	A
0.2	A	A	R	A	A	R	R	A
0.1	A	R	R	A	A	R	R	A
0.05	A	R	R	A	A	R	R	A
0.02	A	R	R	A	A	R	R	A
0.01	A	R	R	A	A	R	R	R
0.005	A	R	R	R	A	R	R	R
0.001	A	R	R	R	A	R	R	R

Note: R = model is rejected, A = model remains adequate.
Source: Kalogerakis (1984).

Table 12.8 Chemostat Kinetics: Results from Model Adequacy Tests Assuming σ_ε is Unknown (Bartlett's χ^2-test) Performed at α=0.01 Level of Significance

σ_ε	Constant Inputs				Square Wave Input			
	Model 1	Model 2	Model 3	Model 4	Model 1	Model 2	Model 3	Model 4
0.5	A	A	R	A	A	A	R	A
0.2	A	A	R	A	A	A	R	A
0.1	A	A	R	A	A	R	R	A
0.05	A	R	R	A	A	R	R	A
0.02	A	R	R	A	A	R	R	A
0.01	A	R	R	A	A	R	R	A
0.005	A	R	R	A	A	R	R	R
0.001	A	R	R	R	A	R	R	R

Note: R = model is rejected, A = model remains adequate.
Source: Kalogerakis (1984).

13

Recursive Parameter Estimation

In this chapter we present very briefly the basic algorithm for recursive least squares estimation and some of its variations for single input - single output systems. These techniques are routinely used for *on-line parameter estimation* in data acquisition systems. They are presented in this chapter without any proof for the sake of completeness and with the aim to provide the reader with a quick overview. For a thorough presentation of the material the reader may look at any of the following references: Söderström et al. (1978), Ljung and Söderström (1983) Shanmugan and Breipohl, (1988), Wellstead and Zarrop (1991). The notation that will be used in this chapter is different from the one we have used up to now. Instead we shall follow the notation typically encountered in the analysis and control of sampled data systems.

13.1 DISCRETE INPUT-OUTPUT MODELS

Recursive estimation methods are routinely used in many applications where process measurements become available continuously and we wish to re-estimate or better *update on-line* the various process or controller parameters as the data become available. Let us consider the linear discrete-time model having the general structure:

$$A(z^{-1})y_n = B(z^{-1})u_{n-k} + e_n \tag{13.1}$$

where z^{-1} is the backward shift operator (i.e., $y_{n-1} = z^{-1}y_n$, $y_{n-2} = z^{-2}y_n$, etc.) and $A(\cdot)$ and $B(\cdot)$ are polynomials of z^{-1}. The input variable is $u_n \equiv u(t_n)$ and the output variable is $y_n \equiv y(t_n)$. The system has a delay of k sampling intervals ($k \geq 1$). In expanded form the system equation becomes

$$(1 + a_1 z^{-1} + a_2 z^{-2} + \ldots + a_p z^{-p}) y_n = (b_0 + b_1 z^{-1} + b_2 z^{-2} + \ldots + b_q z^{-q}) u_{n-k} + e_n \qquad (13.2)$$

or

$$y_n = - a_1 y_{n-1} - a_2 y_{n-2} - \ldots - a_p y_{n-p} + b_0 u_{n-k} + b_1 u_{n-k-1} + b_2 u_{n-k-2} + \ldots$$
$$+ b_n u_{t-k-m} + e_n \qquad (13.3)$$

We shall present three *recursive* estimation methods for the estimation of the process parameters ($a_1, \ldots, a_p$, b_0, $b_1, \ldots, b_q$) that should be employed according to the statistical characteristics of the error term sequence e_n's (the stochastic disturbance).

13.2 RECURSIVE LEAST SQUARES (RLS)

In this case we assume that ε_n is white noise, ξ_n, i.e., the e_n's are identically and independently distributed normally with zero mean and a constant variance σ^2. Thus, the model equation can be rewritten as

$$\begin{aligned} y_n = & -a_1 y_{n-1} - a_2 y_{n-2} - \ldots - a_p y_{p-n} + \\ & b_0 u_{n-k} + b_1 u_{n-k-1} + b_2 u_{n-k-2} + \ldots + b_q u_{n-k-q} + \xi_n \end{aligned} \qquad (13.4)$$

which is of the form

$$y_n = \psi_{n-1}^T \theta_{n-1} + \xi_n \qquad (13.5)$$

where

$$\psi_{n-1} = \left[- y_{n-1}, -y_{n-2}, \ldots, -y_{n-p}, u_{n-k}, u_{n-k-1}, \ldots, u_{n-k-q}\right]^T \qquad (13.6a)$$

and

$$\theta_{n-1} = \left[a_1, a_2, \ldots, a_p, b_0, b_1, b_2, \ldots, b_q\right]^T \qquad (13.6b)$$

Whenever a new measurement, y_n, becomes available, the parameter vector is updated to $\boldsymbol{\theta}_n$ by the formula

$$\theta_n = \theta_{n-1} + \mathbf{K}_n \left[y_n - \psi_{n-1}^T \theta_{n-1}\right] \qquad (13.7)$$

where

$$\mathbf{K}_n = \frac{\mathbf{P}_{n-1}\psi_n}{1 + \psi_n^T \mathbf{P}_{n-1} y_n} \tag{13.8}$$

The new estimate of the normalized parameter covariance matrix, $\mathbf{P}_n$, is obtained from

$$\mathbf{P}_n = \left[\mathbf{I} - \mathbf{K}_n \psi_n^T\right] \mathbf{P}_{n-1} \tag{13.9}$$

The updated quantities θ_n and $\mathbf{P}_n$ represent our best estimates of the unknown parameters and their covariance matrix with information up to and including time t_n. Matrix $\mathbf{P}_n$ represents an estimate of the parameter covariance matrix since,

$$COV(\theta_n) = \sigma^2 \mathbf{P}_n. \tag{13.10}$$

The above equations are developed using the theory of least squares and making use of the matrix inversion lemma

$$(\mathbf{A} + \mathbf{x}\mathbf{x}^T)^{-1} = \mathbf{A}^{-1} + \mathbf{A}^{-1}\mathbf{x}[\mathbf{x}^T\mathbf{A}^{-1}\mathbf{x} + 1]^{-1}\mathbf{x}^T\mathbf{A}^{-1} \tag{13.11}$$

where $\mathbf{A}^{-1}$=$\mathbf{P}_n$, and $\mathbf{x}$=ψ_n is used.

To initialize the algorithm we start with our best initial guess of the parameters, θ_0. Our initial estimate of the covariance matrix $\mathbf{P}_0$ is often set proportional to the identity matrix (i.e., $\mathbf{P}_0 = \gamma^2\mathbf{I}$). If very little information is available about the parameter values, a large value for γ^2 should be chosen.

Of course, as time proceeds and data keeps on accumulating, $\mathbf{P}_n$ becomes smaller and smaller and hence, the algorithm becomes insensitive to process parameter variations.

The standard way of eliminating this problem is by introducing a *forgetting factor* λ. Namely, instead of minimizing

$$S(\theta) = \sum_{i=1}^{n} e_i^2 \tag{13.12}$$

we estimate the parameters by minimizing the weighted least squares objective function

$$S(\boldsymbol{\theta}) = \sum_{i=1}^{n} \lambda^{n-i} e_i^2 \tag{13.13}$$

with $0 < \lambda < 1$. Typically, our selection for λ is in the range $0.90 \leq \lambda \leq 0.9999$). This choice of weighting factors ensures that more attention is paid to recently collected data. It can be shown (Ljung and Söderström, 1983) that in this case the parameters are obtained recursively by the following equations:

$$\boldsymbol{\theta}_n = \boldsymbol{\theta}_{n-1} + \mathbf{K}_n \left[y_n - \boldsymbol{\psi}_{n-1}^T \boldsymbol{\theta}_{n-1}^T \right] \tag{13.14}$$

$$\mathbf{P}_n = \frac{1}{\lambda} \left[\mathbf{I} - \mathbf{K}_n \boldsymbol{\psi}_n^T \right] \mathbf{P}_{n-1} \tag{13.15}$$

$$\mathbf{K}_n = \frac{\mathbf{P}_{n-1} \boldsymbol{\psi}_n}{\lambda + \boldsymbol{\psi}_n^T \mathbf{P}_{n-1} \boldsymbol{\psi}_n} \tag{13.16}$$

Equations 13.14 to 13.16 constitute the well known *recursive least squares* (RLS) algorithm. It is the simplest and most widely used recursive estimation method. It should be noted that it is computationally very efficient as it does not require a matrix inversion at each sampling interval. Several researchers have introduced a *variable forgetting factor* to allow a more precise estimation of $\boldsymbol{\theta}$ when the process is not "sensed" to change.

13.3 RECURSIVE EXTENDED LEAST SQUARES (RELS)

In this case we assume the disturbance term, e_n, is not white noise, rather it is related to ξ_n through the following transfer function (noise filter)

$$e_n = C(z^{-1})\, \xi_n \tag{13.17}$$

where

$$C(z^{-1}) = 1 + c_1 z^{-1} + c_2 z^{-2} + \ldots + c_r z^{-r} \tag{13.18}$$

Thus, the model equation can be rewritten as

$$\begin{aligned} y_t = {} & -a_1 y_{t-1} - a_2 y_{t-2} - \ldots - a_p y_{t-p} + \\ & b_0 u_{t-k} + b_1 u_{t-k-1} + b_2 u_{t-k-2} + \ldots + b_q u_{t-k-q} + \\ & + c_1 \xi_{t-1} + c_2 \xi_{t-2} + c_3 \xi_{t-3} + \ldots + c_r \xi_{t-r} + \xi_t \end{aligned} \tag{13.19}$$

which is again of the form

$$y_n = \boldsymbol{\psi}_{n-1}^T \boldsymbol{\theta}_{n-1} + \xi_n \tag{13.20}$$

where

$$\boldsymbol{\psi}_{n-1} = \Big[-y_{n-1}, -y_{n-2}, \ldots, -y_{n-p}, \\ u_{n-k}, u_{n-k-1}, \ldots, u_{n-k-q}, \xi_{n-1}, \xi_{n-2}, \ldots, \xi_{n-r}\Big]^T \tag{13.21a}$$

and

$$\boldsymbol{\theta}_{n-1} = \big[a_1, a_2, \ldots, a_p, b_0, b_1, b_2, \ldots, bq, c_1, c_2, \ldots, c_r\big]^T \tag{13.21b}$$

This model is of the same form as the one used for RLS and hence the updating equations for $\boldsymbol{\theta}_n$ and $\mathbf{P}_n$ are the same as in the previous section. However, the disturbance terms, ξ_{n-1}, $\xi_{n-2}, \ldots, \xi_{n-r}$, and hence the updating equations cannot be implemented. The usual approach to overcome this difficulty is by using the one-step predictor error, namely,

$$\widetilde{\boldsymbol{\psi}}_{n-1} = \Big[-y_{n-1}, -y_{n-2}, \ldots, -y_{n-p}, \\ u_{n-k}, u_{n-k-1}, \ldots, u_{n-k-q}, \widetilde{\xi}_{n-1}, \widetilde{\xi}_{n-2}, \ldots, \widetilde{\xi}_{n-r}\Big]^T \tag{13.22}$$

where

$$\left.\begin{aligned} \widetilde{\xi}_{n-1} &= y_{n-1} - \widetilde{y}_{n-1|n-2} = y_{n-1} - \widetilde{\boldsymbol{\psi}}_{n-2}^T \boldsymbol{\theta}_{n-2} \\ \widetilde{\xi}_{n-2} &= y_{n-2} - \widetilde{y}_{n-2|n-3} = y_{n-2} - \widetilde{\boldsymbol{\psi}}_{n-3}^T \boldsymbol{\theta}_{n-3} \\ &\vdots \\ \widetilde{\xi}_{n-r} &= y_{n-r} - \widetilde{y}_{n-r|n-r-1} = y_{n-r} - \widetilde{\boldsymbol{\psi}}_{n-r-1}^T \boldsymbol{\theta}_{n-r-1} \end{aligned}\right\} \tag{13.23}$$

where we have denoted with $\widetilde{y}_{n|n-1}$ the one-step ahead predictor of y_n. Therefore the *recursive extended least squares* (RELS) algorithm is given by the following equations:

$$\boldsymbol{\theta}_n = \boldsymbol{\theta}_{n-1} + \mathbf{K}_n\Big[y_n - \widetilde{\boldsymbol{\psi}}_{n-1}^T \boldsymbol{\theta}_{n-1}^T\Big] \tag{13.24}$$

$$\mathbf{P}_n = \frac{1}{\lambda}\left[\mathbf{I} - \mathbf{K}_n \tilde{\psi}_n^T\right] \mathbf{P}_{n-1} \tag{13.25}$$

$$\mathbf{K}_n = \frac{\mathbf{P}_{n-1}\tilde{\psi}_n}{\lambda + \psi_n^T \mathbf{P}_{n-1}\tilde{\psi}_n} \tag{13.26}$$

13.4 RECURSIVE GENERALIZED LEAST SQUARES (RGLS)

In this case we assume again the disturbance term, e_n, is not white noise, rather it is related to ξ_n through the following transfer function (noise filter)

$$e_n = \frac{1}{C(z^{-1})}\xi_n \tag{13.27}$$

where

$$C(z^{-1}) = 1 + c_1 z^{-1} + c_2 z^{-2} + \ldots + c_r z^{-r} \tag{13.28}$$

The model equations now become

$$A(z^{-1})C(z^{-1})y_n = B(z^{-1})C(z^{-1})u_{n-k} + \xi_n \tag{13.29}$$

or

$$A(z^{-1})\bar{y}_n = B(z^{-1})\bar{u}_{n-k} + \xi_n \tag{13.30}$$

where the transformed input and output variables are given by

$$\bar{y}_n = C(z^{-1})y_n \tag{13.31a}$$

or

$$\bar{y}_n = y_n + c_1 y_{n-1} + c_2 y_{n-2} + \ldots + c_r y_{n-r} \tag{13.31b}$$

and

$$\bar{u}_{n-k} = C(z^{-1})\, u_{n-k} \tag{13.32a}$$

or

$$\bar{u}_{n-k} = u_{n-k} + c_1 u_{n-k-1} + c_2 u_{n-k-2} + \ldots + c_r u_{n-k-r} \tag{13.32b}$$

Equation (13.30) can now be rewritten in our usual form as

$$\bar{y}_n = \bar{\psi}_{n-1}^T \theta_{n-1} + \xi_n \tag{13.33}$$

where

$$\bar{\psi}_{n-1} = \left[-\bar{y}_{n-1}, -\bar{y}_{n-2}, \ldots, -\bar{y}_{n-p}, \bar{u}_{n-k}, \bar{u}_{n-k-1}, \ldots, \bar{u}_{n-k-q}\right]^T \tag{13.34a}$$

and

$$\theta_{n-1} = \left[a_1, a_2, \ldots, a_p, b_0, b_1, b_2, \ldots, bq\right]^T \tag{13.34b}$$

The above equations suggest that the unknown parameters in polynomials A(·) and B(·) can be estimated with RLS with the transformed variables $\bar{y}_n$ and $\bar{u}_{n-k}$. Having polynomials A(·) and B(·) we can go back to Equation 13.1 and obtain an estimate of the error term, e_n, as

$$\tilde{e}_n = \tilde{A}(z^{-1}) y_n - \tilde{B}(z^{-1}) u_{n-k} \tag{13.35}$$

where by we have denoted with $\tilde{A}(\cdot)$ and $\tilde{B}(\cdot)$ the current estimates of polynomials A(·) and B(·). The unknown parameters in polynomial C(·) can be obtained next by considering the noise filter transfer function. Namely, Equation 13.27 is written as

$$e_n = -c_1 e_{n-1} - c_2 e_{n-2} - \ldots - c_r e_{n-r} + \xi_n \tag{13.36}$$

The above equation cannot be used directly for RLS estimation. Instead of the true error terms ,e_n, we must use the estimated values from Equation 13.35. Therefore, the recursive generalized least squares (RGLS) algorithm can be implemented as a two-step estimation procedure:

Step 1. Compute the transformed variables $\bar{y}_n$ and $\bar{u}_{n-k}$ based on our knowledge of polynomial C(·) at time t_{n-1}.

Apply RLS on equation $A(z^{-1})\bar{y}_n = B(z^{-1})\bar{u}_{n-k} + \xi_n$ based on information up to time t_n. Namely, obtain the updated value for the parameter vector (i.e. the coefficients of the polynomials A and B), θ_n

Step 2. Having θ_n, estimate $\tilde{e}_n$ as $\tilde{A}(z^{-1}) y_n - \tilde{B}(z^{-1}) u_{n-k}$.

Apply RLS to equation: $C(z^{-1})\tilde{e}_n = \xi_n$ to get the coefficients of $C(z^{-1})$ which are used for the computation of the transformed variables $\bar{y}_n$ and $\bar{u}_{n-k}$ at the next sampling interval.

Essentially, the idea behind the above RGLS algorithm is to apply the ordinary RLS algorithm twice. The method is easy to apply, however, it may have multiple convergence points (Ljung and Söderström, 1983).

The previously presented recursive algorithms (RSL, RELS and RGLS) utilize the same algorithm (i.e., the same subroutine can be used) the only difference being the set up of the state and parameter vectors. That is the primary reason for our selection of the recursive algorithms presented here. Other recursive algorithms (including recursive maximum likelihood) as well as an exhaustive presentation of the subject material including convergence characteristics of the above algorithms can be found in any standard on-line identification book.

14

Parameter Estimation in Nonlinear Thermodynamic Models: Cubic Equations of State

Thermodynamic models are widely used for the calculation of equilibrium and thermophysical properties of fluid mixtures. Two types of such models will be examined: cubic equations of state and activity coefficient models. In this chapter cubic equations of state models are used. Volumetric equations of state (EoS) are employed for the calculation of fluid phase equilibrium and thermophysical properties required in the design of processes involving non-ideal fluid mixtures in the oil and gas and chemical industries. It is well known that the introduction of empirical parameters in equation of state mixing rules enhances the ability of a given EoS as a tool for process design although the number of interaction parameters should be as small as possible. In general, the phase equilibrium calculations with an EoS are very sensitive to the values of the binary interaction parameters.

14.1 EQUATIONS OF STATE

An equation of state is an algebraic relationship that relates the intensive thermodynamic variables (T, P, v, x_1, x_2,..., x_{Nc}) for systems in thermodynamic equilibrium. In this relationship, T is the temperature, P is the pressure, v is the molar volume and x_i are the mole fractions of component i with $i=1,...,N_C$, N_C being the number of species present in the system.

The EoS may be written in many forms. It may be implicit

$$F(T, P, v, x_1, x_2, ..., x_{N_c}) = 0 \tag{14.1}$$

or explicit in any of the variables, such as pressure. In a system of one component (pure fluid) the above equation becomes

$$F(T,P,v) = 0 \quad \text{or} \quad P = P(T,v) \tag{14.2}$$

The above equations describe a three-dimensional T-P-v surface which is a representation of all the experimental or statistical mechanical information concerning the EoS of the system. It should also be noted that it is observed experimentally that

$$\lim_{P \to 0} \frac{Pv}{RT} = 1$$

The quantity Pv/RT is called compressibility factor. The PvT or volumetric behavior of a system depends on the intermolecular forces. Sizes, shapes and structures of molecules determine the forces between them (Tassios, 1993).

Volumetric data are needed to calculate thermodynamic properties (enthalpy, entropy). They are also used for the metering of fluids, the sizing of vessels and in natural gas and oil reservoir calculations.

14.1.1 Cubic Equations of State

These equations originate from van der Waals cubic equation of state that was proposed more than 100 years ago. The EoS has the form $(P+a/v^2)(v-b)=RT$ where the constants a and b are specific for each substance. This was a significant improvement over the ideal gas law because it could describe the volumetric behavior of both gases and liquids. Subsequently, improvements in the accuracy of the calculations, especially for liquids, were made. One of the most widely used cubic equations of state in industry is the *Peng-Robinson EoS* (Peng and Robinson, 1976). The mathematical form of this equation was given in Chapter 1 as Equation 1.7 and it is given again below:

$$P = \frac{RT}{v-b} - \frac{a(T)}{v(v+b)+b(v-b)} \tag{14.3}$$

where P is the pressure, T the temperature and v the molar volume. When the above equation is applied to a single fluid, for example gaseous or liquid CO2, the parameters a and b are specific to CO2. Values for these parameters can be readily computed using Equations 14.5a - 14.5d. Whenever the EoS (Equation 14.3) is applied to a mixture, parameters a and b are now mixture parameters. The way these mixture parameters are defined is important because it affects the ability of the EoS to represent fluid phase behavior. The so-called van der Waals combining rules given below have been widely used. The reader is advised to consult the thermodynamics literature for other combining rules.

$$a = \sum_{i=1}^{N_C} \sum_{j=1}^{N_C} x_i x_j \left(1 - k_{ij}\right) \sqrt{a_i a_j} \tag{14.4a}$$

$$b = \sum_{i=1}^{N_C} x_i b_i \tag{14.4b}$$

where a_i, and b_i are parameters specific for the individual component i and k_{ij} is an empirical interaction parameter characterizing the binary formed by component i and component j. The individual parameters for component i are given by

$$a_i = 0.45724 \frac{R^2 T_{c,i}^2}{P_{c,i}} \left[1 + K_i\left(1 - \sqrt{T_{r,i}}\right)\right]^2 \tag{14.5a}$$

$$K_i = 0.37464 + 1.54226\omega_i - 0.26992\omega_i^2 \tag{14.5b}$$

$$b_i = 0.07780 \frac{R T_{c,i}}{P_{c,i}} \tag{14.5c}$$

$$T_{r,i} = \frac{T}{T_{c,i}} \tag{14.5d}$$

In the above equations, ω is the acentric factor, and T_c, P_c are the critical temperature and pressure respectively. These quantities are readily available for most components.

The *Trebble-Bishnoi EoS* is a cubic equation that may utilize up to four binary interaction parameters, $\mathbf{k}=[k_a, k_b, k_c, k_d]^T$. This equation with its quadratic combining rules is presented next (Trebble and Bishnoi, 1987; 1988).

$$P = \frac{RT}{v - b_m} - \frac{a_m}{v^2 + (b_m + c_m)v - (b_m c_m + d_m^2)} \tag{14.6}$$

where

$$a_m = \sum_{i=1}^{N_C} \sum_{j=1}^{N_C} x_i x_j \sqrt{a_i a_j}\,(1 - k_a) \tag{14.7a}$$

$$b_m = \sum_{i=1}^{N_C} \sum_{j=1}^{N_C} x_i x_j \left(\frac{b_i + b_j}{2}\right)(1 - k_b) \tag{14.7b}$$

$$c_m = \sum_{i=1}^{N_C} \sum_{j=1}^{N_C} x_i x_j \left(\frac{c_i + c_j}{2}\right)(1 - k_c) \tag{14.7c}$$

$$d_m = \sum_{i=1}^{N_C} \sum_{j=1}^{N_C} x_i x_j \left(\frac{d_i + d_j}{2}\right)(1 - k_d) \tag{14.7d}$$

and N_C is the number of components in the fluid mixture. The quantities a_i, b_i,c_i,d_i, $i=1,..,N_C$ are the individual component parameters.

14.1.2 Estimation of Interaction Parameters

As we mentioned earlier, a volumetric EoS expresses the relationship among pressure, P, molar volume, v, temperature, T and composition **z** for a fluid mixture. This relationship for a pressure-explicit EoS is of the form

$$P = P(v, T; \mathbf{z}; \mathbf{u}; \mathbf{k}) \tag{14.8}$$

where the p-dimensional vector **k** represents the unknown binary interaction parameters and **z** is the composition (mole fractions) vector. The vector **u** is the set of EoS parameters which are assumed to be precisely known e.g., pure component critical properties.

Given an EoS, the objective of the parameter estimation problem is to compute optimal values for the interaction parameter vector, **k**, in a statistically correct and computationally efficient manner. Those values are expected to enhance the correlational ability of the EoS without compromising its ability to predict the correct phase behavior.

14.1.3 Fugacity Expressions Using the Peng-Robinson EoS

Any cubic equation of state can give an expression for the fugacity of species i in a gaseous or in liquid mixture. For example, the expression for the fugacity of a component i in a gas mixture, f_i^V, based on the Peng-Robinson EoS is the following

$$ln\left(\frac{f_i^V}{y_iP}\right) = \frac{b_i}{b}(z^V - 1) - ln\left(z^V - \frac{bP}{RT}\right) - \frac{a}{2\sqrt{2}\,bRT}\left(\frac{2\sum_j y_j a_{ij}}{a} - \frac{b_i}{b}\right) ln\left(\frac{z^V + (1+\sqrt{2})\frac{bP}{RT}}{z^V + (1-\sqrt{2})\frac{bP}{RT}}\right) \qquad (14.9)$$

To calculate the fugacity of each species in a gaseous mixture using the above equation at specified T, P, and mole fractions of all components $y_1, y_2, \ldots$, the following procedure is used

1. Obtain the parameters a_i and b_i for each pure species (component) of the mixture from Equation 14.5.
2. Compute a and b parameters for the mixture from Equation 14.4.
3. Write the Peng-Robinson equation as follows

$$z^3 - (1-B)z^2 + (A - 3B^2 - 2B)z - (AB - B^2 - B^3) = 0 \qquad (14.10)$$

where

$$A = \frac{aP}{(RT)^2} \qquad (14.11a)$$

$$B = \frac{bP}{RT} \qquad (14.11b)$$

$$z = \frac{Pv}{RT} \qquad (14.11c)$$

4. Solve the above cubic equation for the compressibility factor, z^V (corresponding to largest root of the cubic EoS).
5. Use the value of z^V to compute the vapor phase fugacity for each species from Equation 14.9.

The expression for the fugacity of component i in a liquid mixture is as follows

$$ln\left(\frac{f_i^L}{x_i P}\right) = \frac{b_i}{b}(z^L - 1) - ln\left(z^L - \frac{bP}{RT}\right)$$

$$-\frac{a}{2\sqrt{2}\,bRT}\left[\frac{2\sum_j x_j a_{ij}}{a} - \frac{b_i}{b}\right] \ln\left[\frac{z^L + (1+\sqrt{2})\dfrac{bP}{RT}}{z^L + (1-\sqrt{2})\dfrac{bP}{RT}}\right] \tag{14.12}$$

where A and B are as before and z^L is the compressibility factor of the liquid phase (corresponding to smallest root of the cubic EoS). It is also noted that in Equations 14.4a and 14.4b the mixture parameters a and b are computed using the liquid phase mole fractions.

14.1.4 Fugacity Expressions Using the Trebble-Bishnoi EoS

The expression for the fugacity of a component j in a gas or liquid mixture, f_j, based on the Trebble-Bishnoi EoS is available in the literature (Trebble and Bishnoi, 1988). This expression is given in Appendix 1. In addition the partial derivative, $(\partial ln f_j/\partial x_j)_{T,P}$, for a binary mixture is also provided. This expression is very useful in the parameter estimation methods that will be presented in this chapter.

14.2 PARAMETER ESTIMATION USING BINARY VLE DATA

Traditionally, the binary interaction parameters such as the k_a, k_b, k_c, k_d in the Trebble-Bishnoi EoS have been estimated from the regression of binary vapor-liquid equilibrium (VLE) data. It is assumed that a set of N experiments have been performed and that at each of these experiments, four state variables were measured. These variables are the temperature (T), pressure (P), liquid (x) and vapor (y) phase mole fractions of one of the components. The measurements of these variables are related to the "true" but unknown values of the state variables by the equations given next

$$\hat{T}_i = T_i + e_{T,i} \quad i = 1,2,\ldots,N \tag{14.13a}$$

$$\hat{P}_i = P_i + e_{P,i} \quad i = 1,2,\ldots,N \tag{14.13b}$$

$$\hat{x}_i = x_i + e_{x,i} \quad i = 1,2,\ldots, N \tag{14.13c}$$

$$\hat{y}_i = y_i + e_{y,i} \quad i = 1,2,\ldots, N \tag{14.13d}$$

where $e_{T,i}$, $e_{P,i}$, $e_{x,i}$ and $e_{y,i}$, are the corresponding errors in the measurements. Given the model described by Equation 14.8 as well as the above experimental information, the objective is to estimate the parameters (k_a, k_b, k_c and k_d) by matching the data with the EoS-based calculated values.

Least squares (LS) or maximum likelihood (ML) estimation methods can be used for the estimation of the parameters. Both methods involve the minimization of an objective function which consists of a weighted sum of squares of deviations (residuals). The objective function is a measure of the correlational ability of the EoS. Depending on how the residuals are formulated we have explicit or implicit estimation methods (Englezos et al. 1990a). In explicit formulations, the differences between the measured values and the EoS (model) based predictions constitute the residuals. At each iteration of the minimization algorithm, explicit formulations involve phase equilibrium calculations at each experimental point. Explicit methods often fail to converge at "difficult" points (e.g. at high pressures). As a consequence, these data points are usually ignored in the regression, with resulting inferior matching ability of the model (Michelsen, 1993). On the other hand, implicit estimation has the advantage that one avoids the iterative phase equilibrium calculations and thus has a parameter estimation method which is robust, and computationally efficient (Englezos et al. 1990a; Peneloux et al. 1990).

It should be kept in mind that an objective function which does not require any phase equilibrium calculations during each minimization step is the basis for a robust and efficient estimation method. The development of implicit objective functions is based on the phase equilibrium criteria (Englezos et al. 1990a). Finally, it should be noted that one important underlying assumption in applying ML estimation is that the model is capable of representing the data without any systematic deviation. Cubic equations of state compute equilibrium properties of fluid mixtures with a variable degree of success and hence the ML method should be used with caution.

14.2.1 Maximum Likelihood Parameter and State Estimation

Over the years two ML estimation approaches have evolved: (a) parameter estimation based an implicit formulation of the objective function; and (b) parameter and state estimation or "error in variables" method based on an explicit formulation of the objective function. In the first approach only the parameters are estimated whereas in the second the true values of the state variables as well as the values of the parameters are estimated. In this section, we are concerned with the latter approach.

Only two of the four state variables measured in a binary VLE experiment are independent. Hence, one can arbitrarily select two as the independent variables and use the EoS and the phase equilibrium criteria to calculate values for the other two (dependent variables). Let ζ_{ij} (i=1,2,...,N and j=1,2) be the independent variables. Then the dependent ones, ξ_{ij} , can be obtained from the phase equilibrium relationships (Modell and Reid, 1983) using the EoS. The relationship between the independent and dependent variables is nonlinear and is written as follows

$$\xi_{ij} = h_j(\zeta_{ij};\mathbf{k};\mathbf{u}); \quad j=1,2 \quad \text{and} \quad i=1,2,...,N \tag{14.14}$$

In this case, the ML parameter estimates are obtained by minimizing the following quadratic optimality criterion (Anderson et al., 1978; Salazar-Sotelo et al. 1986)

$$S(\mathbf{k}) = \sum_{i=1}^{N}\sum_{j=1}^{2}\left[\frac{(\hat{\zeta}_{ij}-\zeta_{ij})^2}{\sigma^2_{\zeta_{ij}}} + \frac{(\hat{\xi}_{ij}-h_j(\zeta_{ij};\mathbf{k};\mathbf{u}))^2}{\sigma^2_{\xi_{ij}}}\right] \tag{14.15}$$

This is the so-called *error in variables method.* The formulation of the above optimality criterion was based on the following assumptions:

(i) The EoS is capable of representing the data without any systematic deviation.
(ii) The experiments are independent.
(iii) The errors in the measurements are normally distributed with zero mean and a *known* covariance matrix *diag*($\sigma^2_{T,i}$, $\sigma^2_{P,i}$, $\sigma^2_{x,i}$ $\sigma^2_{y,i}$).

Unless very few experimental data are available, the dimensionality of the problem is extremely large and hence difficult to treat with standard nonlinear minimization methods. Schwetlick and Tiller (1985), Salazar-Sotelo et al. (1986) and Valko and Vajda (1987) have exploited the structure of the problem and proposed computationally efficient algorithms.

14.2.2 Explicit Least Squares Estimation

The *error in variables method* can be simplified to weighted least squares estimation if the independent variables are assumed to be known precisely or if they have a negligible error variance compared to those of the dependent variables. In practice however, the VLE behavior of the binary system dictates the choice of the pairs (T,x) or (T,P) as independent variables. In systems with a sparingly soluble component in the liquid phase the (T,P) pair should be chosen as independent variables. On the other hand, in systems with azeotropic behavior the (T,x) pair should be chosen.

Assuming that the variance of the errors in the measurement of each dependent variable is known, the following *explicit* LS objective functions may be formulated:

$$S_{Tx}(\mathbf{k}) = \sum_{i=1}^{N}\left[\frac{\left(P_i^{calc} - \hat{P}_i\right)^2}{\sigma_{P,i}^2} + \frac{\left(y_i^{calc} - \hat{y}_i\right)^2}{\sigma_{y,i}^2}\right] \tag{14.16a}$$

$$S_{TP}(\mathbf{k}) = \sum_{i=1}^{N}\left[\frac{\left(x_i^{calc} - \hat{x}_i\right)^2}{\sigma_{x,i}^2} + \frac{\left(y_i^{calc} - \hat{y}_i\right)^2}{\sigma_{y,i}^2}\right] \tag{14.16b}$$

The calculation of y and P in Equation 14.16a is achieved by *bubble point pressure*-type calculations whereas that of x and y in Equation 14.16b is by *isothermal-isobaric flash*-type calculations. These calculations have to be performed during each iteration of the minimization procedure using the current estimates of the parameters. Given that both the bubble point and the flash calculations are iterative in nature the overall computational requirements are significant. Furthermore, convergence problems in the thermodynamic calculations could also be encountered when the parameter values are away from their optimal values.

14.2.3 Implicit Maximum Likelihood Parameter Estimation

Implicit estimation offers the opportunity to avoid the computationally demanding state estimation by formulating a suitable optimality criterion. The penalty one pays is that additional distributional assumptions must be made. Implicit formulation is based on residuals that are implicit functions of the state variables as opposed to the explicit estimation where the residuals are the errors in the state variables. The assumptions that are made are the following:

(i) The EoS is capable of representing the data without any systematic deviation.

(ii) The experiments are independent.

(iii) The covariance matrix of the errors in the measurements is known and it is often of the form: *diag*($\sigma_{T,i}^2, \sigma_{P,i}^2, \sigma_{x,i}^2\ \sigma_{y,i}^2$).

(iv) The residuals employed in the optimality criterion are normally distributed.

The formulation of the residuals to be used in the objective function is based on the phase equilibrium criterion

$$f_{ij}^{V} = f_{ij}^{L} \quad ; \ i=1,2,\ldots,N \text{ and } j=1,2 \tag{14.17}$$

where f is the fugacity of component 1 or 2 in the vapor or liquid phase. The above equation may be written in the following form

$$g_j(T_i, P_i, x_i, y_i; \mathbf{k}; \mathbf{u}) \equiv \left(\ln f_{ij}^{L} - \ln f_{ij}^{V} \right) = 0 \ ; \ i=1,2,\ldots N; \ j=1,2 \tag{14.18}$$

The above equations hold at equilibrium. However, when the measurements of the temperature, pressure and mole fractions are introduced into these expressions the resulting values are not zero even if the EoS were perfect. The reason is the random experimental error associated with each measurement of the state variables. Thus, Equation 14.18 is written as follows

$$g_j(\hat{T}_i, \hat{P}_i, \hat{x}_i, \hat{y}_i; \mathbf{k}; \mathbf{u}) = \left(\ln f_{ij}^{L} - \ln f_{ij}^{V} \right) \equiv r_{ij} \quad ; \ i=1,2,\ldots N; \ j=1,2 \tag{14.19}$$

The estimation of the parameters is now accomplished by minimizing the implicit ML objective function

$$S_{ML}(\mathbf{k}) = [r_{i1}, r_{i2}] \begin{bmatrix} \sigma_{i1}^2 & \sigma_{i12} \\ \sigma_{i12} & \sigma_{i2}^2 \end{bmatrix}^{-1} [r_{i1}, r_{i2}]^T \tag{14.20}$$

where σ_{ij}^2 (j=1,2) is computed by a first order variance approximation also known as the *error propagation law* (Bevington and Robinson, 1992) as follows

$$\sigma_{ij}^2 = \left(\frac{\partial g_j}{\partial T} \right)^2 \sigma_{T,i}^2 + \left(\frac{\partial g_j}{\partial P} \right)^2 \sigma_{P,i}^2 + \left(\frac{\partial g_j}{\partial x_1} \right)^2 \sigma_{x,i}^2 + \left(\frac{\partial g_j}{\partial y_1} \right)^2 \sigma_{y,i}^2 \tag{14.21}$$

$$\sigma_{i12} = \left(\frac{\partial g_1}{\partial T} \right) \left(\frac{\partial g_2}{\partial T} \right) \sigma_{T,i}^2 + \left(\frac{\partial g_1}{\partial P} \right) \left(\frac{\partial g_2}{\partial P} \right) \sigma_{P,i}^2 + \left(\frac{\partial g_1}{\partial x_1} \right) \left(\frac{\partial g_2}{\partial x_1} \right) \sigma_{x,i}^2 + \left(\frac{\partial g_1}{\partial y_1} \right) \left(\frac{\partial g_2}{\partial y_1} \right) \sigma_{y,i}^2 \tag{14.22}$$

where all the derivatives are evaluated at $\hat{T}_i, \hat{P}_i, \hat{x}_i, \hat{y}_i$. Thus, the errors in the measurement of all four state variables are taken into account.

The above form of the residuals was selected based on the following considerations:

(i) The residuals r_{ij} should be approximately normally distributed.
(ii) The residuals should be such that no iterative calculations such as bubble point or flash-type calculations are needed during each step of the minimization.
(iii) The residuals are chosen in a away to avoid numerical instabilities, e.g. exponent overflow.

Because the calculation of these residuals does not require any iterative calculations, the overall computational requirements are significantly less than for the explicit estimation method using Equation 14.15 and the explicit LS estimation method using Equations 14.16a and b (Englezos et al. 1990a).

14.2.4 Implicit Least Squares Estimation

It is well known that cubic equations of state have inherent limitations in describing accurately the fluid phase behavior. Thus our objective is often restricted to the determination of a set of interaction parameters that will yield an "acceptable fit" of the binary VLE data. The following implicit least squares objective function is suitable for this purpose

$$S_{ILS}(\mathbf{k}) = \sum_{i=1}^{N}\sum_{j=1}^{2} Q_{ij}\left(\ln f_{ij}^{L} - \ln f_{ij}^{V}\right)^2 \tag{14.23}$$

where Q_{ij} are user-defined weighting factors so that equal attention is paid to all data points. If the $\ln f_{ij}$ are within the same order of magnitude, Q_{ij}, can be set equal to one.

14.2.5 Constrained Least Squares Estimation

It is well known that cubic equations of state may predict erroneous binary vapor liquid equilibria when using interaction parameter estimates from an unconstrained regression of binary VLE data (Schwartzentruber et al., 1987; Englezos et al. 1989). In other words, the liquid phase stability criterion is violated. Modell and Reid (1983) discuss extensively the phase stability criteria. A general method to alleviate the problem is to perform the least squares estimation subject to satisfying the liquid phase stability criterion. In other words, parameters which yield the optimal fit of the VLE data subject to the liquid phase stability criterion are sought. One such method which ensures that the stability constraint is satisfied over the entire T-P-x surface has been proposed by Englezos et al. (1989).

Given a set of N binary VLE (T-P-x-y) data and an EoS, an efficient method to estimate the EoS interaction parameters subject to the liquid phase stability criterion is accomplished by solving the following problem

$$\textit{Minimize} \quad S_{ILS}(\mathbf{k}) = \sum_{i=1}^{N} \begin{bmatrix} \ln f_1^L - \ln f_1^V \\ \ln f_2^L - \ln f_2^V \end{bmatrix}_i^T \mathbf{Q}_i \begin{bmatrix} \ln f_1^L - \ln f_1^V \\ \ln f_2^L - \ln f_2^V \end{bmatrix}_i \tag{14.24}$$

$$\textit{subject to} \quad \left(\frac{\partial \ln f_1^L}{\partial x_1} \right)_{T,P} \equiv \varphi(T,P,x;\mathbf{k}) > 0, \qquad (T,P,x) \in \Omega \tag{14.25}$$

where $\mathbf{Q}_i$ is a user-defined weighting matrix (the identity matrix is often used), $\mathbf{k}$ is the unknown parameter vector, f_i^L is the fugacity of component i in the liquid phase, f_i^V is the fugacity of component i in the vapor phase, φ is the stability function and Ω is the EoS computed T-P-x surface over which the stability constraint should be satisfied. This is because the EoS-user expects and demands from the equation of state to calculate the correct phase behavior at any specified condition. This is not a typical constrained minimization problem because the constraint is a function of not only the unknown parameters, $\mathbf{k}$, but also of the state variables. In other words, although the objective function is calculated at a finite number of data points, the constraint should be satisfied over the entire feasible range of T, P and x.

14.2.5.1 Simplified Constrained Least Squares Estimation

Solution of the above constrained least squares problem requires the repeated computation of the equilibrium surface at each iteration of the parameter search. This can be avoided by using the equilibrium surface defined by the experimental VLE data points rather than the EoS computed ones in the calculation of the stability function. The above minimization problem can be further simplified by satisfying the constraint only at the given experimental data points (Englezos et al. 1989). In this case, the constraint (Equation 14.25) is replaced by

$$\textit{subject to} \qquad \varphi(T_i,P_i,x_i;\mathbf{k}) \geq \xi > 0 \tag{14.26}$$

where ξ is a small positive number given by

$$\xi = 1.96\sqrt{\left(\frac{\partial\varphi}{\partial T}\right)^2 \sigma_T^2 + \left(\frac{\partial\varphi}{\partial P}\right)^2 \sigma_P^2 + \left(\frac{\partial\varphi}{\partial x_1}\right)^2 \sigma_x^2} \tag{14.27}$$

In Equation 14.27, σ_T, σ_P and σ_x are the standard deviations of the measurements of T, P and x respectively. All the derivatives are evaluated at the point where the stability function φ has its lowest value. We call the minimization of Equation 14.24 subject to the above constraint *simplified Constrained Least Squares* (simplified CLS) estimation.

14.2.5.2 A Potential Problem with Sparse or Not Well Distributed Data

The problem with the above simplified procedure is that it may yield parameters that result in erroneous phase separation at conditions other than the given experimental ones. This problem arises when the given data are sparse or not well distributed. Therefore, we need a procedure that extends the region over which the stability constraint is satisfied.

The objective here is to construct the equilibrium surface in the T-P-x space from a set of available experimental VLE data. In general, this can be accomplished by using a suitable three-dimensional interpolation method. However, if a sufficient number of well distributed data is not available, this interpolation should be avoided as it may misrepresent the real phase behavior of the system.

In practice, VLE data are available as sets of isothermal measurements. The number of isotherms is usually small (typically 1 to 5). Hence, we are often left with limited information to perform interpolation with respect to temperature. On the contrary, one can readily interpolate within an isotherm (two-dimensional interpolation). In particular, for systems with a sparingly soluble component, at each isotherm one interpolates the liquid mole fraction values for a desired pressure range. For any other binary system (e.g., azeotropic), at each isotherm, one interpolates the pressure for a given range of liquid phase mole fraction, typically 0 to 1.

For simplicity and in order to avoid potential misrepresentation of the experimental equilibrium surface, we recommend the use of 2-D interpolation. Extrapolation of the experimental data should generally be avoided. It should be kept in mind that, if prediction of complete miscibility is demanded from the EoS at conditions where no data points are available, a strong prior is imposed on the parameter estimation from a Bayesian point of view.

Location of φ_{min}

The set of points over which the minimum of φ is sought has now been expanded with the addition of the interpolated points to the experimental ones. In this expanded set, instead of using any gradient method for the location of the minimum of the stability function, we advocate the use of direct search. The rationale behind this choice is that first we avoid any local minima and second the computational requirements for a direct search over the interpolated and the given experimental data are rather negligible. Hence, the minimization of Equation 14.24 should be performed subject to the following constraint

$$\varphi_0 = \varphi(T_0, P_0, x_0; \mathbf{k}) \geq \xi > 0. \tag{14.28}$$

where (T_0, P_0, x_0) is the point on the experimental equilibrium surface (interpolated or experimental) where the stability function is minimum. We call the minimization of Equation 14.24 subject to the above constraint (Equation 14.28) *Constrained Least Squares* (CLS) estimation.

It is noted that ξ should be a small positive constant with a value approximately equal to the difference in the value of φ when evaluated at the experimental equilibrium surface and the EoS predicted one in the vicinity of the (T_0, P_0, x_0). Namely, if we have a system with a sparingly soluble component then ξ is given by

$$\xi = \left| \left(\frac{\partial \varphi_0}{\partial x_1} \right)^T \left(x_0^{exp} - x_0^{EoS} \right) \right| \tag{14.29a}$$

For all other systems, ξ can be computed by

$$\xi = \left| \left(\frac{\partial \varphi_0}{\partial P} \right)^T \left(P_0^{exp} - P_0^{EoS} \right) \right| \tag{14.29b}$$

In the above two equations, the superscripts "exp" and "EoS" indicate that the state variable has been obtained from the experimental equilibrium surface or by EoS calculations respectively. The value of ξ which is used is the maximum between the Equations 14.27 and 14.29a or 14.27 and 14.29b. Equation 14.27 accounts for the uncertainty in the measurement of the experimental data, whereas Equations 14.29a and 14.29b account for the deviation between the model prediction and the experimental equilibrium surface. The minimum of φ is computed by using the current estimates of the parameters during the minimization.

The problem of minimizing Equation 14.24 subject to the constraint given by Equation 14.26 or 14.28 is transformed into an unconstrained one by introducing the Lagrange multiplier, ω, and augmenting the LS objective function, $S_{LS}(\mathbf{k})$, to yield

$$S_{LG}(\mathbf{k},\omega) = S_{LS}(\mathbf{k}) + \omega(\varphi_0 - \xi) \tag{14.30}$$

The minimization of $S_{LG}(\mathbf{k},\omega)$ can now be accomplished by applying the Gauss-Newton method with Marquardt's modification and a step-size policy as described in earlier chapters.

14.2.5.3 Constrained Gauss-Newton Method for Regression of Binary VLE Data

As we mentioned earlier, this is not a typical constrained minimization problem although the development of the solution method is very similar to the material presented in Chapter 9. If we assume that an estimate $\mathbf{k}^{(j)}$ is available at the jth iteration, a better estimate, $\mathbf{k}^{(j+1)}$, of the parameter vector is obtained as follows.

Linearization of the residual vector $\mathbf{e} = \left[ln\, f_1^L - ln\, f_1^V,\ ln\, f_2^L - ln\, f_2^V\right]^T$ around $\mathbf{k}^{(j)}$ at the ith data point yields

$$\mathbf{e}_i(\mathbf{k}^{(j+1)}) = \mathbf{e}_i(\mathbf{k}^{(j)}) + \left(\frac{\partial \mathbf{e}^T}{\partial \mathbf{k}}\right)_i^T \Delta\mathbf{k}^{(j+1)} \tag{14.31a}$$

Furthermore, linearization of the stability function yields

$$\varphi_0(\mathbf{k}^{(j+1)}) = \varphi_0(\mathbf{k}^{(j)}) + \left(\frac{\partial \varphi_0}{\partial \mathbf{k}}\right)_i^T \Delta\mathbf{k}^{(j+1)} \tag{14.31b}$$

Substitution of Equations 14.31a and b into the objective function $S_{LG}(\mathbf{k},\omega)$ and use of the stationary criteria $\frac{\partial S_{LG}(\mathbf{k}^{(j+1)},\omega)}{\partial \mathbf{k}^{(j+1)}} = \mathbf{0}$ and $\frac{\partial S_{LG}(\mathbf{k}^{(j+1)},\omega)}{\partial \omega} = 0$ yield the following system of linear equations:

$$\mathbf{A}\,\Delta\mathbf{k}^{(j+1)} = \mathbf{b} - \frac{\omega}{2}\mathbf{c} \tag{14.32a}$$

and

$$\varphi_0(\mathbf{k}^{(j)}) + \mathbf{c}^T \Delta\mathbf{k}^{(j+1)} = \xi \tag{14.32b}$$

where

$$\mathbf{A} = \sum_{i=1}^{N} \mathbf{G}_i^T \mathbf{Q}_i \mathbf{G}_i \tag{14.33a}$$

and

$$\mathbf{b} = -\sum_{i=1}^{N} \mathbf{G}_i^T \mathbf{Q}_i \mathbf{e}_i \tag{14.33b}$$

where $\mathbf{Q}_i$ is a user-defined weighting matrix (often taken as the identity matrix), $\mathbf{G}_i$ is the *(2×p)* sensitivity matrix $\left(\frac{\partial \mathbf{e}^T}{\partial \mathbf{k}}\right)_i^T$ evaluated at $\mathbf{x}_i$ and $\mathbf{k}^{(j)}$ and $\mathbf{c}$ is $\frac{\partial \varphi_0}{\partial \mathbf{k}}$.

Equation 14.32a is solved for $\Delta\mathbf{k}^{(j+1)}$ and the result is substituted in Equation 14.32b which is then solved for ω to yield

$$\omega = 2\left[\frac{\varphi_0 - \xi + \mathbf{c}^T \mathbf{A}^{-1}\mathbf{b}}{\mathbf{c}^T \mathbf{A}^{-1}\mathbf{c}}\right] \tag{14.34a}$$

Substituting the above expression for the Lagrange multiplier into Equation 14.32a we arrive at the following linear equation for $\Delta\mathbf{k}^{(j+1)}$,

$$\mathbf{A}\Delta\mathbf{k}^{(j+1)} = \mathbf{b} - \left(\frac{\varphi_0 - \xi + \mathbf{c}^T \mathbf{A}^{-1}\mathbf{b}}{\mathbf{c}^T \mathbf{A}^{-1}\mathbf{c}}\right)\mathbf{c} \tag{14.34b}$$

Assuming that an estimate of the parameter vector, $\mathbf{k}^{(j)}$, is available at the jth iteration, one can obtain the estimate for the next iteration by following the next steps:

(i) Compute sensitivity matrix $\mathbf{G}_i$, residual vector $\mathbf{e}_i$ and stability function φ at each data point (experimental or interpolated).

(ii) Using direct search determine the experimental or interpolated point (T_0, P_0, x_0) where φ is minimum.

(iii) Set up matrix **A**, vector **b** and compute vector **c** at (T_0, P_0, x_0).

(iv) Decompose matrix **A** using *eigenvalue decomposition* and compute ω from Equation 14.34a.

(v) Compute $\Delta\mathbf{k}^{(j+1)}$ from Equation 14.32b and update the parameter vector by setting $\mathbf{k}^{(j+1)} = \mathbf{k}^{(j)} + \mu\Delta\mathbf{k}^{(j+1)}$ where a stepping parameter, μ ($0<\mu\leq 1$), is used to avoid the problem of overstepping. The bisection rule can be used to arrive at an acceptable value for μ.

14.2.6 A Systematic Approach for Regression of Binary VLE Data

As it was explained earlier the use of the objective function described by Equation 14.15 is not advocated due to the heavy computational requirements expected and the potential convergence problems. Calculations by using the explicit LS objective function given by Equations 14.16a and 14.16b are not advocated for the same reasons. In fact, Englezos et al. (1990a) have shown that the computational time can be as high as two orders of magnitude compared to implicit LS estimation using Equation 14.23. The computational time was also found to be at least twice the time required for ML estimation using Equation 14.20.

The implicit LS, ML and Constrained LS (CLS) estimation methods are now used to synthesize a systematic approach for the parameter estimation problem when no prior knowledge regarding the adequacy of the thermodynamic model is available. Given the availability of methods to estimate the interaction parameters in equations of state there is a need to follow a systematic and computationally efficient approach to deal with all possible cases that could be encountered during the regression of binary VLE data. The following step by step systematic approach is proposed (Englezos et al. 1993)

Step 1. Data Compilation and EOS Selection:

A set of N VLE experimental data points have been made available. These data are the measurements of the state variables (T, P, x, y) at each of the N performed experiments. Prior to the estimation, one should plot the data and look for potential outliers as discussed in Chapter 8. In addition, a suitable EoS with the corresponding mixing rules should be selected.

Step 2. Best Set of Interaction Parameters:

The first task of the estimation procedure is to quickly and efficiently screen all possible sets of interaction parameters that could be used. For example if the Trebble-Bishnoi EoS were to be employed which can utilize up to four binary interaction parameters, the number of possible combinations that should be examined is 15. The implicit LS estimation procedure provides the most efficient means to determine the best set of interaction parameters. The best set is the one that results in the smallest value of the LS objective function after convergence of the minimization algorithm has been achieved. One should not readily accept a set that corresponds to a marginally smaller LS function if it utilizes more interaction parameters.

Step 3. Computation of VLE Phase Equilibria:

Once the best set of interaction parameters has been found, these parameters should be used with the EoS to perform the VLE calculations. The computed values should be plotted together with the data. A comparison of the data with the EoS based calculated phase behavior reveals whether correct or incorrect phase behavior (erroneous liquid phase splitting) is obtained.

CASE A: Correct Phase Behavior:

If the correct phase behavior i.e. absence of erroneous liquid phase splits is predicted by the EoS then the overall fit should be examined and it should be judged whether it is "excellent". If the fit is simply acceptable rather than "excellent", then the previously computed LS parameter estimates should suffice. This was found to be the case for the n-pentane–acetone and the methane–acetone systems presented later in this chapter.

Step 4a. Implicit ML Estimation:

When the fit is judged to be “excellent” the statistically best interaction parameters can be efficiently obtained by performing implicit ML estimation. This was found to be the case with the methane–methanol and the nitrogen–ethane systems presented later in this chapter.

CASE B: Erroneous Phase Behavior

Step 4b. CLS Estimation:

If incorrect phase behavior is predicted by the EOS then constrained least squares (CLS) estimation should be performed and new parameter estimates be obtained. Subsequently, the phase behavior should be computed again and if the fit is found to be acceptable for the intended applications, then the CLS estimates should suffice. This was found to be the case for the carbon dioxide–n-hexane system presented later in this chapter.

Step 5b. Modify EoS/Mixing Rules:

In several occasions the overall fit obtained by the CLS estimation could simply be found unacceptable despite the fact that the predictions of erroneous phase separation have been suppressed. In such cases, one should proceed and either modify the employed mixing rules or use a different EoS all together. Of course, the estimation should start from Step 1 once the new thermodynamic model has been chosen. Calculations using the data for the propane–methanol system illustrate this case as discussed next.

14.2.7 Numerical Results

In this section we consider typical examples. They cover all possible cases that could be encountered during the regression of binary VLE data. Illustration of the methods is done with the Trebble-Bishnoi (Trebble and Bishnoi, 1988) EoS with quadratic mixing rules and temperature-independent interaction parameters. It is noted, however, that the methods are not restricted to any particular EoS/mixing rule.

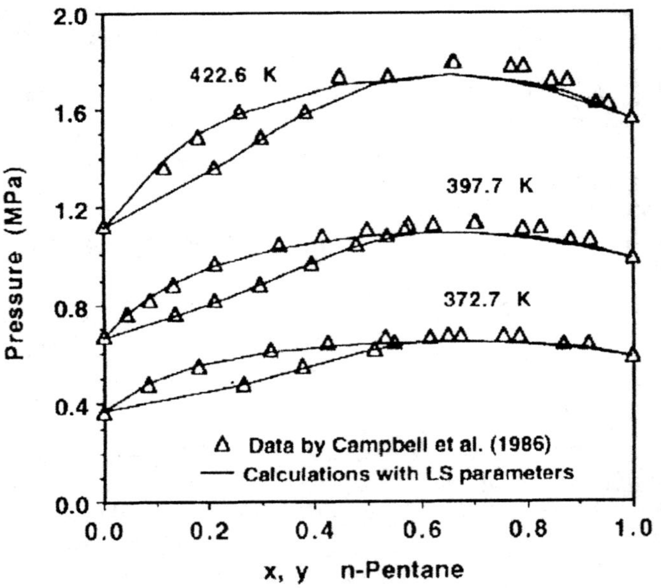

Figure 14.1 Vapor–liquid equilibrium data and calculated values for the n-pentane–acetone system. x and y are the mole fractions in the liquid and vapor phase respectively [reproduced with permission from Canadian Journal of Chemical Engineering].

14.2.7.1 The n-Pentane–Acetone System

Experimental data are available for this system at three temperatures by Campbell et al. (1986). Interaction parameters were estimated by Englezos et al. (1993). It was found that the Trebble-Bishnoi EoS is able to represent the correct phase behavior with an accuracy that does not warrant subsequent use of ML estimation. The best set of interaction parameters was found by implicit LS estimation to be (k_a=0.0977, k_b=0, k_c=0, k_d=0). The standard deviation for k_a was found to be equal to 0.0020. For this system, the use of more than one interaction parameter did not result in any improvement of the overall fit. The deviations between the calculated and the experimental values were found to be larger than the experimental errors. This is attributed to systematic deviations due to model inadequacy. Hence, one should not attempt to perform ML estimation. In this case, the LS parameter estimates suffice. Figure 14.1 shows the calculated phase behavior using the LS parameter estimate. The plot shows the pressure versus the liquid and vapor phase mole fractions. This plot is known as partial phase diagram.

14.2.7.2 The Methane–Acetone System

Data for the methane-acetone system are available by Yokoyama et al. (1985). The implicit LS estimates were computed and found to be sufficient to describe the phase behavior. These estimates are (k_a=0.0447, k_b=0, k_c=0, k_d=0). The standard deviation for k_a was found to be equal to 0.0079.

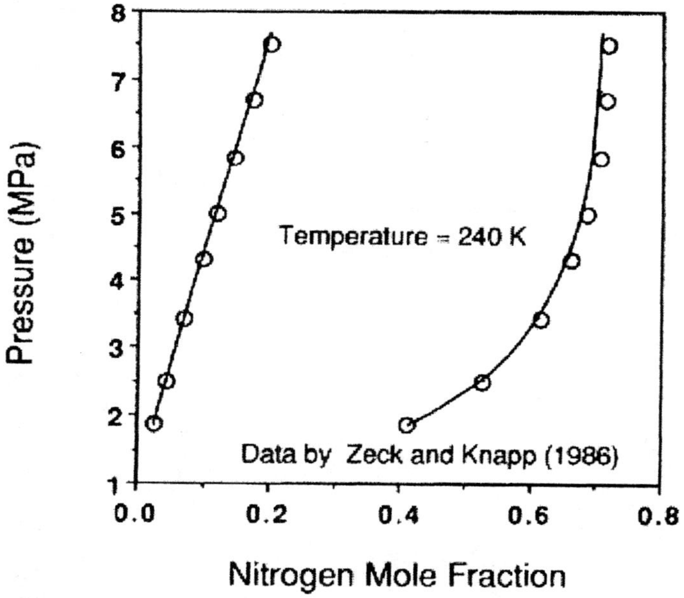

Figure 14.2 *Vapor-liquid equilibrium data and calculated values for the nitrogen-ethane system [reprinted from the Canadian Journal of Chemical Engineering with permission].*

14.2.7.3 The Nitrogen–Ethane System

Data at two temperatures were obtained from Zeck and Knapp (1986) for the nitrogen–ethane system. The implicit LS estimates of the binary interaction parameters are k_a=0, k_b=0, k_c=0 and k_d=0.0460. The standard deviation of k_d was found to be equal to 0.0040. The vapor liquid phase equilibrium was computed and the fit was found to be excellent (Englezos et al. 1993). Subsequently, implicit ML calculations were performed and a parameter value of k_d=0.0493 with a standard deviation equal to 0.0070 was computed. Figure 14.2 shows the experimental phase diagram as well as the calculated one using the implicit ML parameter estimate.

14.2.7.4 The Methane–Methanol System

The methane–methanol binary is another system where the EoS is also capable of matching the experimental data very well and hence, use of ML estimation to obtain the statistically best estimates of the parameters is justified. Data for this system are available from Hong et al. (1987). Using these data, the binary interaction parameters were estimated and together with their standard deviations are shown in Table 14.1. The values of the parameters not shown in the table (i.e., k_a, k_b, k_c) are zero.

Table 14.1 Parameter Estimates for the Methane-Methanol System

Parameter Value	Standard Deviation	Objective Function
k_d=-0.1903	0.0284	Implicit LS (Equation 14.23)
k_d=-0.2317	0.0070	Implicit ML (Equation 14.20)
k_d=-0.2515	0.0025	Explicit LS (T,P) (Equation 14.16b)

Source: Englezos et al. (1990a).

14.2.7.5 Carbon Dioxide–Methanol System

Data for the carbon dioxide-methanol binary are available from Hong and Kobayashi (1988). The parameter values and their standard deviations estimated from the regression of these data are shown in Table 14.2.

Table 14.2 Parameter Estimates for the Carbon Dioxide-Methanol System

Parameter Value	Standard Deviation	Objective Function
k_a=0.0605	0.0058	Implicit LS
k_d=-0.1137	0.0180	(Equation 14.23)
k_a=0.0504	0.0016	Explicit LS (T,x)
k_d=-0.0631	0.0074	(Equation 14.16a)
k_a=0.0566	0.0001	Implicit ML
k_d=-0.2238	0.0003	(Equation 14.20)

Note: Explicit LS (Equation 14.16a) did not converge with a zero value for Marquardt's directional parameter.
Source: Englezos et al. (1990a).

Table 14.3 Parameter Estimates for the Carbon Dioxide-n-Hexane System

Parameter Value	Standard Deviation	Objective Function
k_b= 0.1977	0.0334	Implicit LS
k_c= -1.0699	0.1648	(Equation 14.23)
k_b= 0.1345	0.0375	Simplified CLS
k_c= -0.7686	0.1775	(Eq. 14.24 & 14.26)

14.2.7.6 The Carbon Dioxide-n-Hexane System

This system illustrates the use of simplified constrained least squares (CLS) estimation. In Figure 14.3, the experimental data by Li et al. (1981) together with the calculated phase diagram for the system carbon dioxide–n-hexane are shown. The calculations were done by using the best set of interaction parameter values obtained by implicit LS estimation. These parameter values together with standard deviations are given in Table 14.3. The values of the other parameters (k_a, k_d) were equal to zero. As seen from Figure 14.3, erroneous liquid phase separation is predicted by the EoS in the high pressure region. Subsequently, constrained least squares estimation (CLS) was performed by minimizing Equation 14.24 subject to the constraint of Equation 14.26. In other words by satisfying the liquid phase stability criterion at all experimental points. The new parameter estimates are also given in Table 14.3. These estimates were used for the re-calculation of the phase diagram, which is also shown in Figure 14.3. As seen, the EoS no longer predicts erroneous liquid phase splitting in the high pressure region. The parameter estimates obtained by the constrained LS estimation should suffice for engineering type calculations since the EoS can now adequately represent the correct phase behavior of the system.

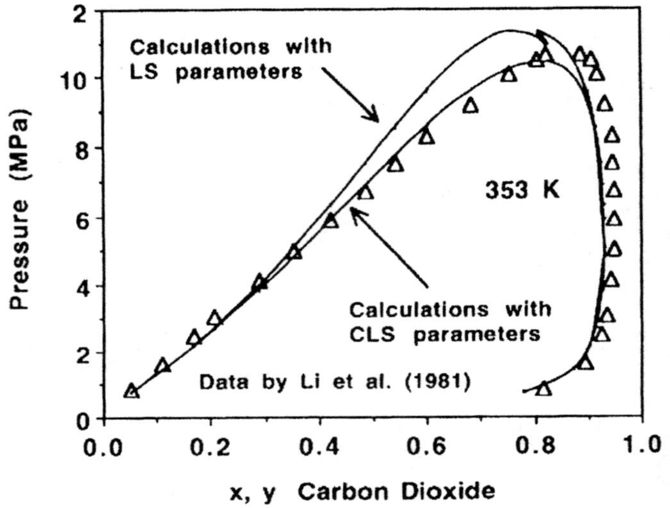

Figure 14.3 Vapor-liquid equilibrium data and calculated values for the carbon dioxide-n-hexane system. Calculations were done using interaction parameters from implicit and constrained least squares (LS) estimation. x and y are the mole fractions in the liquid and vapor phase respectively [reprinted from the Canadian Journal of Chemical Engineering with permission].

14.2.7.7 The Propane–Methanol System

This system represents the case where the structural inadequacy of the thermodynamic model (EoS) is such that the overall fit is simply unacceptable when the EoS is forced to predict the correct phase behavior by using the CLS parameter estimates. In particular, it has been reported by Englezos et al. (1990a) that when the Trebble-Bishnoi EoS is utilized to model the phase behavior of the propane-methanol system (data by Galivel-Solastiuk et al. 1986) erroneous liquid phase splitting is predicted. Those calculations were performed using the best set of LS parameter estimates, which are given in Table 14.4. Following the steps outlined in the systematic approach presented in an earlier section, simplified CLS estimation was performed (Englezos et al. 1993; Englezos and Kalogerakis, 1993). The CLS estimates are also given in Table 14.4.

Table 14.4 Parameter Estimates for the Propane-Methanol System

Parameter Value	Standard Deviation	Objective Function
$k_a = 0.1531$ $k_d = -0.2994$	0.0113 0.0280	Implicit LS (Equation 14.23)
$k_a = 0.1193$ $k_d = -0.3196$	0.0305 0.0731	Simplified CLS (Equation 14.24 & 14.26)

Source: Englezos et al. (1993).

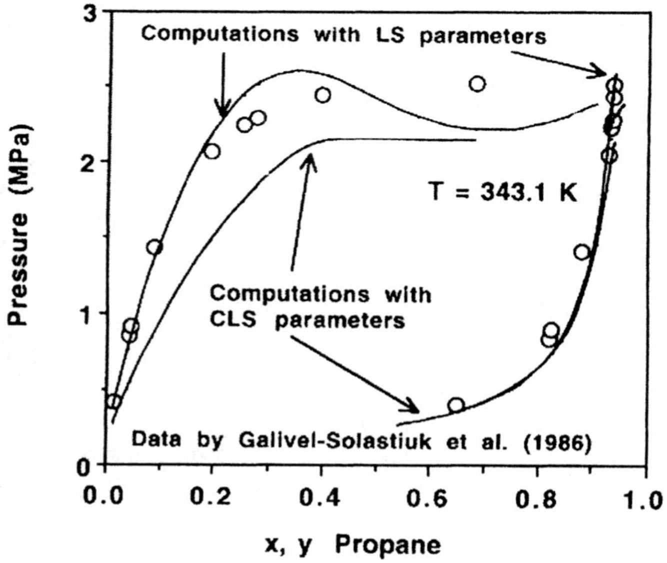

Figure 14.4 *Vapor-liquid equilibrium data and calculated values for the propane-methanol system [reprinted from the Canadian Journal of Chemical Engineering with permission].*

Next, the VLE was calculated using these parameters and the results together with the experimental data are shown in Figure 14.4. The erroneous phase behavior has been suppressed. However, the deviations between the experimental data and the EoS-based calculated phase behavior are excessively large. In this case, the overall fit is judged to be unacceptable and one should proceed and search for more suitable mixing rules. Schwartzentruber et al. (1987) also modeled this system and encountered the same problem.

Table 14.5 Parameter Estimates for the Diethylamine–Water System

Parameter Values	Standard Deviation	Covariance	Objective Function
k_a = -0.2094 k_d = -0.2665	0.0183 0.0279	0.0005	Implicit LS (Equation 14.23)
k_a = -0.8744 k_d = 0.9808	0.0814 0.4846	0.0364	Simplified CLS (Equation 14.24 & 14.26)
k_a = -0.7626 k_d = 0.6952	0.1143 0.4923	0.0551	Constrained LS (Equation 14.24 & 14.28)

Source: Englezos and Kalogerakis (1993).

14.2.7.8 The Diethylamine–Water System

Copp and Everet (1953) have presented 33 experimental VLE data points at three temperatures. The diethylamine-water system demonstrates the problem that may arise when using the simplified constrained least squares estimation due to inadequate number of data. In such case there is a need to interpolate the data points and to perform the minimization subject to constraint of Equation 14.28 instead of Equation 14.26 (Englezos and Kalogerakis, 1993). First, unconstrained LS estimation was performed by using the objective function defined by Equation 14.23. The parameter values together with their standard deviations that were obtained are shown in Table 14.5. The covariances are also given in the table. The other parameter values are zero.

Using the values (-0.2094, -0.2665) for the parameters k_a and k_d in the EoS, the stability function, φ, was calculated at each experimental VLE point. The minima of the stability function at each isotherm are shown in Figure 14.5. The stability function at 311.5 K is shown in Figure 14.6. As seen, φ becomes negative. This indicates that the EoS predicts the existence of two liquid phases. This is also evident in Fig.14.7 where the EoS-based VLE calculations at 311.5 K are shown.

Subsequently, constrained LS estimation was performed by minimizing the objective function given by Equation 14.24 subject to the constraint described in Equation 14.26. The calculated parameters are also shown in Table 14.5. The minima of the stability function were also calculated and they are shown in Figure 14.8. As seen, φ is positive at all the experimental conditions. However, the new VLE calculations indicate in Figure 14.9 that the EoS predicts erroneous liquid phase separation. This result prompted the calculation of the stability function at pressures near the vapor pressure of water where it was found that φ becomes negative. Figure 14.8 shows the stability function becoming negative.

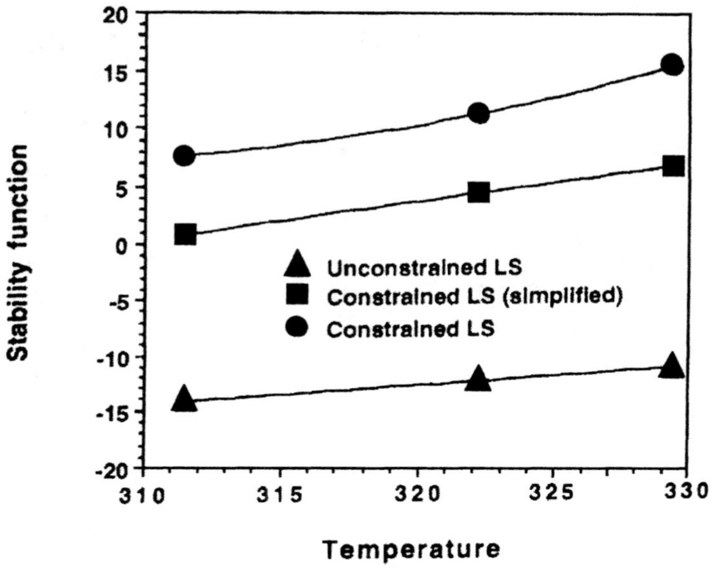

Figure 14.5 *The minima of the stability function at the experimental temperatures for the diethylamine-water system [reprinted from Computers & Chemical Engineering with permission from Elsevier Science].*

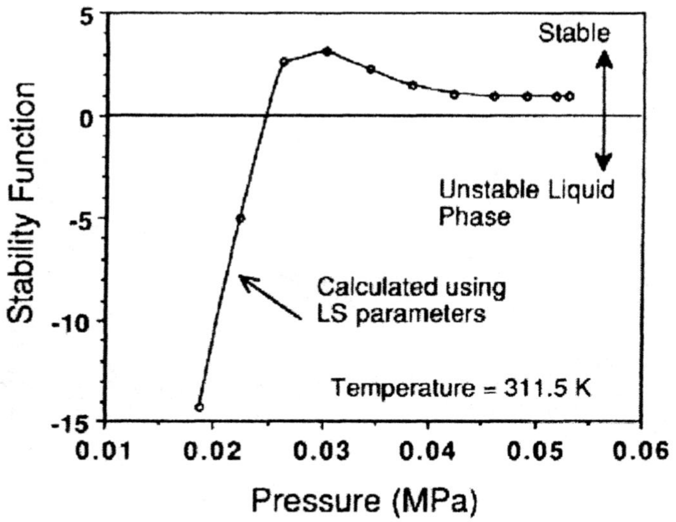

Figure 14.6 *The stability function calculated with interaction parameters from unconstrained least squares estimation.*

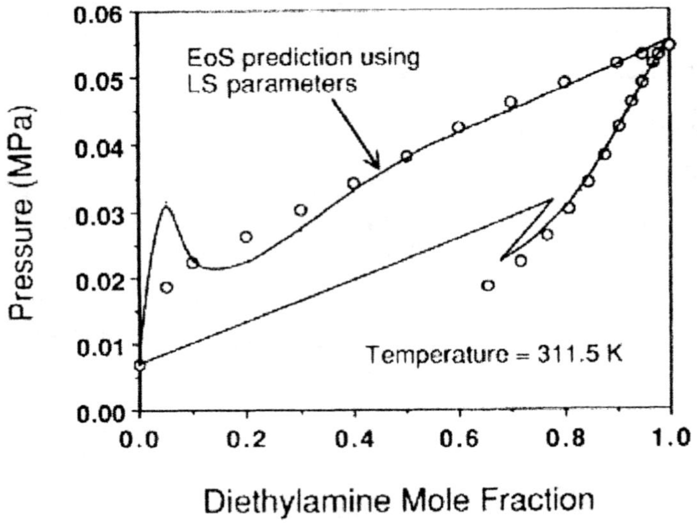

Figure 14.7 *Vapor-liquid equilibrium data and calculated values for the diethylamine-water system. Calculations were done using parameters from unconstrained LS estimation [reprinted from Computers & Chemical Engineering with permission from Elsevier Science].*

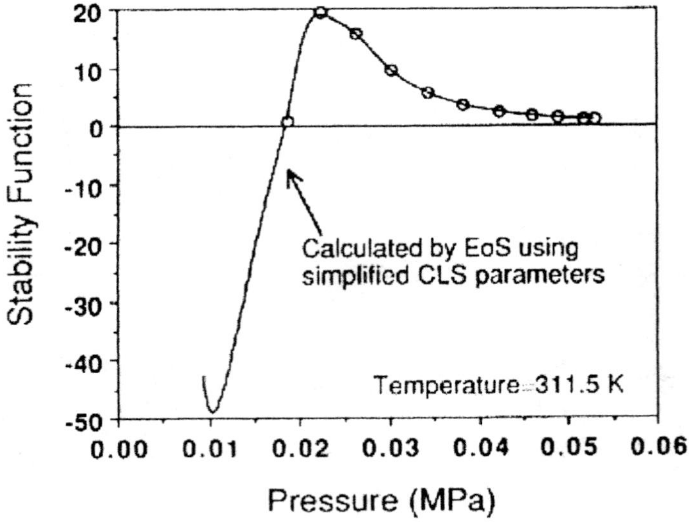

Figure 14.8 *The stability function calculated with interaction parameters from simplified constrained LS estimation.*

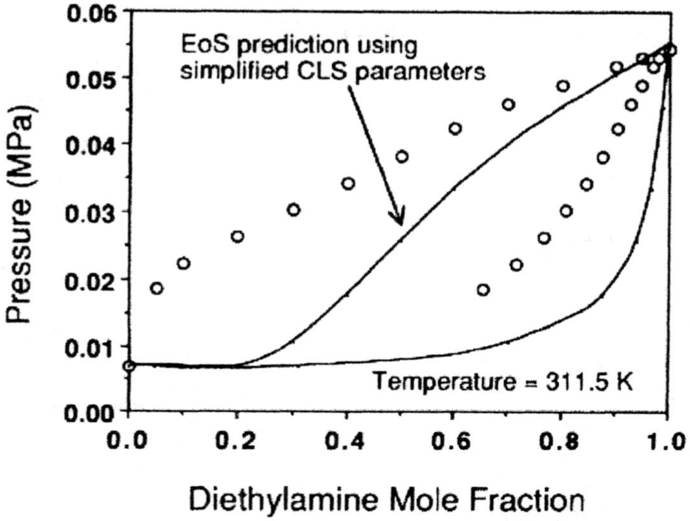

Figure 14.9 Vapor-liquid equilibrium data and calculated values for the diethylamine-water system. Calculations were done using parameters from simplified CLS estimation [reprinted from Computers & Chemical Engineering with permission from Elsevier Science].

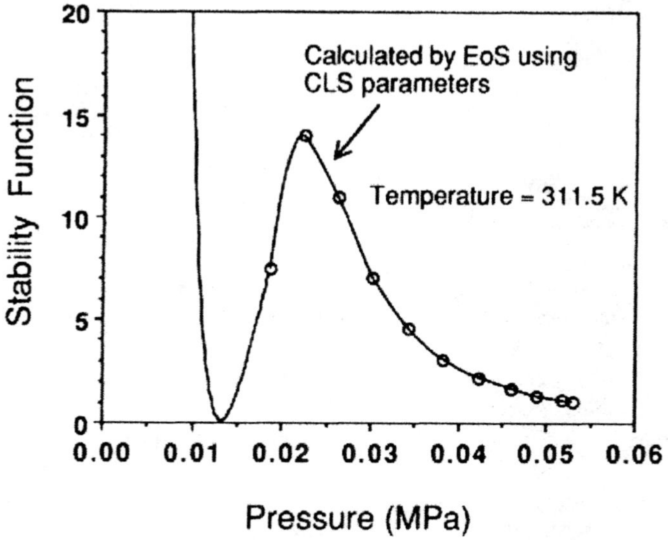

Figure 14.10 The stability function calculated with interaction parameters from constrained LS estimation [reprinted from Computers & Chemical Engineering with permission from Elsevier Science].

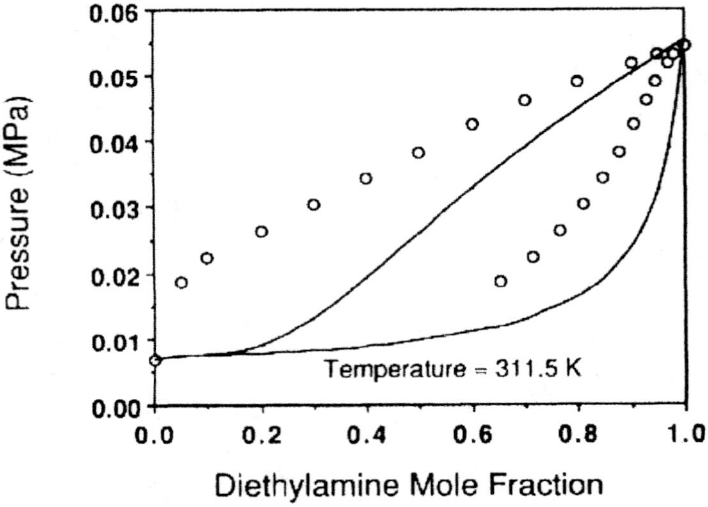

Figure 14.11 Vapor-liquid equilibrium data and calculated values for the diethylamine-water system. Calculations were done using parameters from CLS estimation [reprinted from Computers & Chemical Engineering with permission from Elsevier Science].

Therefore, although the stability function was found to be positive at all the experimental conditions it becomes negative at mole fractions between 0 and the first measured data point. Obviously, if there were additional data available in this region, the simplified constrained LS method that was followed above would have yielded interaction parameters that do not result in prediction of false liquid phase splitting.

Following the procedure described in Section 14.2.5.2 the region over which the stability function is examined is extended as follows. At each experimental temperature, a large number of mole fraction values are selected and then at each T, x point the values of the pressure, P, corresponding to VLE, are calculated. This calculation is performed by using cubic spline interpolation of the experimental pressures. For this purpose the subroutine ICSSCU from the IMSL library was used. The interpolation was performed only once and prior to the parameter estimation by using 100 mole fraction values selected between 0. and 0.05 to ensure a close examination of the stability function. During the minimization, these values of T, P and x together with the given VLE data were used to compute the stability function and find its minimum value. The current estimates of the parameters were employed at each minimization step.

New interaction parameter values were obtained and they are also shown in Table 14.5. These values are used in the EoS to calculate the stability function and the calculated results at 311.5 K are shown in Figure 14.10.

As seen from the figure, the stability function does not become negative at any pressure when the hydrogen sulfide mole fraction lies anywhere between 0 and 1. The phase diagram calculations at 311.5 K are shown in Figure 14.11. As seen, the correct phase behavior is now predicted by the EoS.

The improved method guarantees that the EoS will calculate the correct VLE not only at the experimental data but also at any other point that belongs to the same isotherm. The question that arises is what happens at temperatures different than the experimental. As seen in Figure 14.10 the minima of the stability function increase monotonically with respect to temperature. Hence, it is safe to assume that at any temperature between the lowest and the highest one, the EoS predicts the correct behavior of the system. However, below the minimum experimental temperature, it is likely that the EoS will predict erroneous liquid phase separation.

The above observation provides useful information to the experimenter when investigating systems that exhibit vapor-liquid-liquid equilibrium. In particular, it is desirable to obtain VLE measurements at a temperature near the one where the third phase appears. Then by performing CLS estimation, it is guaranteed that the EoS predicts complete miscibility everywhere in the actual two phase region. It should be noted, however, that in general the minima of the stability function at each temperature might not change monotonically. This is the case with the CO_2-n-Hexane system where it is risky to interpolate for intermediate temperatures. Hence, VLE data should also be collected at intermediate temperatures too.

14.3 PARAMETER ESTIMATION USING THE ENTIRE BINARY PHASE EQUILIBRIUM DATA

It is known that in five of the six principal types of binary fluid phase equilibrium diagrams, data other than VLE may also be available for a particular binary (van Konynenburg and Scott, 1980). Thus, the entire database may also contain VL_2E, VL_1E, VL_1L_2E, and L_1L_2E data. In this section, a systematic approach to utilize the entire phase equilibrium database is presented. The material is based on the work of Englezos et al. (1990b; 1998)

14.3.1 The Objective Function

Let us assume that for a binary system there are available N_1 VL_1E, N_2 VL_2E, N_3 L_1L_2E and N_4 VL_1L_2E data points. The *light* liquid phase is L_1 and L_2 is the *heavy* one. Thus, the total number of available data is $N=N_1+N_2+N_3+N_4$. Gas-gas equilibrium type of data are not included in the analysis because they are beyond the range of most practical applications. An implicit objective function that provides an appropriate measure of the ability of the EoS to represent all these types of equilibrium data is the following

$$S(\mathbf{k}) = \sum_{i=1}^{N} \mathbf{r}_i^T \mathbf{R}_i \mathbf{r}_i \tag{14.35}$$

where the vectors $\mathbf{r}_i$ are the residuals and $\mathbf{R}_i$ are suitable weighting matrices. The residuals are based on the *iso-fugacity phase equilibrium* criterion. When two phase data (VL_1E, or VL_2E or L_1L_2E) are considered, the residual vector takes the form

$$\mathbf{r}_i = \begin{pmatrix} ln\, f_1^{\alpha} - ln\, f_1^{\beta} \\ ln\, f_2^{\alpha} - ln\, f_2^{\beta} \end{pmatrix}_i \tag{14.36}$$

where f_j^a and f_j^{β} are the fugacities of component j (j=1 or 2) in phase α (Vapor or Liquid$_1$) and phase β (Liquid$_1$ or Liquid$_2$) respectively for the ith experiment. When three phase data (VL_1L_2E) are used, the residuals become

$$\mathbf{r}_i = \begin{pmatrix} \ln f_1^V - \ln f_1^{L1} \\ \ln f_2^V - \ln f_2^{L1} \\ \ln f_1^V - \ln f_1^{L2} \\ \ln f_2^V - \ln f_2^{L2} \end{pmatrix} \tag{14.37}$$

The residuals are functions of temperature, pressure, composition and the interaction parameters. These functions can easily be derived analytically for any equation of state. At equilibrium the value of these residuals should be equal to zero. However, when the measurements of the temperature, pressure and mole fractions are introduced into these expressions the resulting values are not zero even if the EoS were perfect. The reason is the random experimental error associated with each measurement of the state variables.

The values of the elements of the weighting matrices $\mathbf{R}_i$ depend on the type of estimation method being used. When the residuals in the above equations can be assumed to be independent, normally distributed with zero mean and the same constant variance, Least Squares (LS) estimation should be performed. In this case, the weighting matrices in Equation 14.35 are replaced by the identity matrix **I**. Maximum likelihood (ML) estimation should be applied when the EoS is capable of calculating the correct phase behavior of the system within the experimental error. Its application requires the knowledge of the measurement errors for each variable (i.e. temperature, pressure, and mole fractions) at each experimental data point. Each of the weighting matrices is replaced by the inverse of the covariance matrix of the residuals. The elements of the covariance matrices are computed by a first order variance approximation of the residuals and from the variances of the errors in the measurements of the T, P and composition (**z**) in the coexisting phases (Englezos et al. 1990a;b).

The optimal parameter values are found by the Gauss-Newton method using Marquardt's modification. The estimation problem is of considerable magnitude, especially for a multi-parameter EoS, when the entire fluid phase equilibrium database is used. Usually, there is no prior information about the ability of the EoS to represent the phase behavior of a particular system. It is possible that convergence of the minimization of the objective function given by Eq. 14.35 will be difficult to achieve due to structural inadequacies of the EoS. Based on the systematic approach for the treatment of VLE data of Englezos et al. (1993) the following step-wise methodology is advocated for the treatment of the entire phase database:

- Consider each type of data separately and estimate the best set of interaction parameters by Least Squares.
- If the estimated best set of interaction parameters is found to be different for each type of data then it is rather meaningless to correlate the entire database simultaneously. One may proceed, however, to find the parameter set that correlates the maximum number of data types.
- If the estimated best set of interaction parameters is found to be the same for each type of data then use the entire database and perform least squares estimation.
- Compute the phase behavior of the system and compare with the data. If the fit is excellent then proceed to maximum likelihood parameter estimation.
- If the fit is simply acceptable then the LS estimates suffice.

14.3.2 Covariance Matrix of the Parameters

The elements of the covariance matrix of the parameter estimates are calculated when the minimization algorithm has converged with a zero value for the Marquardt's directional parameter. The covariance matrix of the parameters *COV*(**k**) is determined using Equation 11.1 where the degrees of freedom used in the calculation of $\hat{\sigma}_\varepsilon^2$ are

$$(\text{d.f.}) = 2(N_1 + N_2 + N_3) + 4N_4 - p \tag{14.38}$$

14.3.3 Numerical Results

Data for the hydrogen sulfide-water and the methane-n-hexane binary systems were considered. The first is a type III system in the binary phase diagram classification scheme of van Konynenburg and Scott. Experimental data from Selleck et al. (1952) were used. Carroll and Mather (1989a; b) presented a new interpretation of these data and also new three phase data. In this work, only those VLE data from Selleck et al. (1952) that are consistent with the new data were used. Data for the methane-n-hexane system are available from Poston and McKetta (1966) and Lin et al. (1977). This is a type V system.

It was shown by Englezos et al. (1998) that use of the entire database can be a stringent test of the correlational ability of the EoS and/or the mixing rules. An additional benefit of using all types of phase equilibrium data in the parameter estimation database is the fact that the statistical properties of the estimated parameter values are usually improved in terms of their standard deviation.

14.3.3.1 The Hydrogen Sulfide–Water System

The same two interaction parameters (k_a, k_d) were found to be adequate to correlate the VLE, LLE and VLLE data of the H_2S-H_2O system. Each data set was used separately to estimate the parameters by implicit least squares (LS).

Table 14.6 Parameter Estimates for the Hydrogen Sulfide-Water System

Parameter Values	Standard Deviation	Database	Objective Function
$k_a = 0.2209$ $k_d = -0.5862$	0.0120 0.0306	VL_2E	Implicit LS (Eqn 14.35 with $\mathbf{R}_i = \mathbf{I}$)
$k_a = 0.1929$ $k_d = -0.5455$	0.0149 0.0366	L_1L_2E	Implicit LS (Eqn 14.35 with $\mathbf{R}_i = \mathbf{I}$)
$k_a = 0.2324$ $k_d = -0.6054$	0.0232 0.0581	VL_1L_2E	Implicit LS (Eqn 14.35 with $\mathbf{R}_i = \mathbf{I}$)
$k_a = 0.2340$ $k_d = -0.6179$	0.0132 0.0333	VL_2E & L_1L_2E & VL_1L_2E	Implicit LS (Eqn 14.35 with $\mathbf{R}_i = \mathbf{I}$)
$k_a = 0.2108$ $k_d = -0.5633$	0.0003 0.0013	VL_2E & L_1L_2E & VL_1L_2E	Implicit ML (Equation 14.35 with $\mathbf{R}_i = \mathbf{COV}^{-1}(\mathbf{e}_i)$)

Note: L_1 is the light liquid phase and L_2 the heavy.
Source: Englezos et al. (1998).

The values of these parameter estimates together with the standard deviations are shown in Table 14.6. Subsequently, these two interaction parameters were recalculated using the entire database. The algorithm easily converged with a zero value for Marquardt's directional parameter and the estimated values are also shown in the table. The phase behavior of the system was then computed using these interaction parameter values and it was found to agree well with the experimental data. This allows maximum likelihood (ML) estimation to be performed. The ML estimates that are also shown in Table 14.6 were utilized in the EoS to illustrate the computed phase behavior. The calculated pressure-composition diagrams corresponding to VLE and LLE data at 344.26 K are shown in Figures 14.12 and 14.13. There is an excellent agreement between the values computed by the EoS and the experimental data. The Trebble-Bishnoi EoS was also found capable of representing well the three phase equilibrium data (Englezos et al. 1998).

14.3.3.2 The Methane–n-Hexane System

The proposed methodology was also followed and the best parameter estimates for the various types of data are shown in Table 14.7 for the methane-n-hexane system. As seen, the parameter set (k_a, k_d) was found to be the best to correlate the VL_2E, the L_1L_2E and the VL_2L_1E data and another (k_a, k_b) for the VL_1E data.

Table 14.7 Parameter Estimates for the Methane–n-Hexane System

Parameter Values	Standard Deviation	Used Data	Objective Function
k_a = -0.0793 k_d = 0.0695	0.0102 0.0120	VL_2E	Implicit LS (Eqn 14.35 with $\mathbf{R}_i$=$\mathbf{I}$)
k_a = -0.1503 k_b = -0.2385	0.0277 0.0403	VL_1E	Implicit LS (Eqn 14.35 with $\mathbf{R}_i$=$\mathbf{I}$)
k_a = -0.0061 k_d = 2236	0.0018 0.0116	L_1L_2E	Implicit LS (Eqn 14.35 with $\mathbf{R}_i$=$\mathbf{I}$)
k_a = 0.0113 k_d = 3185	0.0084 0.0254	VL_1L_2E	Implicit LS (Eqn 14.35 with $\mathbf{R}_i$=$\mathbf{I}$)
k_a = -0.0587 k_d = 0.0632	0.0075 0.0104	VL_1E & L_1L_2E & VL_1L_2E	Implicit LS (Eqn 14.35 with $\mathbf{R}_i$=$\mathbf{I}$)

Source: Englezos et al. (1998).

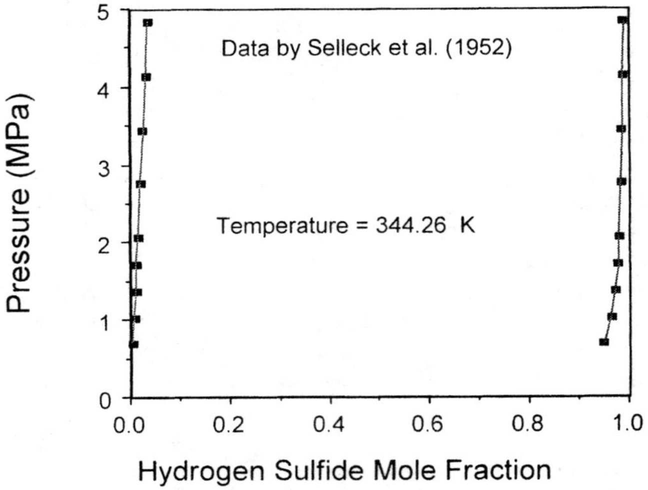

Figure 14.12 *VLE data and calculated phase diagram for the hydrogen sulfide-water system [reprinted from Industrial Engineering Chemistry Research with permission from the American Chemical Society].*

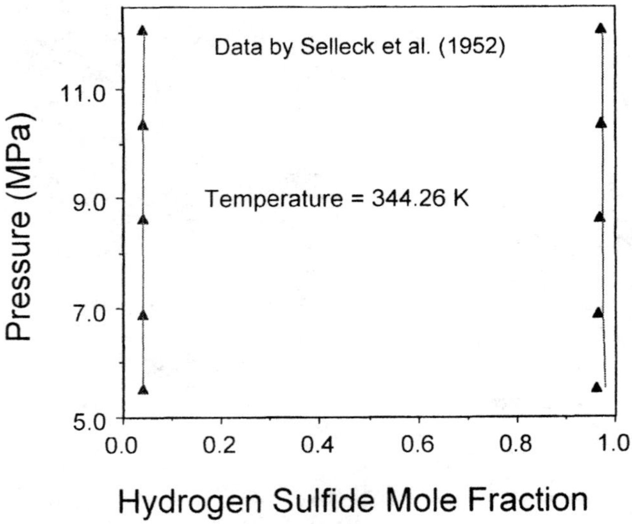

Figure 14.13 *LLE data and calculated phase diagram for the hydrogen sulfide-water system [reprinted from Industrial Engineering Chemistry Research with permission from the American Chemical Society].*

Phase equilibrium calculations using the best set of interaction parameter estimates for each type of data revealed that the EoS is able to represent well only the VL_2E data. The EoS was not found to be able to represent the L_2L_1E and the VL_2L_1E data. Although the EoS is capable of representing only part of the fluid phase equilibrium diagram, a single set of values for the best set of interaction parameters was found by using all but the VL_1E data. The values for this set (k_a, k_d) were calculated by least squares and they are also given in Table 14.7.

Using the estimated interaction parameters phase equilibrium computations were performed. It was found that the EoS is able to represent the VL_2E behavior of the methane-n-hexane system in the temperature range of 198.05 to 444.25 K reasonably well. Typical results together with the experimental data at 273.16 and 444.25 K are shown in Figures 14.14 and 14.15 respectively. However, the EoS was found to be unable to correlate the entire phase behavior in the temperature range of 195.91 K (Upper Critical Solution Temperature) and 182.46K (Lower Critical Solution Temperature).

14.4 PARAMETER ESTIMATION USING BINARY CRITICAL POINT DATA

Given the inherent robustness and efficiency of implicit methods, their use was extended to include critical point data (Englezos et al. 1998). Illustration of the method was done with the Trebble-Bishnoi EoS (Trebble and Bishnoi, 1988) and the Peng-Robinson EoS (Peng and Robinson, 1976) with quadratic mixing rules and temperature-independent interaction parameters. The methodology, however, is not restricted to any particular EoS/mixing rule.

Prior work on the use of critical point data to estimate binary interaction parameters employed the minimization of a summation of squared differences between experimental and calculated critical temperature and/or pressure (Equation 14.39). During that minimization the EoS uses the current parameter estimates in order to compute the critical pressure and/or the critical temperature. However, the initial estimates are often away from the optimum and as a consequence, such iterative computations are difficult to converge and the overall computational requirements are significant.

14.4.1 The Objective Function

It is assumed that there are available N_{CP} experimental binary critical point data. These data include values of the pressure, P_c, the temperature, T_c, and the mole fraction, x_c, of one of the components at each of the critical points for the binary mixture. The vector **k** of interaction parameters is determined by fitting the EoS to the critical data. In explicit formulations the interaction parameters are obtained by the minimization of the following least squares objective function:

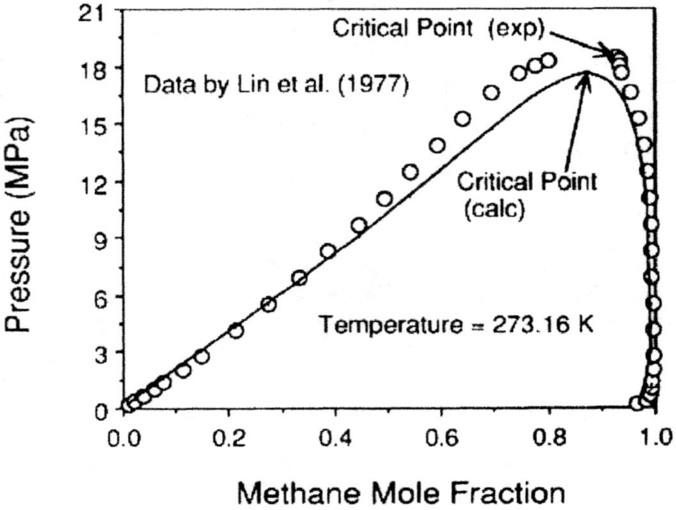

Figure 14.14 *VLE data and calculated phase diagram for the methane–n-hexane system [reprinted from Industrial Engineering Chemistry Research with permission from the American Chemical Society].*

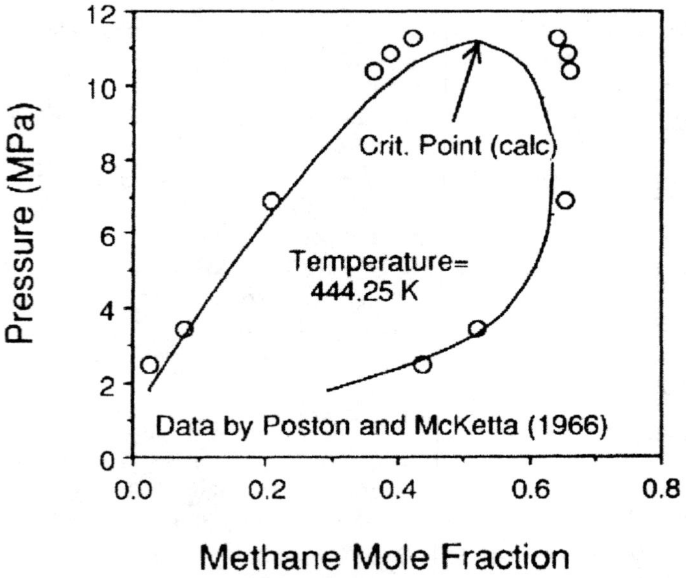

Figure 14.15 *VLE data and calculated phase diagram for the methane–n-hexane system.*

$$S_{LS}(\mathbf{k}) = q_1 \sum_{i=1}^{N_{CP}} (P_{c,i}^{exp} - P_{c,i}^{calc})^2 + q_2 \sum_{i=1}^{N_{CP}} (T_{c,i}^{exp} - T_{c,i}^{calc})^2 \quad (14.39)$$

The values of q_1 and q_2 are usually taken equal to one. Hissong and Kay (1970) assigned a value of four for the ratio of q_2 over q_1.

The critical temperature and/or the critical pressure are calculated by solving the equations that define the critical point given below (Modell and Reid, 1983)

$$\left(\frac{\partial \ln f_1}{\partial x_1}\right)_{T,P} = 0. \quad (14.40a)$$

$$\left(\frac{\partial^2 \ln f_1}{\partial x_1^2}\right)_{T,P} = 0. \quad (14.40b)$$

Because the current estimates of the interaction parameters are used when solving the above equations, convergence problems are often encountered when these estimates are far from their optimal values. It is therefore desirable to have, especially for multi-parameter equations of state, an efficient and robust estimation procedure. Such a procedure is presented next.

At each point on the critical locus Equations 40a and b are satisfied when the true values of the binary interaction parameters and the state variables, T_c, P_c and x_c are used. As a result, following an implicit formulation, one may attempt to minimize the following residuals.

$$\left(\frac{\partial \ln f_1}{\partial x_1}\right)_{T,P} = r_1. \quad (14.41a)$$

$$\left(\frac{\partial^2 \ln f_1}{\partial x_1^2}\right)_{T,P} = r_2 \quad (14.41b)$$

Furthermore, in order to avoid any iterative computations for each critical point, we use the experimental measurements of the state variables instead of their unknown true values. In the above equations r_1 and r_2 are residuals which can be easily calculated for any equation of state.

Based on the above residuals the following objective function is formulated:

$$S(\mathbf{k}) = \sum_{i=1}^{N_{CP}} \mathbf{r}_i^{\,T} \mathbf{R}_i \mathbf{r}_i \tag{14.42}$$

where $\mathbf{r}=(r_1, r_2)^T$. The optimal parameter values are found by using the Gauss-Newton-Marquardt minimization algorithm. For LS estimation, $\mathbf{R}_i=\mathbf{I}$ is used, however, for ML estimation, the weighting matrices need to be computed. This is done using a first order variance approximation.

The elements of the covariance matrix of the parameter estimates are calculated when the minimization algorithm has converged with a zero value for the Marquardt's directional parameter. The covariance matrix of the parameters $COV(\mathbf{k})$ is determined using Equation 11.1 where the degrees of freedom used in the calculation of $\hat{\sigma}_\varepsilon^2$ are

$$(d.f.) = 2N_{CP} - p \tag{14.43}$$

14.4.2 Numerical Results

Five critical points for the methane-n-hexane system in the temperature range of 198 to 273 K measured by Lin et al. (1977) are available. By employing the Trebble-Bishnoi EoS in our critical point regression least squares estimation method, the parameter set (k_a, k_b) was found to be the optimal one. Convergence from an initial guess of (k_a,k_b=0.001, -0.001) was achieved in six iterations. The estimated values are given in Table 14.8.

As seen, coincidentally, this is the set that was optimal for the VL_1E data. The same data were also used with the implicit least squares optimization procedure but by using the Peng-Robinson EoS. The results are also shown in Table 14.8 together with results for the CH_4-C_3H_8 system. One interaction parameter in the quadratic mixing rule for the attractive parameter was used. Temperature-dependent values estimated from VLE data have also been reported by Ohe (1990) and are also given in the table. As seen from the table, we did not obtain optimal parameters values same as those obtained from the phase equilibrium data. This is expected because the EoS is a semi-empirical model and different data are used for the regression. A perfect thermodynamic model would be expected to correlate both sets of data equally well. Thus, regression of the critical point data can provide additional information about the correlational and predictive capability of the EoS/mixing rule.

In principle, one may combine equilibrium and critical data in one database for the parameter estimation. From a numerical implementation point of view this can easily be done with the proposed estimation methods. However, it was not done because it puts a tremendous demand in the correlational ability of the EoS to describe all the data and it will be simply a computational exercise.

Table 14.8 Interaction Parameter Values from Binary Critical Point Data

System	Reference	Temperature Range (K)	Parameters Values Using		EoS
			Critical Point Data	VLE Data	
CH_4–nC_6H_{14}	Lin et al. (1977)	198 - 273	k_a=-0.1397 ±0.015 k_b=-0.5251 ±0.035	see Table 14.7	Trebble-Bishnoi
CH_4–nC_6H_{14}	Lin et al. (1977)	198 - 273	k_{ij}= -0.0507 ±0.014	k_{ij}= 0.0269 k_{ij}=0.0462[b]	Peng-Robinson
CH_4–C_3H_8	Reamer et al. (1950)	277.6-327.6	k_{ij}= 0.1062 ±0.013	k_{ij}=0.0249 k_{ij}=0.0575[b]	Peng-Robinson

[b]Interaction parameter values reported by Ohe (1990).
Source: Englezos et al. (1998).

Table 14.9 Vapor-Liquid Equilibrium Data for the Methanol (1)-Isobutane (2) System at 133.0 °C

Pressure (MPa)	Liquid Phase Mole Fraction (x_2)	Vapor Phase Mole Fraction (y_2)
0.886	0.0000	0.0000
1.136	0.0112	0.1790
1.633	0.0343	0.3815
2.586	0.1229	0.6169
3.662	0.4107	0.7334
3.723	0.4671	0.7567
3.920	0.7244	0.7868
3.959	0.8048[a]	0.8048
3.951	0.8465[b]	0.8465
3.816	0.9130	0.9130
3.742	0.9452	0.9369
3.651	0.9782	0.9711
3.591	0.9897	0.9833
3.529	1.0000	1.0000

[a] This is an azeotropic composition.
[b] This is a critical point composition.
Source: Leu and Robinson (1992).

14.5 PROBLEMS

There are numerous sources of phase equilibrium data available that serve as a database to those developing or improving equations of state. References to these databases are widely available. In addition, new data are added mainly through the Journal of Chemical and Engineering Data and the journal Fluid Phase Equilibria. Next, we give data for two systems so that the reader may practice the estimation methods discussed in this chapter.

14.5.1 Data for the Methanol–Isobutane System

Leu and Robinson (1992) reported data for this binary system. The data were obtained at temperatures of 0.0, 50.0, 100.0, 125.0, 133.0 and 150.0 °C. At each temperature the vapor and liquid phase mole fractions of isobutane were measured at different pressures. The data at 133.0 and 150.0 are given in Tables 14.9 and 14.10 respectively. The reader should test if the Peng-Robinson and the Trebble-Bishnoi equations of state are capable of describing the observed phase behaviour. First, each isothermal data set should be examined separately.

Table 14.10 Vapor–Liquid Equilibrium Data for the Methanol (1)–Isobutane (2) System at 150.0 °C

Pressure (MPa)	Liquid Phase Mole Fraction (x_2)	Vapor Phase Mole Fraction (y_2)
1.336	0.000	0.0000
1.520	0.0092	0.1209
1.982	0.0285	0.2819
2.448	0.0566	0.4169
3.555	0.1550	0.6196
4.424	0.3279	0.6601
4.542	0.4876	0.6489
4.623	0.5724	0.6400
4.679	0.6254[a]	0.6254

[a] This is a critical point composition.
Source: Leu and Robinson (1992).

14.5.2 Data for the Carbon Dioxide–Cyclohexane System

Shibata and Sandler (1989) reported VLE data for the CO_2–cyclohexane and N_2–cyclohexane systems. The data for the CO_2–cyclohexane system are given in Tables 14.11 and 14.12. The reader should test if the Peng-

Robinson and the Trebble-Bishnoi equations of state are capable of describing the observed phase behavior. First, each isothermal data set should be examined separately.

Table 14.11 Vapor–Liquid Equilibrium Data for the Carbon Dioxide (1) –Cyclohexane (2) System at 366.5 K

Pressure (MPa)	Liquid Phase Mole Fraction (x_1)	Vapor Phase Mole Fraction (y_1)
0.171	0.0000	0.0000
1.752	0.0959	0.8984
3.469	0.1681	0.9374
5.191	0.2657	0.9495
6.886	0.3484	0.9515
8.637	0.4598	0.9500
10.360	0.5698	0.9402
12.511	0.7526	0.8977
12.800	0.7901	0.8732

Source: Shibata and Sandler (1989).

Table 14.12 Vapor–Liquid Equilibrium Data for the Carbon Dioxide (1) –Cyclohexane (2) System at 410.9 K

Pressure (MPa)	Liquid Phase Mole Fraction (x_1)	Vapor Phase Mole Fraction (y_1)
0.443	0.0000	0.0000
1.718	0.0671	0.7166
3.444	0.1470	0.8295
5.151	0.2050	0.8593
6.928	0.2887	0.8850
8.651	0.3578	0.8861
10.402	0.4329	0.8777
12.056	0.5165	0.8636
13.814	0.6107	0.8320
14.441	0.6850	0.7846
14.510	0.7448	0.7671

Source: Shibata and Sandler (1989).

15

Parameter Estimation in Nonlinear Thermodynamic Models: Activity Coefficients

Activity coefficient models offer an alternative approach to equations of state for the calculation of fugacities in liquid solutions (Prausnitz et al. 1986; Tassios, 1993). These models are also mechanistic and contain adjustable parameters to enhance their correlational ability. The parameters are estimated by matching the thermodynamic model to available equilibrium data. In this chapter, we consider the estimation of parameters in activity coefficient models for electrolyte and non-electrolyte solutions.

15.1 ELECTROLYTE SOLUTIONS

We consider Pitzer's model for the calculation of activity coefficients in aqueous electrolyte solutions (Pitzer, 1991). It is the most widely used thermodynamic model for electrolyte solutions.

15.1.1 Pitzer's Model Parameters for Aqueous Na_2SiO_3 Solutions

Osmotic coefficient data measured by Park (Park and Englezos, 1998; Park, 1999) are used for the estimation of the model parameters. There are 16 osmotic coefficient data available for the Na_2SiO_3 aqueous solution. The data are given in Table 15.1. Based on these measurements the following parameters in Pitzer's activity coefficient model can be calculated: $\beta^{(0)}$, $\beta^{(1)}$, and C^{φ} for Na_2SiO_3. The three binary parameters may be determined by minimizing the following least squares objective function (Park and Englezos, 1998).

$$S(\beta^{(0)},\beta^{(1)},C^{\varphi}) = \sum_{j=1}^{N} \frac{(\varphi_j^{calc} - \varphi_j^{exp})^2}{\sigma_{\varphi j}^2} \tag{15.1}$$

where φ^{calc} is the calculated and φ^{exp} is the measured osmotic coefficient and σ_{φ} is the uncertainty.

Table 15.1 Osmotic Coefficient Data for the Aqueous Na_2SiO_3 Solution

Molality	Osmotic Coefficient (φ^{exp})	Standard Deviation (σ_{φ})
0.0603	0.8923	0.0096
0.0603	0.8926	0.0096
0.3674	0.8339	0.0091
0.3690	0.8304	0.0090
0.5313	0.8188	0.0088
0.5313	0.8188	0.0088
0.8637	0.7790	0.0083
0.8629	0.7797	0.0083
1.2063	0.7614	0.0080
1.2059	0.7617	0.0080
1.4928	0.7607	0.0078
1.4927	0.7608	0.0078
1.8213	0.7685	0.0078
1.8241	0.7674	0.0077
2.3745	0.8021	0.0078
2.3725	0.8028	0.0078

Source: Park and Englezos (1998).

The calculated osmotic coefficient is obtained by the next equation

$$\varphi^{calc} = 1 + |z_M z_X| f^{\varphi} + m(2\nu_M \nu_X/\nu) B_{MX}^{\varphi} + m^2[2(\nu_M \nu_X)^{3/2}/\nu] C_{MX}^{\varphi} \tag{15.2}$$

where

z	is the charge
M	denotes a cation
X	denotes an anion
f^{φ}	is equal to $-A_{\varphi}I^{1/2}/(1 + bI^{1/2})$
A_{φ}	is the Debye-Hückel osmotic coefficient parameter
m	is the molality of solute
I	is the ionic strength ($I=\frac{1}{2}\sum_i m_i z_i^2$)
b	is a universal parameter with the value of 1.2 $(kg.mol)^{1/2}$
v	is the number of ions produced by 1 mole of the solute, and
B^{φ}_{MX}	is $\beta^{(0)}_{MX}+\beta^{(1)}_{MX}\exp(-\alpha_1 I^{1/2})+\beta^{(2)}_{MX}\exp(-\alpha_2 I^{1/2})$

The parameters $\beta^{(0)}_{MX}$, $\beta^{(1)}_{MX}$, C^{φ}_{MX} are tabulated binary parameters specific to the electrolyte MX. The $\beta^{(2)}_{MX}$ is a parameter to account for the ion pairing effect of 2-2 electrolytes. When either cation M or anion X is univalent, α_1 = 2.0. For 2-2, or higher valence pairs, α_1 = 1.4. The constant α_2 is equal to 12. The parameter vector to be estimated is $\mathbf{k}=[\beta^{(0)}, \beta^{(1)}, C^{\varphi}]^T$.

15.1.2 Pitzer's Model Parameters for Aqueous Na_2SiO_3–NaOH Solutions

There are 26 experimental osmotic coefficient data and they are given in Table 15.2 (Park and Englezos, 1999; Park, 1999). Two sets of the binary parameters for the NaOH and Na_2SiO_3 systems and two mixing parameters, $\theta_{OH^-SiO_3^{2-}}$ and $\Psi_{Na^+OH^-SiO_3^{2-}}$ are required to model this system. The binary parameters for the Na_2SiO_3 solution were obtained previously and those for the NaOH system, $\beta^{(0)}_{NaOH}$ = 0.0864, $\beta^{(1)}_{NaOH}$ = 0.253, and C^{φ}_{NaOH} = 0.0044 at 298.15 K are available in the literature. The remaining two Pitzer's mixing parameters, $\theta_{OH^-SiO_3^{2-}}$ and $\Psi_{Na^+OH^-SiO_3^{2-}}$ were determined by the least squares optimization method using the 26 osmotic coefficient data of Table 15.2.

The Gauss-Newton method may be used to minimize the following LS objective function (Park and Englezos, 1999):

$$S(\theta_{OH^-SiO_3^{2-}}, \Psi_{Na^+OH^-SiO_3^{2-}}) = \sum_{j=1}^{N} \frac{(\varphi_j^{calc} - \varphi_j^{exp})^2}{\sigma^2_{\varphi j}} \tag{15.3}$$

where φ^{exp} is the measured osmotic coefficient, σ_φ is the uncertainty computed using the error propagation law (Park, 1999) and φ^{calc} is calculated by the following equation,

$$\varphi^{calc} = 1 + (2/\sum_i m_i)[f^\varphi I + \sum_c \sum_a m_c m_a (B^\varphi_{ca} + ZC_{ca}) + \sum_c \sum_{c'} m_c m_{c'} (\Phi^\varphi_{cc'} + \sum_a m_a \Psi_{cc'a}) + \sum_a \sum_{a'} m_a m_{a'} (\Phi^\varphi_{aa'} + \sum_c m_c \Psi_{caa'}) \tag{15.4}$$

where

m	is the molality of the solute
f^φ	is equal to $-A_\varphi I^{1/2}/(1 + bI^{1/2})$
A_φ	is the Debye-Hückel osmotic coefficient parameter
b	is a universal parameter with the value of 1.2 $(kg.mol)^{1/2}$
I	is the ionic strength ($I=\frac{1}{2}\sum_i m_i z_i^2$)
c	is a cation
a	is an anion
B^φ_{ca}	is equal to $\beta^{(0)}_{ca}+\beta^{(1)}_{ca}\exp(-\alpha_1 I^{1/2})+\beta^{(2)}_{ca}\exp(-\alpha_2 I^{1/2})$
Z	is equal to $\sum_i m_i \mid z_i \mid$
z	is the charge
C_{ca}	is equal to $\dfrac{C^\varphi}{2\lvert z_c z_a\rvert^{1/2}}$
Φ^φ_{ij}	is equal to $\Phi_{ij} + I\Phi'_{ij}$
Φ_{ij}	is equal to $\theta_{ij} + {}^E\theta_{ij}(I)$
Φ'_{ij}	is equal to ${}^E\theta'_{ij}(I)$

θ_{ij} are tabulated mixing parameters specific to the cation-cation or anion-anion pairs. The Ψ_{ijk} is also tabulated mixing parameter specific to the cation-anion-anion or anion-cation-cation pairs.

Table 15.2 Osmotic Coefficient Data for the Aqueous Na_2SiO_3-NaOH Solution

Na^+, molality	OH^- molality	SiO_3^{2-} molality	(φ^{exp})	(σ_φ)
0.1397	0.0466	0.0466	0.9625	0.0103
01405	0.0468	0.0468	0.9572	0.0101
0.2911	0.0970	0.0970	0.9267	0.0100
0.2920	0.0973	0.0973	0.9239	0.0096
0.4442	0.1481	0.1481	0.8816	0.0097
0.4395	0.1465	0.1465	0.8910	0.0094
0.7005	0.2335	0.2335	0.8624	0.0093
0.7015	0.2338	0.2338	0.8611	0.0090
1.0670	0.3557	0.3557	0.8352	0.0091
1.0558	0.3519	0.3519	0.8440	0.0088
1.4192	0.4731	0.4731	0.8253	0.0088
1.4221	0.4740	0.4740	0.8236	0.0087
1.5759	0.5263	0.5263	0.8216	0.0087
1.5868	0.5289	0.5289	0.8175	0.0085
2.3577	0.7859	0.7859	0.8222	0.0085
2.3537	0.7846	0.7846	0.8236	0.0084
2.8797	0.9599	0.9599	0.8323	0.0084
2.8747	0.9582	0.9582	0.8338	0.0085
3.0557	1.0186	1.0186	0.8408	0.0085
3.0542	1.0181	1.0181	0.8412	0.0085
3.9807	1.3269	1.3269	0.8843	0.0085
3.9739	1.3246	1.3246	0.8858	0.0086
4.4861	1.4954	1.4954	0.9027	0.0085
4.4901	1.4967	1.4967	0.9019	0.0085
4.9904	1.6635	1.6635	0.9326	0.0086
4.9888	1.6629	1.6629	0.9329	0.0086

Source: Park and Englezos (1998).

The parameters ${}^{E}\theta_{ij}(I)$ and ${}^{E}\theta'_{ij}(I)$ represent the effects of asymmetrical mixing. These values are significant only for 3-1 or higher electrolytes (Pitzer, 1975). The 2-dimensional parameter vector to be estimated is simply $\mathbf{k}=[\theta_{OH^-SiO_3^{2-}}, \Psi_{Na^+OH^-SiO_3^{2-}}]^T$.

15.1.3 Numerical Results

First the values for the parameter vector $k=[\beta^{(0)}, \beta^{(1)}, C^{\varphi}]^T$ were obtained by using the Na_2SiO_3 data and minimizing Equation 15.1. The estimated parameter values are shown in Table 15.3 together with their standard deviation.

Subsequently, the parameter vector $\mathbf{k}=[\theta_{OH^-SiO_3^{2-}}, \Psi_{Na^+OH^-SiO_3^{2-}}]^T$ was estimated by using the data for the Na_2SiO_3–NaOH system and minimizing Equation 15.3. The estimated parameter values and their standard errors of estimate are also given in Table 15.3. It is noted that for the minimization of Equation 15.3 knowledge of the binary parameters for NaOH is needed. These parameter values are available in the literature (Park and Englezos, 1998).

Table 15.3 Calculated Pitzer's Model Parameters for Na_2SiO_3 and Na_2SiO_3-NaOH Systems

Parameter Value	Standard Deviation
$\beta^{(0)} = 0.0577$	0.0039
$\beta^{(1)} = 2.8965$	0.0559
$C^{\varphi} = 0.00977$	0.00176
$\theta_{OH^-SiO_3^{2-}} = -0.2703$	0.0384
$\Psi_{Na^+OH^-SiO_3^{2-}} = 0.0233$	0.0095

Source: Park and Englezos (1998).

The calculated osmotic coefficients using Pitzer's parameters were compared with the experimentally obtained values and found to have an average percent error of 0.33 for Na_2SiO_3 and 1.74 for the Na_2SiO_3–NaOH system respectively (Park, 1999; Park and Englezos, 1998). Figure 15.1 shows the experimental and calculated osmotic coefficients of Na_2SiO_3 and Figure 15.2 those for the Na_2SiO_3–NaOH system respectively. As seen the agreement between calculated and experimental values is excellent. There are some minor inflections on the calculated curves at molalities below 0.1 mol/kg H_2O. Similar inflections were also observed in other systems in the literature (Park and Englezos, 1998). It is noted that it is known that the isopiestic method does not give reliable results below 0.1 *mol/kg* H_2O.

Park has also obtained osmotic coefficient data for the aqueous solutions of NaOH–NaCl– $NaAl(OH)_4$ at 25°C employing the isopiestic method (Park and Englezos, 1999; Park, 1999). The solutions were prepared by dissolving $AlCl_3 \cdot 6H_2O$ in aqueous NaOH solutions. The osmotic coefficient data were then used to evaluate the unknown Pitzer's binary and mixing parameters for the NaOH–NaCl–$NaAl(OH)_4$–H_2O system. The binary Pitzer's parameters, $\beta^{(0)}$, $\beta^{(1)}$, and C^{φ}, for $NaAl(OH)_4$ were found to be -0.0083, 0.0710, and 0.00184 respectively. These binary parameters were obtained from the data on the ternary system because it was not possible to prepare a single ($NaAl(OH)_4$) solution.

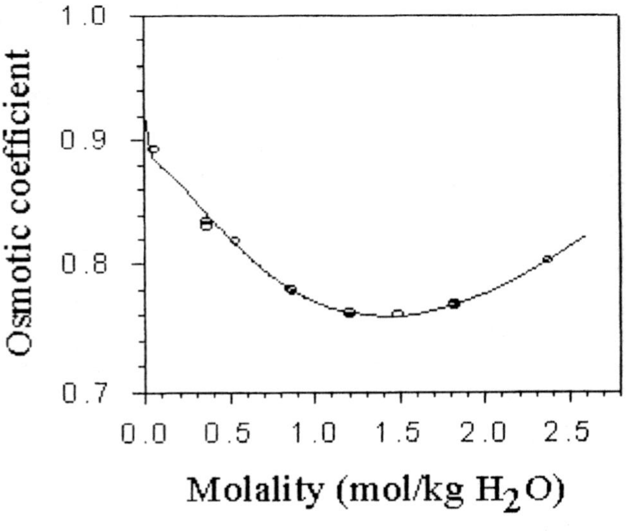

Figure 15.1 *Calculated and experimental osmotic coefficients for Na_2SiO_3. The line represents the calculated values.*

15.2 NON-ELECTROLYTE SOLUTIONS

Activity coefficient models are functions of temperature, composition and to a very small extent pressure. They offer the possibility of expressing the fugacity of a chemical j, f_j^L, in a liquid solution as follows (Prausnitz et al. 1986; Tassios, 1993)

$$\hat{f}_j^L = x_j \gamma_j f_j \tag{15.5}$$

where x_j is the mole fraction, γ_j is the activity coefficient and f_j is the fugacity of chemical species j.

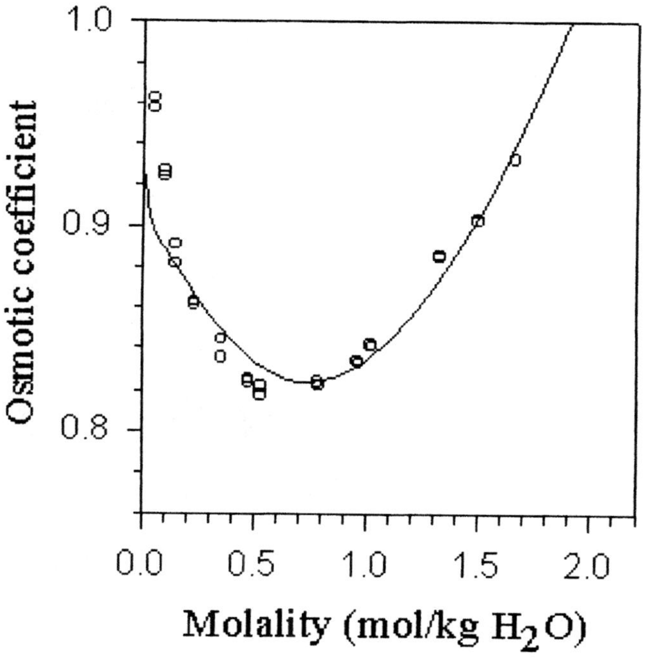

Figure 15.2 Calculated and experimental osmotic coefficients for the Na_2SiO_3 –NaOH system. The line represents the calculated values.

Several activity coefficient models are available for industrial use. They are presented extensively in the thermodynamics literature (Prausnitz et al., 1986). Here we will give the equations for the activity coefficients of each component in a binary mixture. These equations can be used to regress binary parameters from binary experimental vapor-liquid equilibrium data.

15.2.1 The Two-Parameter Wilson Model

The activity coefficients are given by the following equations

$$ln\gamma_1 = -ln(x_1 + \Lambda_{12}x_2) + x_2\left[\frac{\Lambda_{12}}{x_1 + \Lambda_{12}x_2} - \frac{\Lambda_{21}}{\Lambda_{21}x_1 + x_2}\right] \tag{15.6a}$$

$$ln\gamma_2 = -ln(x_2 + \Lambda_{21}x_1) - x_1\left[\frac{\Lambda_{12}}{x_1 + \Lambda_{12}x_2} - \frac{\Lambda_{21}}{\Lambda_{21}x_1 + x_2}\right] \tag{15.6b}$$

The adjustable parameters are related to pure component molar volumes and to characteristic energy differences as following

$$\Lambda_{12} \equiv \frac{v_2}{v_1} exp\left[-\frac{\lambda_{12} - \lambda_{11}}{RT}\right] \tag{15.7a}$$

$$\Lambda_{21} \equiv \frac{v_1}{v_2} exp\left[-\frac{\lambda_{12} - \lambda_{22}}{RT}\right] \tag{15.7b}$$

where v_1 and v_2 are the liquid molar volumes of components 1 and 2 and the λ's are energies of interaction between the molecules designated in the subscripts. The temperature dependence of the quantities $(\lambda_{12}\text{-}\lambda_{11})$ and $(\lambda_{12}\text{-}\lambda_{22})$ can be neglected without serious error.

15.2.2 The Three-Parameter NRTL Model

Renon used the concept of local composition to develop a non-random, two-liquid (NRTL) three parameter (α_{12}, τ_{12}, τ_{21}) equations given below (Prausnitz et al., 1986).

$$\ln\gamma_1 = x_2^2\left[\tau_{21}\left(\frac{G_{21}}{x_1 + x_2G_{21}}\right)^2 + \frac{\tau_{12}G_{12}}{(x_2 + x_1G_{12})^2}\right] \tag{15.8a}$$

$$ln\gamma_2 = x_1^2\left[\tau_{12}\left(\frac{G_{12}}{x_2 + x_1G_{12}}\right)^2 + \frac{\tau_{21}G_{21}}{(x_1 + x_2G_{21})^2}\right] \tag{15.8b}$$

where

$$G_{12} = exp(-\alpha_{12}\tau_{12}) \tag{15.8c}$$

$$G_{21} = exp(-\alpha_{12}\tau_{21}) \tag{15.8d}$$

$$\tau_{12} = \frac{g_{12} - g_{22}}{RT} \tag{15.8e}$$

$$\tau_{21} = \frac{g_{21} - g_{11}}{RT} \tag{15.8f}$$

The parameter g_{ij} is an energy parameter characteristic of the i-j interaction. The parameter α_{12} is related to the non-randomness in the mixture. The NRTL model contains three parameters which are independent of temperature and composition. However, experience has shown that for a large number of binary systems the parameter α_{12} varies from about 0.20 to 0.47. Typically, the value of 0.3 is set.

15.2.3 The Two-Parameter UNIQUAC Model

The Universal Quasichemical (UNIQUAC) is a two-parameter (τ_{12}, τ_{21}) model based on statistical mechanical theory. Activity coefficients are obtained by

$$\ln\gamma_1 = \ln\frac{\varphi_1}{x_1} + \frac{z}{2}q_1 \ln\frac{\theta_1}{\varphi_1} + \varphi_2(l_1 - \frac{r_1}{r_2}l_2) - q_1' \ln(\theta_1' + \theta_2'\tau_{21}) + \theta_2' q_1'\left(\frac{\tau_{21}}{\theta_1' + \theta_2'\tau_{21}} - \frac{\tau_{12}}{\theta_2' + \theta_1'\tau_{12}}\right) \tag{15.9a}$$

$$\ln\gamma_2 = \ln\frac{\varphi_2}{x_2} + \frac{z}{2}q_2 \ln\frac{\theta_2}{\varphi_2} + \varphi_1(l_2 - \frac{r_2}{r_1}l_1) - q_2' \ln(\theta_2' + \theta_1'\tau_{12}) + \theta_1' q_2'\left(\frac{\tau_{12}}{\theta_2' + \theta_1'\tau_{12}} - \frac{\tau_{21}}{\theta_1' + \theta_2'\tau_{21}}\right) \tag{15.9b}$$

where

$$l_1 = \frac{z}{2}(r_1 - q_1) - (r_1 - 1) \tag{15.9c}$$

$$l_2 = \frac{z}{2}(r_2 - q_2) - (r_2 - 1) \tag{15.9d}$$

Segment or volume fractions, φ, and area fractions, θ and θ', are given by

$$\varphi_1 = \frac{x_1 r_1}{x_1 r_1 + x_2 r_2}, \qquad \varphi_2 = \frac{x_2 r_2}{x_1 r_1 + x_2 r_2} \tag{15.9e}$$

$$\theta_1 = \frac{x_1 q_1}{x_1 q_1 + x_2 q_2}, \qquad \theta_2 = \frac{x_2 q_2}{x_1 q_1 + x_2 q_2} \tag{15.9f}$$

$$\theta_1' = \frac{x_1 q_1'}{x_1 q_1' + x_2 q_2'}, \qquad \theta_2' = \frac{x_2 q_2'}{x_1 q_1' + x_2 q_2'} \tag{15.9g}$$

Parameters r, q and q' are pure component molecular-structure constants depending on molecular size and external surface areas. For fluids other than water or lower alcohols, $q = q'$.

For each binary mixture there are two adjustable parameters, τ_{12} and τ_{21}. These in turn, are given in terms of characteristic energies $\Delta u_{12} = u_{12} - u_{22}$ and $\Delta u_{21} = u_{21} - u_{11}$ given by

$$\ln \tau_{12} = -\frac{\Delta u_{12}}{RT} \equiv -\frac{u_{12} - u_{22}}{RT} \tag{15.10a}$$

$$\ln \tau_{21} = -\frac{\Delta u_{21}}{RT} \equiv -\frac{u_{21} - u_{11}}{RT} \tag{15.10b}$$

Characteristic energies, Δu_{12} and Δu_{21} are often only weakly dependent on temperature. The UNIQUAC equation is applicable to a wide variety of nonelectrolyte liquid mixtures containing nonpolar or polar fluids such as hydrocarbons, alcohols, nitriles, ketones, aldehydes, organic acids, etc. and water, including partially miscible mixtures. The main advantages are its relative simplicity using only two adjustable parameters and its wide range of applicability.

15.2.4 Parameter Estimation: The Objective Function

According to Tassios (1993) a suitable objective function to be minimized in such cases is the following

$$S(\mathbf{k}) = \sum_{i=1}^{N}\sum_{j=1}^{2}\left(\frac{\gamma_j^{calc} - \gamma_j^{exp}}{\gamma_j^{exp}}\right)_i^2 \tag{15.11}$$

This is equivalent to assuming that the standard error in the measurement of γ_j is proportional to its value.

Experimental values for the activity coefficients for components 1 and 2 are obtained from the vapor-liquid equilibrium data. During an experiment, the following information is obtained: Pressure (P), temperature (T), liquid phase mole fraction (x_1 and x_2=1$-x_1$) and vapor phase mole fraction (y_1 and y_2=1$-y_1$).

The activity coefficients are evaluated from the above phase equilibrium data by procedures widely available in the thermodynamics literature (Tassios, 1993; Prausnitz et al. 1986). Since the objective in this book is parameter estimation we will provide evaluated values of the activity coefficients based on the phase equilibrium data and we will call these values experimental. These γ_j^{exp} values can then be employed in Equation 15.11.

Alternatively, one may use implicit LS estimation, e.g., minimize Equation 14.23 where liquid phase fugacities are computed by Equation 15.5 whereas vapor phase fugacities are computed by an EoS or any other available method (Prausnitz et al., 1986).

15.3 PROBLEMS

A number of problems formulated with data from the literature are given next as exercises. In addition, to the objective function given by Equation 15.11 the reader who is familiar with thermodynamic computations may explore the use of implicit objective functions based on fugacity calculations.

15.3.1 Osmotic Coefficients for Aqueous Solutions of KCl Obtained by the Isopiestic Method

Thiessen and Wilson (1987) presented a modified isopiestic apparatus and obtained osmotic coefficient data for KCl solutions using NaCl as reference solution. The data are given in Table 15.4. Subsequently, they employed Pitzer's method to correlate the data. They obtained the following values for three Pitzer's model parameters: $\beta_{mx}^{(0)} = 0.05041176, \beta_{mx}^{(1)} = 0.195522, C_{mx}^{\varphi} = 0.001355442$.

Using a constant error for the measurement of the osmotic coefficient, estimate Pitzer's parameters as well as the standard error of the parameter estimates by minimizing the objective function given by Equation 15.1 and compare the results with the reported parameters.

Table 15.4 Osmotic Coefficients for Aqueous KCl Solutions

Molality of KCl	Osmotic Coefficient (φ)
0.09872	0.9325
0.09893	0.9265
0.5274	0.8946
0.9634	0.8944
1.043	0.8981
1.157	0.9009
1.929	0.9120
2.919	0.9351
4.148	0.9675

Source: Thiessen and Wilson (1987).

15.3.2 Osmotic Coefficients for Aqueous Solutions of High-Purity $NiCl_2$

Rard (1992) reported the results of isopiestic vapor-pressure measurements for the aqueous solution of high-purity $NiCl_2$ solution form 1.4382 to 5.7199 *mol/kg* at 298.15±0.005 K. Based on these measurements he calculated the osmotic coefficient of aqueous $NiCl_2$ solutions. He also evaluated other data from the literature and finally presented a set of smoothed osmotic coefficient and activity of water data (see Table IV in original reference).

Rard also employed Pitzer's electrolyte activity coefficient model to correlate the data. It was found that the quality of the fit depended on the range of molalities that were used. In particular, the fit was very good when the molalities were less than 3 *mol/kg*.

Estimate Pitzer's electrolyte activity coefficient model by minimizing the objective function given by Equation 15.1 and using the following osmotic coefficient data from Rard (1992) given in Table 15.5. First, use the data for molalities less than 3 *mol/kg* and then all the data together. Compare your estimated values with those reported by Rard (1992). Use a constant value for σ_φ in Equation 15.1.

Table 15.5 Osmotic Coefficients for Aqueous $NiCl_2$ Solutions at 298.15 K

Molality (mol/kg)	Osmotic Coefficient (φ)	Molality (mol/kg)	Osmotic Coefficient (φ)
0.1	0.8556	1.8	1.3659
0.2	0.8656	2.0	1.4415
0.3	0.8842	2.2	1.5171
0.4	0.9064	2.4	1.5919
0.5	0.9312	2.5	1.6288
0.6	0.9580	2.6	1.6653
0.7	0.9864	2.8	1.7364
0.8	1.0163	3.0	1.8048
0.9	1.0475	3.2	1.8700
1.0	1.0798	3.4	1.9316
1.2	1.1473	3.5	1.9610
1.4	1.2180	3.6	1.9894
1.5	1.2543	3.8	2.0433
1.6	1.2911	4.0	2.0933

Source: Rard (1992).

15.3.3 The Benzene (1)–i-Propyl Alcohol (2) System

Calculate the binary parameters for the UNIQUAC equation by using the vapour-liquid equilibrium data for benzene(1)–i-propyl alcohol (2) at 760 *mmHg* (Tassios, 1993). The following values for other UNIQUAC parameters are available from Tassios (1993): r_1=3.19, q_1=2.40, r_2=2.78, q_2=2.51. The data are given in Table 15.6.

Tassios (1993) also reported the following parameter estimates

$$-\frac{\Delta u_{12}}{R} = -231.5 \tag{15.12a}$$

$$-\frac{\Delta u_{21}}{R} = 10.6 \tag{15.12b}$$

The objective function to be minimized is given by Equation 15.11. The experimental values for the activity coefficients are also given in Table 15.5.

Table 15.6 Vapor-Liquid Equilibrium Data and Activity Coefficients for Benzene(1)–i-Propyl Alcohol at 760 mmHg

Temperature (°C)	x_1	y_1	γ_1	γ_2
79.9	0.053	0.140	2.7187	0.9944
78.5	0.084	0.208	2.6494	1.0009
77.1	0.126	0.276	2.4376	1.0145
75.3	0.199	0.371	2.1834	1.0356
74.4	0.240	0.410	2.0540	1.0628
73.6	0.291	0.451	1.9074	1.0965
73.0	0.357	0.493	1.7291	1.1458
72.4	0.440	0.535	1.5492	1.2386
72.2	0.556	0.583	1.3424	1.4152
72.0	0.624	0.612	1.2625	1.5698
72.1	0.685	0.638	1.1944	1.7429
72.4	0.762	0.673	1.1210	2.0614
73.8	0.887	0.760	1.0393	3.0212
77.5	0.972	0.901	1.0025	4.3630

Source: Tassios (1993).

15.3.4 Vapor-Liquid Equilibria of Coal-Derived Liquids: Binary Systems with Tetralin

Blanco et al. (1994) presented VLE data at 26.66±0.03 *kPa* for binary systems of tetralin with p-xylene, g-picoline, piperidine, and pyridine. The data for the pyridine (1)–tetralin (2) binary are given in Table 15.7.

Blanco et al. have also correlated the results with the van Laar, Wilson, NRTL and UNIQUAC activity coefficient models and found all of them able to describe the observed phase behavior. The value of the parameter a_{12} in the NRTL model was set equal to 0.3. The estimated parameters were reported in Table 10 of the above reference. Using the data of Table 15.7 estimate the binary parameters in the Wislon, NRTL and UNIQUAC models. The objective function to be minimized is given by Equation 15.11.

*Table 15.7 Vapor-Liquid Equilibrium Data and Activity Coefficients for Pyridine (1)-Tetralin (2) at 26.66 kPa**

Temperature (K)	Liquid Phase Mole Fraction of Pyridine (x_1)	Vapor Phase Mole Fraction of Tetralin (y_1)	Activity Coefficient of Pyridine (γ_1)	Activity Coefficient of Tetralin (γ_2)
430.15	0.000	0.000		1.003
417.85	0.025	0.332	1.549	1.010
416.65	0.030	0.348	1.395	1.048
411.30	0.050	0.494	1.366	0.975
389.85	0.160	0.795	1.252	0.962
385.0	0.196	0.840	1.248	0.946
380.9	0.237	0.873	1.216	0.931
376.7	0.287	0.900	1.181	0.930
369.55	0.378	0.932	1.170	0.979
364.10	0.495	0.955	1.099	1.013
360.50	0.575	0.967	1.084	1.038
357.7	0.664	0.977	1.046	1.041
355.55	0.720	0.982	1.047	1.081
352.55	0.830	0.988	1.018	1.369
351.05	0.882	0.992	1.017	1.414
350.05	0.926	0.995	1.007	1.479
349.30	0.953	0.996	0.997	2.455
348.85	0.980	0.998	0.998	2.322
348.20	1.000	1.000	1.004	

* The standard deviation of the measured compositions is 0.005. The temperature was measured with a thermometer having 0.01 K divisions (Blanco et al., 1994).

Source: Blanco et al. (1994).

15.3.5 Vapor-Liquid Equilibria of Ethylbenzene (1) – o-Xylene (2) at 26.66 *kPa*

Monton and Llopis (1994) presented VLE data at 6.66 and 26.66 *kPa* for binary systems of ethylbenzene with m-xylene and o-xylene. The accuracy of the temperature measurement was 0.1 K and that of the pressure was 0.01 kPa. The standard deviations of the measured mole fractions were less than 0.001. The data at 26.66 for the ethylbenzene (1) – o-Xylene (2) are given in Table 15.8 and the objective is to estimate the NRTL and UNIQUAC parameters based on these data.

The reader should refer to the original reference for further details and may also use the additional data at 6.66 *kPa* to estimate the parameters.

Table 15.8 Vapor-Liquid Equilibrium Data for Ethylbenzene (1)–o-Xylene (2) at 26.66 kPa

Temperature (K)	Liquid Phase Mole Fraction of Ethylbenzene (x_1)	Vapor Phase Mole Fraction of o-Xylene (y_1)	Activity Coefficient of Ethylbenzene (γ_1)	Activity Coefficient of o-Xylene (γ_2)
373.25	0.000	0.000		
372.85	0.044	0.057	1.109	1.001
372.45	0.091	0.116	1.063	1.004
371.75	0.171	0.214	1.020	1.010
371.15	0.242	0.294	1.003	1.015
370.45	0.328	0.391	0.997	1.017
369.85	0.399	0.468	1.000	1.015
369.25	0.481	0.545	1.007	1.009
368.65	0.559	0.622	1.015	1.001
368.05	0.638	0.698	1.021	0.992
367.45	0.717	0.767	1.022	0.990
366.85	0.803	0.842	1.017	1.006
366.25	0.892	0.914	1.008	1.063
365.95	0.943	0.955	1.003	1.127
365.55	1.000	1.000		

Source: Monton and Llopis (1994).

16

Parameter Estimation in Chemical Engineering Kinetic Models

A number of examples have been presented in Chapters 4 and 6. The solutions to all these problems are given here except for the two numerical problems that were solved in Chapter 4. In addition a number of problems have been included for solution by the reader.

16.1 ALGEBRAIC EQUATION MODELS

16.1.1 Chemical Kinetics: Catalytic Oxidation of 3-Hexanol

Gallot et al. (1998) studied the catalytic oxidation of 3-hexanol with hydrogen peroxide. The data on the effect of the solvent (CH_3OH) on the partial conversion, y, of hydrogen peroxide are given in Table 4.1. The proposed model is:

$$y = k_1 [1 - exp(-k_2 t)] \tag{16.1}$$

As mentioned in Chapter 4, although this is a dynamic experiment where data are collected over time, we consider it as a simple algebraic equation model with two unknown parameters. The data were given for two different conditions: (i) with 0.75 *g* and (ii) with 1.30 *g* of methanol as solvent. An initial guess of k_1=1.0 and k_2=0.01 was used. The method converged in six and seven iterations respectively without the need for Marquardt's modification. Actually, if Marquardt's modification is used, the algorithm slows down somewhat. The estimated parameters are given in Table 16.1 In addition, the model-calculated values are compared with the experimental data in Table 16.2. As seen the agreement is very good in this case. The quadratic convergence of the Gauss-Newton method is shown in Table 16.3 where the reduction of the LS objective function is shown when an initial guess of k_1=100 and k_2=10 was used.

Table 16.1 Catalytic Oxidation of 3-Hexanol: Estimated Parameter Values and Standard Deviations

	Parameter Value		Standard Deviation	
Mass of CH_3OH	k_1	k_2	σ_{k1}	σ_{k2}
0.75 g	0.1776	0.1055	0.0095	0.0158
1.30 g	0.1787	0.0726	0.0129	0.0116

Table 16.2 Catalytic Oxidation of 3-Hexanol: Experimental Data and Model Calculations

Reaction Time	Partial Conversion			
	Run with 0.75 g methanol		Run with 1.30 g methanol	
	Data	Model	Data	Model
3	0.055	0.048	0.140	0.035
6	0.090	0.083	0.070	0.063
13	0.120	0.133	0.100	0.109
18	0.150	0.151	0.130	0.130
26	0.165	0.166	0.150	0.151
28	0.175	0.168	0.160	0.155

Table 16.3 Catalytic Oxidation of 3-Hexanol: Reduction of the LS Objective Function (Data for 0.75 g CH_3OH)

Iteration	Objective function	k_1	k_2
0	59849.1	100	10
1	59849.1	100.0	4.360
2	32775.6	96.88	0.1140
3	.00035242	0.1769	0.1140
4	.00029574	0.1769	0.1064
5	.00029534	0.1776	0.1056
6	.00029534	0.1776	0.1055

16.1.2 Chemical Kinetics: Isomerization of Bicyclo [2,1,1] Hexane

Data on the thermal isomerization of bicyclo [2,1,1] hexane were measured by Srinivasan and Levi (1963). The data are given in Table 4.4. The following nonlinear model was proposed to describe the fraction of original material remaining (y) as a function of time (x_1) and temperature (x_2).

$$y = exp\left\{-k_1 x_1\, exp\left[-k_2\left(\frac{1}{x_2}-\frac{1}{620}\right)\right]\right\} \tag{16.2}$$

This problem was described in Chapter 4 (Problem 4.3.4). An initial guess of $\mathbf{k}^{(0)}$=(0.001, 10000) was used and convergence of the Gauss-Newton method without the need for Marquardt's modification was achieved in five iterations. The reduction in the LS objective function as the iterations proceed is shown in Table 16.4. In this case the initial guess was fairly close the optimum, **k***. As the initial guess is further away from **k***, the number of iterations increases. For example, if we use as initial guess $\mathbf{k}^{(0)}$=(1, 1000000), convergence is achieved in eight iterations. At the optimum, the following parameter values and standard deviations were obtained: k_1=0.0037838±0.000057 and k_2=27643±461.

Table 16.4 Isomerization of Bicyclo [2,1,1] Hexane: Reduction of the LS Objective Function

Iteration	Objective function	$k_1 \times 10^2$	k_2
0	2.2375	0.1	10000
1	0.206668	0.22663	38111
2	0.029579	0.36650	25038
3	0.0102817	0.37380	27677
4	0.0102817	0.37384	27643
5	0.0102817	0.37383	27643

Using the above parameter estimates the fraction of original material of bicyclo [2,1,1] hexane was calculated and is shown together with the data in Table 16.5. As seen the model matches the data well.

This problem was also solved with Matlab™ (Student version 5, The MATH WORKS Inc.) Using as initial guess the values (k_1=0.001, k_2=10000) convergence of the Gauss-Newton method was achieved in 5 iterations to the values (k_1=0.003738, k_2=27643). As expected, the parameter values that were obtained are the same with those obtained with the Fortran program. In addition the same standard errors of the parameter estimates were computed.

Table 16.5 Isomerization of Bicyclo [2,1,1] Hexane: Experimental Data and Model Calculated Values

x_1	x_2	$\hat{y}$ (data)	y (model)	x_1	x_2	$\hat{y}$ (data)	y (model)
120.	600	.900	.9035	60.0	620	.802	.7991
60.0	600	.949	.9505	60.0	620	.802	.7991
60.0	612	.886	.8823	60.0	620	.804	.7991
120.	612	.785	.7784	60.0	620	.794	.7991
120.	612	.791	.7784	60.0	620	.804	.7991
60.0	612	.890	.8823	60.0	620	.799	.7991
60.0	620	.787	.7991	30.0	631	.764	.7835
30.0	620	.877	.8939	45.1	631	.688	.6930
15.0	620	.938	.9455	30.0	631	.717	.7835
60.0	620	.782	.7991	30.0	631	.802	.7835
45.1	620	.827	.8448	45.0	631	.695	.6935
90.0	620	.696	.7143	15.0	639	.808	.8097
150.	620	.582	.5708	30.0	639	.655	.6556
60.0	620	.795	.7991	90.0	639	.309	.2818
60.0	620	.800	.7991	25.0	639	.689	.7034
60.0	620	.790	.7991	60.1	639	.437	.4292
30.0	620	.883	.8939	60.0	639	.425	.4298
90.0	620	.712	.7143	30.0	639	.638	.6556
150.	620	.576	.5708	30.0	639	.659	.6556
90.4	620	.715	.7132	60.0	639	.449	.4298
120.	620	.673	.6385				

16.1.3 Catalytic Reduction of Nitric Oxide

As another example from chemical kinetics, we consider the catalytic reduction of nitric oxide (NO) by hydrogen which was studied using a flow reactor operated differentially at atmospheric pressure (Ayen and Peters, 1962). The following reaction was considered to be important

$$NO + H_2 \longleftrightarrow H_2O + \frac{1}{2}N_2$$

Data were taken at 375°*C*, and 400°*C*, and 425°*C* using nitrogen as the diluent. The reaction rate in *gmol/(min·g-catalyst)* and the total NO conversion were measured at different partial pressures for H_2 and NO.

A Langmuir-Hinshelwood reaction rate model for the reaction between an adsorbed nitric oxide molecule and one adjacently adsorbed hydrogen molecule is described by:

$$r = \frac{k K_{H_2} K_{NO} P_{H_2} P_{NO}}{\left(1 + K_{NO} P_{NO} + K_{H_2} P_{H_2}\right)^2} \tag{16.3}$$

where r is the reaction rate in *gmol/(min·g-catalyst)*, P_{H2} is the partial pressure of hydrogen (*atm*), P_{NO} is the partial pressure of NO (*atm*), $K_{NO}= A_2 exp\{-E_2/RT\}$ atm^{-1} is the adsorption equilibrium constant for NO, $K_{H2}= A_3 exp\{-E_3/RT\}$ atm^{-1} is the adsorption equilibrium constant for H_2 and $k=A_1 exp\{-E_1/RT\}$ *gmol/(min·g-catalyst)* is the forward reaction rate constant for surface reaction. The data for the above problem are given in Table 4.5.

The objective of the estimation procedure is to determine the parameters k, K_{H2} and K_{NO} (if data from one isotherm are only considered) or the parameters A_1, A_2, A_3, E_1, E_2, E_3 (when all data are regressed together). The units of E_1, E_2, E_3 are in *cal/mol* and R is the universal gas constant (1.987 *cal/mol K*).

Kittrell et al. (1965a) considered three models for the description of the reduction of nitric oxide. The one given in Chapter 4 corresponds to a reaction between one adsorbed molecule of nitric oxide and one adsorbed molecule of hydrogen. This was done on the basis of the shape of the curves passing through the plotted data.

In this work, we first regressed the isothermal data. The estimated parameters from the treatment of the isothermal data are given in Table 16.6. An initial guess of (k_1=1.0, k_2=1.0, k_3=1.0) was used for all isotherms and convergence of the Gauss-Newton method without the need for Marquardt's modification was achieved in 13, 16 and 15 iterations for the data at 375, 400, and 425°*C* respectively.

Plotting of lnk_j (j=1,2,3) versus 1/T shows that only k_1 exhibits Arrhenius type of behavior. However, given the large standard deviations of the other two estimated parameters one cannot draw definite conclusions about these two parameters.

Table 16.6 Catalytic reduction of NO: Estimated Model Parameters by the Gauss-Newton Method Using Isothermal Data

Temperature (°C)	$(k_1 \pm \sigma_{k1})\times 10^4$	$k_2 \pm \sigma_{k2}$	$k_3 \pm \sigma_{k3}$
375	5.2 ± 1.2	18.5 ± 3.4	13.2 ± 3.4
400	5.5 ± 3.2	31.5 ± 13.0	35.9 ± 14.0
425	13.5 ± 8.0	25.9 ± 10.3	14.0 ± 8.9

Kittrell et al. (1965a) also performed two types of estimation. First the data at each isotherm were used separately and subsequently all data were regressed simultaneously. The regression of the isothermal data was also done with linear least squares by linearizing the model equation. In Tables 16.7 and 16.8 the reported parameter estimates are given together with the reported standard error. Ayen and Peters (1962) have also reported values for the unknown parameters and they are given here in Table 16.9.

Table 16.7 Catalytic reduction of NO: Estimated Model Parameters by Linear Least Squares Using Isothermal Data

Temperature (°C)	$(k_1 \pm \sigma_{k1})\times 10^4$	$k_2 \pm \sigma_{k2}$	$k_3 \pm \sigma_{k3}$
375	4.9 ± 0.7	18.8 ± 4.6	14.6 ± 2.9
400	5.3± 8.5	38.6 ± 19.6	35.4 ± 11.3
425	8.8 ± 2.3	48.9 ± 31.3	30.9 ± 20.2

Source: Kittrell et al. (1965a).

Table 16.8 Catalytic reduction of NO: Estimated Model Parameters by Nonlinear Least Squares Using Isothermal Data

Temperature (°C)	$(k_1 \pm \sigma_{k1})\times 10^4$	$k_2 \pm \sigma_{k2}$	$k_3 \pm \sigma_{k3}$
375	5.19 ± 0.9	18.5 ± 3.4	13.2 ± 3.4
400	5.51 ± 1.2	31.6 ± 12.9	36.0 ± 13.9
425	10.1 ± 3.0	34.5 ± 15.2	23.1 ± 11.6

Source: Kittrell et al. (1965a).

Table 16.9 Catalytic reduction of NO: Estimated Model Parameters by Nonlinear Least Squares

Temperature (°C)	$k_1\times 10^4$	k_2	k_3
375	4.94	19.00	14.64
400	7.08	30.45	20.96
425	8.79	48.55	30.95

Source: Ayen and Peters (1962).

Table 16.10 Catalytic reduction of NO: Estimated Model Parameters by Nonlinear Least Squares Using Nonisothermal Data

Temperature (°C)	$(k_1 \pm \sigma_{k1})\times 10^4$	$k_2 \pm \sigma_{k2}$	$k_3 \pm \sigma_{k3}$
375	4.92 ± 3.71	15.5 ± 13.4	17.5 ± 11.5
400	6.58 ± 3.94	26.3 ± 18.3	23.8 ± 12.8
425	8.63 ± 3.92	42.9 ± 23.6	31.7 ± 14.6

Source: Kittrell et al. (1965a).

Kittrell et al. (1965a) also used all the data simultaneously to compute the parameter values. These parameter values are reported for each temperature and are given in Table 16.10.

Writing Arrhenius-type expressions, $k_j=A_j exp(-E_j/RT)$, for the kinetic constants, the mathematical model with six unknown parameters (A_1, A_2, A_3, E_1, E_2 and E_3) becomes

$$f(\mathbf{x},\mathbf{k})=\frac{A_1 e^{-\frac{E_1}{RT}} A_2 e^{-\frac{E_2}{RT}} A_3 e^{-\frac{E_3}{RT}} x_1 x_2}{\left(1+A_2 e^{-\frac{E_2}{RT}} x_2 + A_3 e^{-\frac{E_3}{RT}} x_1\right)^2} \tag{16.4}$$

The elements of the (1×6)-dimensional sensitivity coefficient matrix **G** are obtained by evaluating the partial derivatives:

$$G_{11}=\left(\frac{\partial f}{\partial A_1}\right)=\frac{A_2 A_3 e^{-\frac{E_1+E_2+E_3}{RT}} x_1 x_2}{Y^2} \tag{16.5}$$

$$G_{12}=\left(\frac{\partial f}{\partial A_2}\right)=\frac{-2A_1 A_2 A_3 e^{-\frac{E_1+2E_2+E_3}{RT}} x_1 x_2^2}{Y^3} + \frac{A_1 A_3 e^{-\frac{E_1+E_2+E_3}{RT}} x_1 x_2}{Y^2} \tag{16.6}$$

$$G_{13}=\left(\frac{\partial f}{\partial A_3}\right)=\frac{-2A_1 A_2 A_3 e^{-\frac{E_1+E_2+2E_3}{RT}} x_1^2 x_2}{Y^3} + \frac{A_1 A_2 e^{-\frac{E_1+E_2+E_3}{RT}} x_1 x_2}{Y^2} \tag{16.7}$$

$$G_{14}=\left(\frac{\partial f}{\partial E_1}\right)=\frac{-A_1 A_2 A_3 e^{-\frac{E_1+E_2+E_3}{RT}} x_1 x_2}{RTY^2} \tag{16.8}$$

$$G_{15} = \left(\frac{\partial f}{\partial E_2}\right) = \frac{2A_1A_2^2A_3e^{-\frac{E_1+2E_2+E_3}{RT}}x_1x_2^2}{RTY^3} - \frac{A_1A_2A_3e^{-\frac{E_1+E_2+E_3}{RT}}x_1x_2}{RTY^2} \tag{16.9}$$

$$G_{16} = \left(\frac{\partial f}{\partial E_3}\right) = \frac{2A_1A_2A_3^2e^{-\frac{E_1+E_2+2E_3}{RT}}x_1^2x_2}{RTY^3} - \frac{A_1A_2A_3e^{-\frac{E_1+E_2+E_3}{RT}}x_1x_2}{RTY^2} \tag{16.10}$$

where

$$Y = 1 + A_2e^{-\frac{E_2}{RT}}x_2 + A_3e^{-\frac{E_3}{RT}}x_1 \tag{16.11}$$

The results were obtained using three different sets of initial guesses which were given by Kittrell et al. (1965b). None of them was good enough to converge to the global optimum. In particular the first two converged to local optima and the third diverged. The lowest LS objective function was obtained with the first initial guess and it was 0.1464×10^{-6}. The corresponding estimated parameter values were A_1=0.8039±0.3352, $A_2=1.371\times10^5\pm6.798\times10^4$, $A_3=1.768\times10^7\pm8.739\times10^6$, $E_1=9520\pm0.4\times10^{-5}$, $E_2=11500\pm0.9\times10^{-7}$ and $E_3=17{,}900\pm0.9\times10^{-7}$.

In this problem it is very difficult to obtain convergence to the global optimum as the condition number of matrix **A** at the above local optimum is 3×10^{18}. Even if this was the global optimum, a small change in the data would result in widely different parameter estimates since this parameter estimation problem appears to be fairly ill-conditioned.

At this point we should always try and see whether there is anything else that could be done to reduce the ill-conditioning of the problem. Upon reexamination of the structure of the model given by Equation 16.4 we can readily notice that it can be rewritten as

$$f(\mathbf{x},\mathbf{k}) = \frac{A_1^* e^{-\frac{E_1^*}{RT}} x_1 x_2}{\left(1 + A_2 e^{-\frac{E_2}{RT}} x_2 + A_3 e^{-\frac{E_3}{RT}} x_1\right)^2} \tag{16.12}$$

where

$$A_1^* = A_1 A_2 A_3 \tag{16.13}$$

and

$$E_1^* = E_1 + E_2 + E_3 \tag{16.14}$$

The reparameterized model has the same number of unknown parameters (A_1^*, A_2, A_3, E_1^*, E_2 and E_3) as the original problem, however, it has a simpler structure. This often results in much better convergence characteristics of the iterative estimation algorithm. Indeed, convergence to the global optimum was obtained after many iterations using Marquardt's modification. The value of Marquardt's parameter was always kept one order of magnitude greater than the smallest eigenvalue of matrix **A**. At the optimum, a value of zero for Marquardt's parameter was used and convergence was maintained.

The LS objective function was found to be 0.7604×10^{-9}. This value is almost three orders of magnitude smaller than the one found earlier at a local optimum. The estimated parameter values were: A_1=22.672, A_2=132.4, A_3=585320, E_1=13899, E_2=2439.6 and E_3=13506 where parameters A_1 and E_1 were estimated back from A_1^* and E_1^*. With this reparameterization we were able to lessen the ill-conditioning of the problem since the condition number of matrix **A** was now 5.6×10^8.

The model-calculated reaction rates are compared to the experimental data in Table 16.11 where it can be seen that the match is quite satisfactory. Based on the six estimated parameter values, the kinetic constants (k_1, k_2 and k_3) were computed at each temperature and they are shown in Table 16.12.

Having found the optimum, we returned back to the original structure of the problem and used an initial guess fairly close to the global optimum. In this case the parameters converged very close to the optimum where the LS objective function was 0.774×10^{-9}. The condition number of matrix **A** was found to be 1.7×10^{13} which is about 5 orders of magnitude higher that the one for the reparameterized formulation calculated at the same point. In conclusion, reparameterization should be seriously considered for hard to converge problems.

Table 16.11 Catalytic reduction of NO: Experimental Measurements and Model Calculated Values (with **k** *from Table 16.12)*

P_{H2} *(atm)*	P_{NO} *(atm)*	Measured Rate $r\times10^5$ *gmol/(min·g-catalyst)*	Calculated Rate $r\times10^5$ *gmol/(min·g-catalyst)*
T=375°C, Weight of catalyst=2.39 *g*			
0.00922	0.0500	1.60	1.514
0.0136	0.0500	2.56	2.091
0.0197	0.0500	3.27	2.775
0.0280	0.0500	3.64	3.521
0.0291	0.0500	3.48	3.607
0.0389	0.0500	4.46	4.253
0.0485	0.0500	4.75	4.724
0.0500	0.00918	1.47	1.739
0.0500	0.0184	2.48	2.924
0.0500	0.0298	3.45	3.885
0.0500	0.0378	4.06	4.336
0.0500	0.0491	4.75	4.761
T=400°C, Weight of catalyst=1.066 *g*			
0.00659	0.0500	2.52	2.376
0.0113	0.0500	4.21	3.690
0.0228	0.0500	5.41	5.954
0.0311	0.0500	6.61	7.011
0.0402	0.0500	6.86	7.803
0.0500	0.0500	8.79	8.357
0.0500	0.0100	3.64	3.062
0.0500	0.0153	4.77	4.275
0.0500	0.0270	6.61	6.249
0.0500	0.0361	7.94	7.298
0.0500	0.0432	7.82	7.907
T=425°C, Weight of catalyst=1.066 *g*			
0.00474	0.0500	5.02	3.550
0.0136	0.0500	7.23	7.938
0.0290	0.0500	11.35	11.68
0.0400	0.0500	13.00	12.82
0.0500	0.0500	13.91	13.29
0.0500	0.0269	9.29	9.586
0.0500	0.0302	9.75	10.29
0.0500	0.0387	11.89	11.81

Table 16.12 Catalytic reduction of NO: Estimated Model Parameters by Nonlinear Least Squares Using All the Data.

Temperature (°C)	$k_1\times10^4$	k_2	k_3
375	4.65	19.91	16.29
400	6.94	21.36	24.05
425	10.07	22.80	34.53

16.2 PROBLEMS WITH ALGEBRAIC MODELS

The following parameter estimation problems were formulated from research papers available in the literature and are left as exercises.

16.2.1 Catalytic Dehydrogenation of sec-butyl Alcohol

Data for the initial reaction rate for the catalytic dehydrogenation of *sec*-butyl alcohol to methyl ethyl ketone are given in Table 16.13 (Thaller and Thodos, 1960; Shah, 1965). The following two models were considered for the initial rate:

<u>Model A</u>

$$r_{Ai} = R - \sqrt{R^2 - k_H^2} \tag{16.15}$$

where

$$R = k_H + \frac{k_H^2}{2k_R}\frac{(1+K_A p_A)^2}{K_A p_A} \tag{16.16}$$

<u>Model B</u>

$$r_{Ai} = \left[(k_H)^\lambda + \left(\frac{k_R K_A P_A}{(1+K_A P_A)^2}\right)^s\right]^{1/\lambda} \tag{16.17}$$

where λ=–0.7, K_A is the adsorption equilibrium constant for *sec*-butyl alcohol, k_H is the rate coefficient for the rate of hydrogen desorption controlling, k_R is the rate coefficient for surface reaction controlling, P_A is the partial pressure of *sec*-butyl alcohol.

Table 16.13 Data for the Catalytic Dehydrogenation of sec-butyl Alcohol

Temperature (°F)	Pressure (atm)	Feed Rate (lb-moles/h)	Initial Rate (r_{Ai}) lb-moles of alcohol/(h)(lb-catalyst)
600	1.0	0.01359	0.0392
600	7.0	0.01366	0.0416
600	4.0	0.01394	0.0416
600	10.0	0.01367	0.0326
600	14.6	0.01398	0.0247
600	5.5	0.01389	0.0415
600	8.5	0.01384	0.0376
600	3.0	0.01392	0.0420
600	0.22	0.01362	0.0295
600	1.0	0.01390	0.0410
575	1.0	0.01411	0.0227
575	3.0	0.01400	0.0277
575	5.0	0.01401	0.0255
575	7.0	0.01374	0.0217
575	9.6	0.01342	0.0183

Source: Thaller and Thodos (1960); Shah (1965).

Table 16.14 Parameter Estimates for Models A and B

Model	Temperature (°F)	$k_H \times 10^2$ (lbmoles alcohol /(hr lb catalyst)	$k_R \times 10^2$ (lbmoles alcohol /(hr lb catalyst)	$K_A \times 10^2$ (atm^{-1})
A	575	7.65	23.5	44.4
	600	7.89	81.7	53.5
B	575	11.5	20.2	40.7
	600	9.50	62.8	51.5

Source: Shad (1965).

Using the initial rate data given above do the following: (a) Determine the parameters, k_R, k_H and K_A for model-A and model-B and their 95% confidence intervals; and (b) Using the parameter estimates calculate the initial rate and compare it with the data. Shah (1965) reported the parameter estimates given in Table 16.14.

16.2.2 Oxidation of Propylene

The following data given in Tables 16.15, 16.16 and 16.17 on the oxidation of propylene over bismuth molybdate catalyst were obtained at three temperatures, 350, 375, and 390°C (Watts, 1994).

One model proposed for the rate of propylene disappearance, r_p, as a function of the oxygen concentration, C_o, the propylene concentration, C_p, and the stoichiometric number, n, is

$$r_p = \frac{k_o k_p c_o^{0.5} c_p}{k_o c_o^{0.5} + n k_p c_p} \tag{16.18}$$

where k_o and k_p are the rate parameters.

Table 16.15 Data for the Oxidation of Propylene at 350°C.

C_p	C_o	n	r_p
3.05	3.07	0.658	2.73
1.37	3.18	0.439	2.86
3.17	1.24	0.452	3.00
3.02	3.85	0.695	2.64
4.31	3.15	0.635	2.60
2.78	3.89	0.670	2.73
3.11	6.48	0.760	2.56
2.96	3.13	0.642	2.69
2.84	3.14	0.665	2.77
1.46	7.93	0.525	2.91
1.38	7.79	0.483	2.87
1.42	8.03	0.522	2.97
1.49	7.78	0.530	2.93
3.01	3.03	0.635	2.75
1.35	8.00	0.480	2.90
1.52	8.22	0.544	2.94
5.95	6.13	0.893	2.38
1.46	8.41	0.517	2.89
5.68	7.75	0.996	2.41
1.36	3.10	0.416	2.81
1.42	1.25	0.367	2.86
3.18	7.89	0.835	2.59
2.87	3.06	0.609	2.76

Source: Watts (1994).

Table 16.16 Data for the Oxidation of Propylene at 375°C

C_p	C_o	n	r_p
2.94	2.96	1.160	2.37
1.35	3.06	0.680	2.58
3.04	1.19	0.740	2.24
2.90	3.70	1.170	2.19
4.14	3.03	1.390	2.32
2.69	3.76	1.190	2.31
2.99	6.23	1.290	2.16
2.85	3.03	1.130	2.25
5.46	7.46	2.030	1.93
1.39	7.67	0.804	2.63
1.34	1.15	0.630	2.58
2.73	3.02	1.080	2.16
1.46	7.65	0.864	2.64
1.39	7.56	0.772	2.53
1.33	7.49	0.777	2.64
1.37	7.75	0.745	2.51
7.02	2.93	1.310	2.25
2.89	2.91	1.160	2.27
7.30	2.96	1.360	2.22
1.35	7.66	0.741	2.55
3.15	7.52	1.440	2.14
2.75	2.93	1.050	2.15

Source: Watts (1994).

The objective is to determine the parameters and their standard errors by the Gauss-Newton method for each temperature and then check to see if the parameter estimates obey Arrhenius type behavior.

Watts (1994) reported the following parameter estimates at 350°C: k_o=1,334 ± 0.081 *[(mmol L)$^{0.5}$/(g s)]* and k_p=0.611 ± 0.055 *[(L/g s)]*. Similar results were found for the data at the other two temperatures.

The parameter values were then plotted versus the inverse temperature and were found to follow an Arrhenius type relationship

$$k_j = A_j \, exp\left(-\frac{E_j}{RT}\right) \quad ; \quad j = o, p \tag{16.19}$$

where A_j is the pre-exponential factor and E_j the activation energy. Both numbers should be positive.

Table 16.17 Data for the Oxidation of Propylene at 390°C

C_p	C_o	n	r_p
2.62	3.66	1.480	1.95
2.79	2.96	1.510	2.00
3.02	6.12	1.800	1.92
3.07	7.32	1.900	1.96
1.36	7.52	0.990	2.36
1.31	1.12	0.805	2.33
1.42	7.47	0.991	2.26
2.72	3.48	1.520	1.93
6.86	2.86	2.210	2.06
7.13	2.89	2.300	2.10
1.32	7.48	0.936	2.36
7.09	3.27	2.430	2.16
2.88	3.76	1.640	1.85
1.33	7.84	0.975	2.38
7.14	3.22	2.300	2.10
1.37	7.89	0.996	2.39
5.39	7.25	2.760	1.76
1.31	2.90	0.823	2.28
2.74	3.54	1.530	1.84
2.89	7.48	1.790	1.83
5.29	7.23	2.760	1.75

Source: Watts (1994).

Subsequently, Watts performed a parameter estimation by using the data from all temperatures simultaneously and by employing the formulation of the rate constants as in Equation 16.19. The parameter values that they found as well as their standard errors are reported in Table 16.18. It is noted that they found that the residuals from the fit were well behaved except for two at 375°C. These residuals were found to account for 40% of the residual sum of squares of deviations between experimental data and calculated values.

Watts (1994) dealt with the issue of confidence interval estimation when estimating parameters in nonlinear models. He proceeded with the reformulation of Equation 16.19 because the pre-exponential parameter estimates "behaved highly nonlinearly." The rate constants were formulated as follows

$$k_j = A_j^* \, exp\left(-\frac{E}{R}\left(\frac{1}{T}-\frac{1}{T_o}\right)\right) \quad ; \quad j = o, p \tag{16.20}$$

with

$$A_j^* = A_j \, exp\left(-\frac{E_j}{RT_o}\right) \tag{16.21}$$

Table 16.18 Parameter Estimates for the Model for the Oxidation of Propylene (Noncentered Formulation)

Parameter	*Estimated Value*	*Standard Error*
A_o	8.94	9.77
E_o	105.6	5.9
A_p	145.1	209.4
E_p	28.1	7.89

Source: Watts (1994).

Table 16.19 Parameter Estimates for the Model for the Oxidation of Propylene (Centered Formulation)

Parameter	*Estimated Value*	*Standard Error*
A_o^*	2.74	0.071
E_o	105.6	5.9
A_p^*	0.794	0.025
E_p	28.1	7.89

Source: Watts (1994).

Thus, in order to improve the behavior of the parameter estimates, Watts (1994) centers the temperature factor about a reference value T_o which was chosen to be the middle temperature of 375°*C* (648 *K*). The parameters estimates and their standard errors are given in Table 16.19.

You are asked to verify the calculations of Watts (1994) using the Gauss-Newton method. You are also asked to determine by how much the condition number of matrix **A** is improved when the centered formulation is used.

16.2.3 Model Reduction Through Parameter Estimation in the s-Domain

Quite often we are face with the task of reducing the order of a transfer function without losing essential dynamic behavior of the system. Many methods have been proposed for model reduction, however quite often with unsatisfactory results. A reliable method has been suggested by Luus (1980) where the deviations between the reduced model and the original one in the Nyquist plot are minimized.

Consider the following 8th order system

$$f(s) = \frac{194480 + 482964s + 511812s^2 + 278376s^3 + 82402s^4 + 13285s^5 + 1086s^6 + 35s^7}{9600 + 28880s + 37492s^2 + 27470s^3 + 11870s^4 + 3017s^5 + 437s^6 + 33s^7 + s^8} \tag{16.22}$$

The poles of the transfer function (roots of the denominator) are at -1, -1±j, -3, -4, -5, -8 and -10. Let us assume that we seek a third order system that follows as closely as possible the behavior of the high order system. Namely, consider

$$g(s) = \frac{20.2583(1 + k_1 s + k_2 s^2)}{1 + k_3 s + k_4 s^2 + k_5 s^3} \tag{16.23}$$

The constant in the numerator can always be chosen to preserve the steady state gain of the transfer function. As suggested by Luus (1980) the 5 unknown parameters can be obtained by minimizing the following quadratic objective function

$$S(\mathbf{k}) = \sum_{i=1}^{N} \left[Re(f(j\omega_i)) - Re(g(j\omega_i))\right]^2 + \left[Im(f(j\omega_i)) - Im(g(j\omega_i))\right]^2 \tag{16.24}$$

where $j = \sqrt{-1}$, $\omega_{i+1}=1.1\times\omega_i$, $\omega_1=0.01$ and N=100.

In this problem you are asked to determine the unknown parameters using the dominant zeros and poles of the original system as an initial guess. LJ optimization procedure can be used to obtain the best parameter estimates.

Redo the problem but take N=1,000, 10,000 and 100,000.

After the parameters have been estimated, generate the Nyquist plots for the reduced models and the original one. Comment on the result at high frequencies. Is N=100 a wise choice?

Redo this problem. However, this time assume that the reduced model is a fourth order one. Namely, it is of the form

$$g(s) = \frac{20.2583(1 + k_1 s + k_2 s^2 + k_3 s^3)}{1 + k_4 s + k_5 s^2 + k_6 s^3 + k_7 s^4} \tag{16.25}$$

How important is the choice of ω_1=0.01?

16.3 ORDINARY DIFFERENTIAL EQUATION MODELS

The formulation for the next three problems of the parameter estimation problem was given in Chapter 6. These examples were formulated with data from the literature and hence the reader is strongly recommended to read the original papers for a thorough understanding of the relevant physical and chemical phenomena.

16.3.1 A Homogeneous Gas Phase Reaction

Bellman et al. (1967) have considered the estimation of the two rate constants k_1 and k_2 in the Bodenstein-Linder model for the homogeneous gas phase reaction of NO with O_2:

$$2NO + O_2 \leftrightarrow 2NO_2$$

The model is described by the following equation

$$\frac{dx}{dt} = k_1(\alpha - x)(\beta - x)^2 - k_2 x^2 \quad ; \quad x(0) = 0 \tag{16.26}$$

where α=126.2, β=91.9 and x is the concentration of NO_2. The concentration of NO_2 was measured experimentally as a function of time and the data are given in Table 6.1

Bellman et al. (1967) employed the quasilinearization technique and obtained the following parameter estimates: k_1=0.4577×10^{-5} and k_2=0.2797×10^{-3}. Bodenstein and Lidner who had obtained the kinetic data reported slightly different values: k_1=0.53×10^{-5} and k_2=0.41×10^{-3}. The latter values were obtained by a combination of chemical theory and the data. The residual sum of squares of deviations was found to be equal to 0.210×10^{-2}. The corresponding value reported by Bodenstein and Lidner who had obtained the kinetic data is 0.555×10^{-2}. Bellman et al. (1967) stated that the difference does not reflect one set of parameters being better than the other.

Using the computer program Bayes_ODE1 which is given in Appendix 2 the following parameter estimates were obtained: k_1=0.4577×10^{-5} and k_2=0.2796×10^{-3} using as initial guess k_1=0.1×10^{-5} and k_2=0.1×10^{-5}. Convergence was achieved in seven iterations as seen in Table 16.20. The calculated standard deviations for the parameters k_1 and k_2 were 3.3% and 18.8% respectively. Based on these parameter values, the concentration of NO_2 was computed versus time and compared to the experimental data as shown in Figure 16.1. The overall fit is quite satisfactory.

Table 16.20 Homogeneous Gas Phase Reaction: Convergence of the Gauss-Newton Method

Iteration	LS objective function	k_1	k_2
0	4011.7	0.1×10^{-5}	0.1×10^{-5}
1	1739.9	0.33825×10^{-5}	0.25876×10^{-2}
2	774.15	0.35410×10^{-5}	0.11655×10^{-2}
3	123.35	0.39707×10^{-5}	0.40790×10^{-3}
4	22.020	0.45581×10^{-5}	0.26482×10^{-3}
5	21.867	0.45786×10^{-5}	0.28017×10^{-3}
6	21.867	0.45770×10^{-5}	0.27959×10^{-3}
7	21.867	0.45771×10^{-5}	0.27962×10^{-3}

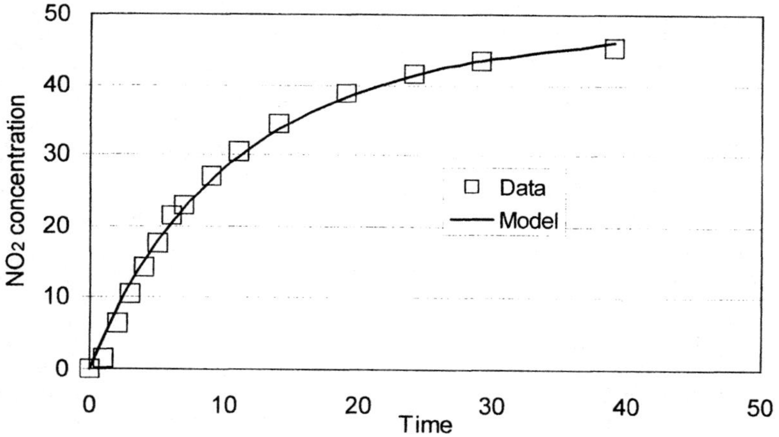

Figure 16.1 Homogeneous Gas Phase Reaction: Experimental data and model calculated values of the NO_2 concentration.

16.3.2 Pyrolytic Dehydrogenation of Benzene to Diphenyl and Triphenyl

Let us now consider the pyrolytic dehydrogenation of benzene to diphenyl and triphenyl (Seinfeld and Gavalas, 1970; Hougen and Watson, 1948):

$$2C_2H_6 \leftrightarrow C_{12}H_{10} + H_2$$

$$C_6H_6 + C_{12}H_{10} \leftrightarrow C_{10}H_{14} + H_2$$

The following kinetic model has been proposed

$$\frac{dx_1}{dt} = -r_1 - r_2 \tag{16.27a}$$

$$\frac{dx_2}{dt} = \frac{r_1}{2} - r_2 \tag{16.27b}$$

where

$$r_1 = k_1\left(x_1^2 - x_2(2 - 2x_1 - x_2)/3K_1\right) \tag{16.28a}$$

$$r_2 = k_2\left(x_1 x_2 - (1 - x_1 - 2x_2)(2 - 2x_1 - x_2)/9K_2\right) \tag{16.28b}$$

and where x_1 denotes *lb-moles* of benzene per *lb-mole* of pure benzene feed and x_2 denotes *lb-moles* of diphenyl per *lb-mole* of pure benzene feed. The parameters k_1 and k_2 are unknown reaction rate constants whereas K_1 and K_2 are equilibrium constants. The data consist of measurements of x_1 and x_2 in a flow reactor at eight values of the reciprocal space velocity t and are given in Table 6.2. The feed to the reactor was pure benzene. The equilibrium constants K_1 and K_2 were determined from the run at the lowest space velocity to be 0.242 and 0.428, respectively.

Seinfeld and Gavalas (1970) employed the quasilinearization method to estimate the parameters for the kinetic model that was proposed by Hougen and Watson (1948). Seinfeld and Gavalas (1970) examined the significance of having a good initial guess. As weighting matrix in the LS objective function they used the identity matrix. It was found that with an initial guess k_1=k_2=300 or k_1=k_2=500 convergence was achieved in four iterations. The corresponding estimated parameters were 347.4, 403.1. The quasilinearization algorithm diverged when the initial guesses were k_1=k_2=100 or k_1=k_2=1000. It is interesting to note that Hougen and Watson (1948) reported the values of k_1=348 and k_2=404 [*(lb-moles)/359 (ft^3)(hr)(atm^2)*]. Seinfeld and Gavalas pointed out that it was a coincidence that the values estimated by Hougen and Watson in a rather "crude" manner were very close to the values estimated by the quasilinearization method.

Subsequently, Seinfeld and Gavalas examined the role of the weighting matrix and the role of using fewer data points. It was found that the estimated values at the global minimum are not affected appreciably by using different weighting factors. They proposed this as a quick test to see whether the global minimum has been reached and as a means to move away from a local minimum of the LS objective function. It was also found that there is some variation in the estimates of k_2 as the number of data points used in the regression is reduced. The estimate of k_1 was found to remain constant. The same problem was also studied by Kalogerakis and Luus (1983) to demonstrate the substantial enlargement of the region of convergence for the Gauss-Newton method through the use of the information index discussed in Chapter 8 (Section 8.7.2.2).

Using the program provided in Appendix 2 and starting with an initial guess far from the optimum ($k_1=k_2=10000$), the Gauss-Newton method converged within nine iterations. Marquardt's parameter was zero at all times. The reduction in the LS objective function is shown in Table 16.21. As weighting matrix, the identity matrix was used. The uncertainty in the parameter estimates is quite small, namely, 0.181% and 0.857% for k_1 and k_2 respectively. The corresponding match between the experimental data and the model-calculated values is shown in Table 16.22.

Table 16.21 Benzene Dehydrogenation: Convergence of the Gauss-Newton Method

Iteration	LS objective function	k_1	k_2
0	0.27501	10000	10000
1	0.27325	4971.1	9518.4
2	0.25841	4254.5	1526.1
3	0.21317	1885.2	1479.5
4	0.65848×10^{-1}	703.35	893.62
5	0.75509×10^{-2}	436.83	515.53
6	0.23063×10^{-3}	343.05	378.92
7	0.71811×10^{-5}	354.32	399.62
8	0.70523×10^{-5}	354.60	400.24
9	0.70523×10^{-5}	354.61	400.23
Standard Deviation		0.642	3.43

Table 16.22 Benzene Dehydrogenation: Experimental Data and Model Calculated Values

Reciprocal Space Velocity $\times10^4$	x_1 (data)	x_1 (model)	x_2 (data)	x_2 (model)
5.63	0.828	0.82833	0.0737	0.0738
11.32	0.704	0.70541	0.1130	0.1122
16.97	0.622	0.62147	0.1322	0.1313
22.62	0.565	0.56433	0.1400	0.1407
34.00	0.499	0.49898	0.1468	0.1473
39.70	0.482	0.48110	0.1477	0.1481
45.20	0.470	0.46934	0.1477	0.1484
169.7	0.443	0.44330	0.1476	0.1477

If instead of the identity matrix, we use $\mathbf{Q}_i = diag(y_1(t_i)^{-2}, y_2(t_i)^{-2})$ as a time varying weighting matrix, we arrive at $\mathbf{k}^* = [355.55, 402.91]^T$ which is quite close to $\mathbf{k}^* = [354.61, 400.23]^T$ obtained with $\mathbf{Q}=\mathbf{I}$. This choice of the weighting matrix assumes that the error in the measured concentration is proportional to the value of the measured variable, whereas the choice of $\mathbf{Q}=\mathbf{I}$ assumes a constant standard error in the measurement. The parameter estimates are essentially the same because the measurements of x_1 and x_2 do not differ by more than one order of magnitude. In general, the computed uncertainty in the parameters is expected to have a higher dependence on the choice of $\mathbf{Q}$. The standard estimation error of k_1 almost doubled from 0.642 to 1.17, while that of k_2 increased marginally from 3.43 to 3.62.

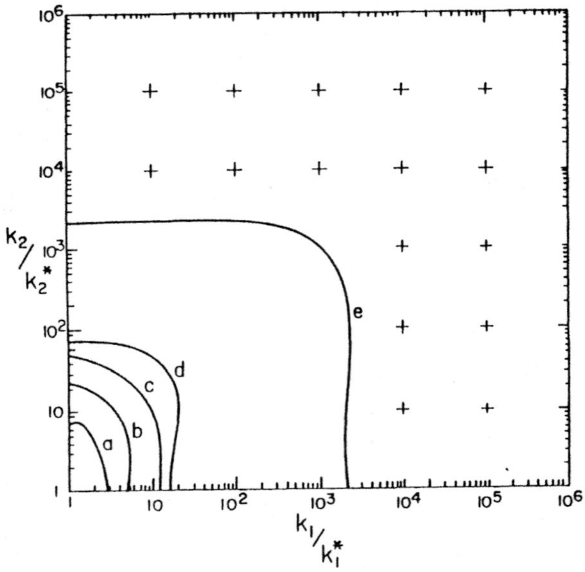

Figure 16.2 Benzene Dehydrogenation: Region of convergence of the Gauss-Newton method (a) standard Gauss-Newton method (4th order Runge-Kutta for integration); (b) G-N method with optimal step-size policy (4th order Runge-Kutta for integration); (c) G-N method with optimal step-size policy (DVERK for integration); (d) G-N method with optimal step-size policy (DGEAR for integration); (e) G-N method with optimal step-size policy and use of information index (DVERK for integration). Test points for G-N method with optimal step-size policy and use of information index (DGEAR for integration) where convergence was obtained are denoted by +. [reprinted from Industrial Engineering Chemistry Fundamentals with permission from the American Chemical Society].

The region of convergence can be substantially enlarged by implementing the simple bisection rule for step-size control or even better by implementing the optimal step-size policy described in Section 8.7.2.1. Furthermore, it is quite important to use a robust integration routine. Kalogerakis and Luus (1983) found that the use of a *stiff differential equation solver* (like IMSL routine DGEAR which uses Gear's method) resulted in a large expansion of the region of convergence compared to IMSL routine DVERK (using J.H. Verner's Runge-Kutta formulas of 5th and 6th order) or to the simple 4th order Runge-Kutta method. Although this problem is nonstiff at the optimum, this may not be the case when the initial parameter estimates far from the optimum and hence, use of a stiff ODE solver is generally beneficial. The effects of the step-size policy, use of a robust integration routine and use of the information index on the region of convergence are shown in Figure 16.2.

As seen in Figure 16.2, the effect of the information index and the integration routine on the region of is very significant. It should be noted that use of DGEAR or any other stiff differential equation solver does not require extra programming effort since an analytical expression for the jacobean $(\partial \mathbf{f}^T/\partial \mathbf{x})^T$ is also required by the Gauss-Newton method. Actually quite often this also results in savings in computer time as the parameters change from iteration to iteration and the system ODEs could become stiff and their integration requires excessive computer effort by nonstiff solvers (Kalogerakis and Luus, 1983). In all the computer programs for ODE models provided in Appendix 2, we have used DIVPAG, the latest IMSL integration routine that employs Gear's method.

16.3.3 Catalytic Hydrogenation of 3-Hydroxypropanal (HPA) to 1,3-Propanediol (PD)

The hydrogenation of 3-hydroxypropanal (HPA) to 1,3-propanediol (PD) over $Ni/SiO_2/Al_2O_3$ catalyst powder was studied by Professor Hoffman's group at the Friedrich-Alexander University in Erlagen, Germany (Zhu et al., 1997). PD is a potentially attractive monomer for polymers like polypropylene terephthalate. They used a batch stirred autoclave. The experimental data were kindly provided by Professor Hoffman and consist of measurements of the concentration of HPA and PD (C_{HPA}, C_{PD}) versus time at various operating temperatures and pressures.

The same group also proposed a reaction scheme and a mathematical model that describe the rates of HPA consumption, PD formation as well as the formation of acrolein (Ac). The model is as follows

$$\frac{dC_{HPA}}{dt} = -[r_1 + r_2]C_k - [r_3 + r_4 - r_{-3}] \qquad (16.29a)$$

$$\frac{dC_{PD}}{dt} = [r_1 - r_2]C_k \tag{16.29b}$$

$$\frac{dC_{Ac}}{dt} = r_3 - r_4 - r_{-3} \tag{16.29c}$$

where C_k is the concentration of the catalyst (10 *g/L*). The initial conditions in all experiments were $C_{HPA}(0)$=1.34953, $C_{PD}(0)$ = 0 and $C_{Ac}(0)$=0.

The kinetic expressions for the reaction rates are given next,

$$r_1 = \frac{k_1 P C_{HPA}}{H\left[1 + \left(\frac{K_1 P}{H}\right)^{0.5} + K_2 C_{HPA}\right]^3} \tag{16.30a}$$

$$r_2 = \frac{k_2 C_{PD} C_{HPA}}{1 + \left(\frac{K_1 P}{H}\right)^{0.5} + K_2 C_{HPA}} \tag{16.30b}$$

$$r_3 = k_3 C_{HPA} \tag{16.30c}$$

$$r_{-3} = k_{-3} C_{Ac} \tag{16.30d}$$

$$r_4 = k_4 C_{Ac} C_{HPA} \tag{16.30e}$$

In the above equations, k_j (j=1, 2, 3, -3, 4) are rate constants *(L/(mol min g))*, K_1 and K_2 are the adsorption equilibrium constants *(L/mol)* for H_2 and HPA respectively; P is the hydrogen pressure *(MPa)* in the reactor and H is the Henry's law constant with a value equal to 1379 *(L bar/mol)* at 298 *K*. The seven parameters (k_1, k_2, k_3, k_{-3}, k_4, K_1 and K_2) are to be determined from the measured concentrations of HPA and PD versus time. In this example, we shall consider only the data gathered at one isotherm (318 *K*) and three pressures 2.6, 4.0 and 5.15 *MPa*. The experimental data are given in Table 16.23.

In this example the number of measured variables is less than the number of state variables. Zhu et al. (1997) minimized an unweighted sum of squares of deviations of calculated and experimental concentrations of HPA and PD. They used Marquardt's modification of the Gauss-Newton method and reported the parameter estimates shown in Table 16.24.

Table 16.23 HPA Hydrogenation: Experimental Data Collected at 318 K and Pressure 2.6, 4.0 and 5.15 MPa

Pressure *(MPa)*	Time *(min)*	C_{HPA} *(mol/L)*	C_{PD} *(mol/L)*
2.6	10	1.37395	0.0
	20	1.25821	0.0197109
	30	1.18707	0.0642576
	40	1.13292	0.136399
	50	1.03556	0.238633
	60	0.961339	0.304599
	80	0.734436	0.492378
	100	0.564551	0.732326
	120	0.374385	0.887254
	140	0.214799	1.04284
	160	0.100976	1.17306
	180	0.0364192	1.25769
	200	0.00530892	1.26032
4.0	10	1.3295	0.00262812
	20	1.31157	0.0525624
	30	1.22828	0.120736
	40	1.087	0.241393
	50	0.994539	0.384888
	60	0.811825	0.4682
	80	0.600962	0.773193
	100	0.386302	0.990802
	120	0.204222	1.14954
	140	0.0782304	1.28
	160	0.0277708	1.29
	180	0.00316296	1.30
	200	0.00210864	1.30
5.15	10	1.36324	0.00262812
	20	1.25882	0.0700394
	30	1.17918	0.184363
	40	0.972102	0.354008
	50	0.825203	0.469777
	60	0.697109	0.607359
	80	0.421451	0.852431
	100	0.232296	1.03535
	120	0.128095	1.16413
	140	0.0289817	1.30053
	160	0.00962368	1.31971

Using the FORTRAN program given in Appendix 2 and starting with the values given by Zhu et al. (1997) as an initial guess, the LS objective function was computed using the identity matrix as a weighting matrix. The LS objective function was 0.26325 and the corresponding condition number of matrix **A** was 0.345×10^{17}. It should be noted that since the parameter values appear to differ by several orders of magnitude, we used the formulation with the scaled matrix **A** discussed in Section 8.1.3. Hence, the magnitude of the computed condition number of matrix **A** is solely due to ill-conditioning of the problem.

Indeed, the ill-conditioning of this problem is quite severe. Using program Bayes_ODE3 and using as initial guess the parameter values reported by Zhu et al. (1997) we were unable to converge. At this point it should be emphasized that a tight test of convergence should be used (NSIG=5 in Equation 4.11), otherwise it may appear that the algorithm has converged. In this test Marquardt's parameter was zero and no prior information was used (i.e., a zero was entered into the computer program for the inverse of the variance of the prior parameter estimates).

In problems like this one which are very difficult to converge, we should use Marquardt's modification first to reduce the LS objective function as much as possible. Then we can approach closer to the global minimum in a *sequential way by letting only one or two parameters to vary at a time*. The estimated standard errors for the parameters provide excellent information on what the next step should be. For example if we use as an initial guess $k_i=10^{-3}$ for all the parameters and a constant Marquardt's parameter $\gamma=10^{-4}$, the Gauss-Newton iterates lead to **k**=[2.6866, 0.108×10^{-6}, 0.672×10^{-3}, 0.68×10^{-5}, 0.0273, 35.56, 2.57]^T. This corresponds to a significant reduction in the LS Objective function from 40.654 to 0.30452. Subsequently, using the last estimates as an initial guess and using a smaller value for Marquardt's directional parameter (any value in the range 10^{-5} to 10^{-14} yielded similar results) we arrive at **k**=[2.6866 , 0.236×10^{-8}, 0.672×10^{-3}, 0.126×10^{-5}, 0.0273, 35.56, 2.57]^T. The corresponding reduction in the LS objective function was rather marginal from 0.30452 to 0.30447. Any further attempt to get closer to the global minimum using Marquardt's modification was unsuccessful. The values for Marquardt's parameter were varied from a value equal to the smallest eigenvalue all the way up to the third largest eigenvalue.

At this point we switched to our *sequential approach*. First we examine the estimated standard errors in the parameters obtained using Marquardt's modification. These were 22.2, 0.37×10^{8}, 352., 0.3×10^{7}, 1820., 30.3 and 14.5 (%) for k_1, k_2, k_3, k_{-3}, k_4, K_1 and K_2 respectively. Since K_2 ,k_1, K_1 and k_3 have the smallest standard errors, these are the parameters we should try to optimize first. Thus, letting only K_2 change, the Gauss-Newton method converged in three iterations to K_2=2.5322 and the LS objective function was reduced from 0.30447 to 0.27807. In these runs, Marquardt's parameter was set to zero. Next we let k_1 to vary. Gauss-Newton method converged to k_1= 2.7397 in three iterations and the LS objective function was further reduced from 0.27807 to 0.26938. Next we optimize K_2 and k_1 together. This step yields K_2=2.7119 and k_1=3.0436 with a corresponding LS objective function of 0.26753. Next we optimize three parameters

K_2, k_1 and K_1. Using as Marquardt's parameter γ=0.5 (since the smallest eigenvalue was 0.362), the Gauss-Newton method yielded the following parameter values K_2=172.25, k_1=13.354 and K_1=4.5435 with a corresponding reduction of the LS objective function from 0.26753 to 0.24357. The corresponding standard errors of estimate were 9.3, 3.1 and 3.6 (%). Next starting from our current best parameter values for K_2, k_1 and K_1, we include another parameter in our search. Based on our earlier calculations this should be k_3. Optimizing simultaneously K_2, k_1, K_1 and k_3, we obtain K_2=191.30, k_1=13.502, K_1=4.3531 and k_3=$.3922\times10^{-3}$ reducing further the LS objective function to 0.21610. At this point we can try all seven parameters to see whether we can reduce the objective function any further. Using Marquardt's modification with γ taking values from 0 all the way up to 0.5, there was no further improvement in the performance index. Similarly, there was no improvement by optimizing five or six parameters. Even when four parameters were tried with k_4 instead of k_3, there was no further improvement. Thus, we conclude that we have reached the global minimum. All the above steps are summarized in Table 16.24. The fact that we have reached the global minimum was verified by starting from widely different initial guesses. For example, starting with the parameter values reported by Zhu et al. (1997) as initial guesses, we arrive at the same value of the LS objective function.

At this point is worthwhile commenting on the computer standard estimation errors of the parameters also shown in Table 16.24. As seen in the last four estimation runs we are at the minimum of the LS objective function. The parameter estimates in the run where we optimized four only parameters (K_2, k_1, K_1 & k_3) have the smallest standard error of estimate. This is due to the fact that in the computation of the standard errors, it is assumed that all other parameters are known precisely. In all subsequent runs by introducing additional parameters the overall uncertainty increases and as a result the standard error of all the parameters increases too.

Finally, in Figures 16.3a, 16.3b and 16.3c we present the experimental data in graphical form as well as the model calculations based on the parameter values reported by Zhu et al. (1997) and from the parameter estimates determined here, namely, $\mathbf{k}^*=[13.502,\ 0.236\times10^{-8},\ 0.3922\times10^{-3},\ 0.126\times10^{-5},\ 0.0273,\ 4.3531,\ 191.30]^T$. As seen, the difference between the two model calculations is very small and all the gains realized in the LS objective function (from 0.26325 to 0.21610) produce a slightly better match of the HPA and PD transients at 5.15 *MPa*.

Zhu et al. (1997) have also reported data at two other temperatures, 333 *K* and 353 *K*. The determination of the parameters at these temperatures is left as an exercise for the reader (see Section 16.4.3 for details).

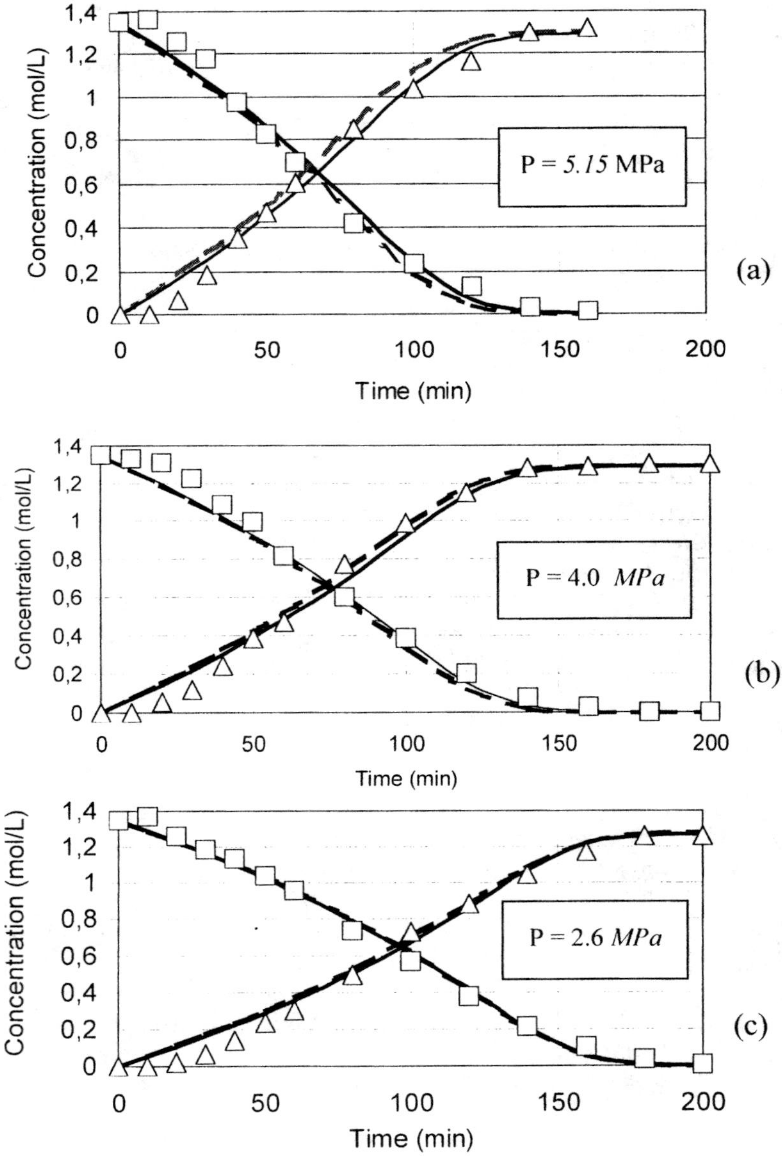

Figure 16.3 HPA Hydrogenation: Experimental data at 318 K of HPA (□) and PD (Δ) concentrations compared to model calculated values (- - - - parameters by Zhu et al., 1997; —— parameters from this work) at (a) 5.15 MPa, (b) 4.0 MPa and (c) 2.6 MPa.

Table 16.24 HPA Hydrogenation: Systematic Estimation of Parameter Values Using the Data Collected at 318 K

	k_1	k_2	k_3	k_{-3}	k_4	K_1	K_2	LS Object. Function	*cond*(**A**)
Initial Guess	10^{-3}	10^{-3}	10^{-3}	10^{-3}	10^{-3}	10^{-3}	10^{-3}	40.654	
G-N Marquardt, $\gamma=10^{-4}$	2.6866	0.108×10^{-6}	0.672×10^{-3}	0.68×10^{-5}	0.0273	35.56	2.57	0.30452	
G-N Marquardt, $\gamma=10^{-10}$	2.6866 ±22.2%	0.236×10^{-8} $\pm0.37\times10^{8}\%$	0.672×10^{-3} ±352%	0.126×10^{-5} $\pm0.3\times10^{7}\%$	0.0273 ±1820%	35.56 ±30.3%	2.57 ±14.5%	0.30447	
Optimizing K_2	*	*	*	*	*	*	2.5322	0.27807	
Optimizing k_1	2.7397	*	*	*	*	*	*	0.26938	
Optimizing K_2 & k_1	3.0436	*	*	*	*	*	2.7119	0.26753	
Optimizing K_2, k_1 & K_1	13.354 ±3.1%	*	*	*	*	4.5435 ±3.6%	172.25 ±9.3%	0.24357	
Optimizing K_2, k_1, K_1 & k_3	13.502 ±8.8%	*	0.3922×10^{-3} ±135%	*	*	4.3531 ±9.8%	191.30 ±8.5%	0.21610	0.62×10^{5}
Optimizing all except k_2,k_{-3}	13.502 ±9.5%	*	0.3922×10^{-3} ±226%	*	0.0273 ±1452%	4.3531 ±10.9%	191.30 ±8.8%	0.21610	0.11×10^{13}
Optimizing all except k_2	13.502 ±21.1%	*	0.3922×10^{-3} ±445%	0.126×10^{-5} $\pm0.6\times10^{6}\%$	0.0273 ±2535%	4.3531 ±12.7%	191.30 ±18.0%	0.21610	0.12×10^{17}
Optimizing all 7 parameters	13.502 ±21.1%	0.236×10^{-8} $\pm0.6\times10^{8}\%$	0.3922×10^{-3} ±547%	0.126×10^{-5} $\pm0.4\times10^{7}\%$	0.0273 ±2857%	4.3531 ±12.7%	191.30 ±19.4%	0.21610	0.12×10^{17}
Parameters reported by Zhu et al. (1997)	6.533 ±0.045	3.048×10^{-4} $\pm1.07\times10^{-4}$	6.233×10^{-6}	7.219 ± 10^{-4}	3.902×10^{-6}	95.00 ±1.28	3.227 ±0.033	0.26325	0.34×10^{17}

* *Parameters with a star are assumed to be known and have a constant value equal to the one determined in the previous run.*

16.3.4 Gas Hydrate Formation Kinetics

Gas hydrates are non-stoichiometric crystals formed by the enclosure of molecules like methane, carbon dioxide and hydrogen sulfide inside cages formed by hydrogen-bonded water molecules. There are more than 100 compounds (guests) that can combine with water (host) and form hydrates. Formation of gas hydrates is a problem in oil and gas operations because it causes plugging of the pipelines and other facilities. On the other hand natural methane hydrate exists in vast quantities in the earth's crust and is regarded as a future energy resource.

A mechanistic model for the kinetics of gas hydrate formation was proposed by Englezos et al. (1987). The model contains one adjustable parameter for each gas hydrate forming substance. The parameters for methane and ethane were determined from experimental data in a semi-batch agitated gas-liquid vessel. During a typical experiment in such a vessel one monitors the rate of methane or ethane gas consumption, the temperature and the pressure. Gas hydrate formation is a crystallization process but the fact that it occurs from a gas-liquid system under pressure makes it difficult to measure and monitor in situ the particle size and particle size distribution as well as the concentration of the methane or ethane in the water phase.

The experiments were conducted at four different temperatures for each gas. At each temperature experiments were performed at different pressures. A total of 14 and 11 experiments were performed for methane and ethane respectively. Based on crystallization theory, and the two film theory for gas-liquid mass transfer Englezos et al. (1987) formulated five differential equations to describe the kinetics of hydrate formation in the vessel and the associate mass transfer rates. The governing ODEs are given next.

$$\frac{dn}{dt} = \left(\frac{D^{*}\gamma A_{(g-1)}}{y_1}\right)\left(\frac{[f_g - f_{eq}]\cosh\gamma - [f_b - f_{eq}]}{\sinh\gamma}\right) \quad ; \quad n(t_0) = n_0 \tag{16.31a}$$

$$\frac{df_b}{dt} = \frac{HD^{*}\gamma a}{c_{w0}\,y_1\sinh\gamma}\Big[[f_g - f_{eq}] - [f_b - f_{eq}]\cosh\gamma\Big] - \frac{4\pi\pi^{*}\mu_2 H[f_b - f_{eq}]}{c_{w0}} \quad ; \quad f_b(t_0) = f_{eq} \tag{16.31b}$$

$$\frac{d\mu_o}{dt} = \alpha_2\mu_2 \quad ; \quad \mu_o(0) = \mu_o^0 \tag{16.31c}$$

$$\frac{d\mu_1}{dt} = G\mu_0 \quad ; \quad \mu_1(0) = \mu_1^0 \tag{16.31d}$$

$$\frac{d\mu_2}{dt} = 2G\mu_1 \quad ; \quad \mu_2(0) = \mu_2^0 \tag{16.31e}$$

The first equation gives the rate of gas consumption as moles of gas (n) versus time. This is the only state variable that is measured. The initial number of moles, n0 is known. The intrinsic rate constant, K* is the only unknown model parameter and it enters the first model equation through the Hatta number γ. The Hatta number is given by the following equation

$$\gamma = y_L \sqrt{4\pi K^* \mu_2 H / D^*} \tag{16.32}$$

The other state variables are the fugacity of dissolved methane in the bulk of the liquid water phase (f_b) and the zero, first and second moment of the particle size distribution (μ_0, μ_1, μ_2). The initial value for the fugacity, f_b^0 is equal to the three phase equilibrium fugacity f_{eq}. The initial number of particles, μ_o^0, or nuclei initially formed was calculated from a mass balance of the amount of gas consumed at the turbidity point. The explanation of the other variables and parameters as well as the initial conditions are described in detail in the reference. The equations are given to illustrate the nature of this parameter estimation problem with five ODEs, one kinetic parameter (K*) and only one measured state variable.

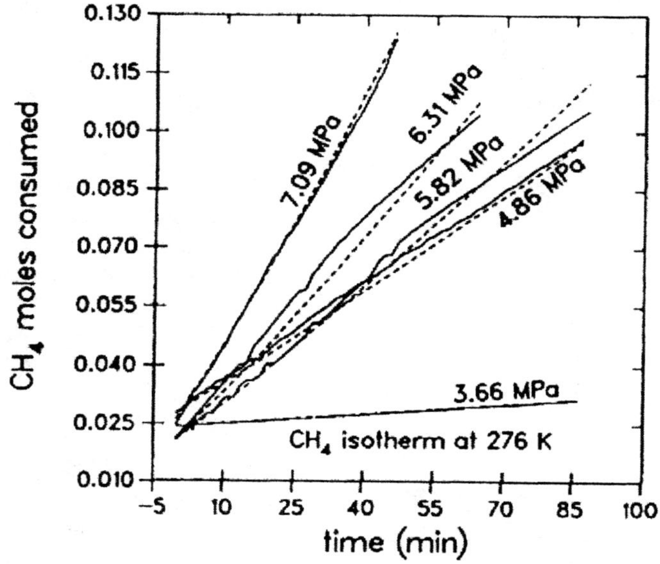

Figure 16.4: Experimental (——) and fitted (- - - - -) curves for the methane hydrate formation at 279 K [reprinted from Chemical Engineering Science with permission from Elsevier Science].

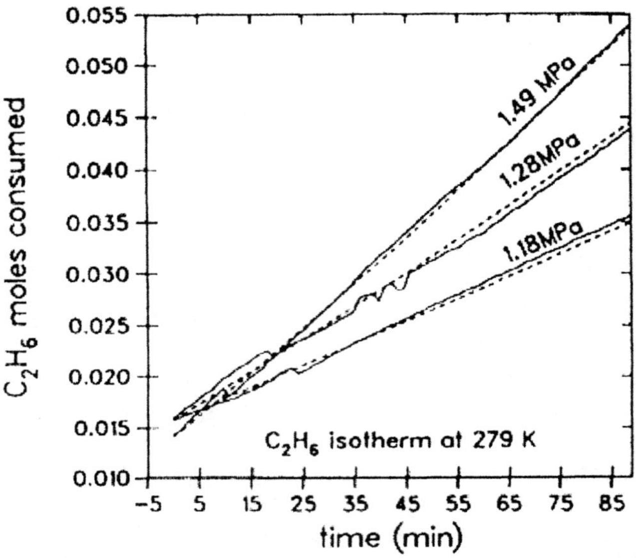

Figure 16.5: Experimental (——) and fitted (- - - - -) curves for the ethane hydrate formation at 279 K [reprinted from Chemical Engineering Science with permission from Elsevier Science].

The Gauss-Newton method with an optimal step-size policy and Marquardt's modification to ensure rapid convergence was used to match the calculated gas consumption curve with the measured one. The five state and five sensitivity equations were integrated using DGEAR (an IMSL routine for integration of stiff differential equations). Initially, parameter estimation was performed for each experiment but it was found that for each isotherm the pressure dependence was not statistically significant. Consequently, each isothermal set of data for each gas was treated simultaneously to obtain the optimal parameter value. The experimental data and the corresponding model calculations for methane and ethane gas hydrate formation are shown in Figures 16.4 and 16.5.

16.4 PROBLEMS WITH ODE MODELS

The following problems were formulated with data from the literature. Although the information provided here is sufficient to solve the parameter estimation problem, the reader is strongly recommended to see the papers in order to fully comprehend the relevant physical and chemical phenomena.

16.4.1 Toluene Hydrogenation

Consider the following reaction scheme

$$A \underset{r_{-1}}{\overset{r_1}{\longleftrightarrow}} B \overset{r_2}{\longrightarrow} C$$

where A is toluene, B is 1-methyl-cyclohexane, C is methyl-cyclohexane, r_1 is the hydrogenation rate (forward reaction) and r_{-1} the disproportionation rate (backward reaction). Data are available from Belohlav et al. (1997).

The proposed kinetic model describing the above system is given next (Belohlav et al. 1997):

$$\frac{dC_A}{dt} = -r_1 + r_{-1} \quad ; \quad C_A(0) = 1 \tag{16.33a}$$

$$\frac{dC_B}{dt} = r_1 - r_{-1} - r_2 \quad ; \quad C_B(0) = 0 \tag{16.33b}$$

$$\frac{dC_C}{dt} = r_2 \quad ; \quad C_C(0) = 0 \tag{16.33c}$$

The rate equations are as follows

$$r_1 = \frac{k_H K_A^{rel} C_A}{K_A^{rel} C_A + C_B + K_C^{rel} C_C} \tag{16.34a}$$

$$r_{-1} = \frac{k_D K_B^{rel} C_B}{K_A^{rel} C_A + C_B + K_C^{rel} C_C} \tag{16.34b}$$

$$r_2 = \frac{k_2 K_B^{rel} C_B}{K_A^{rel} C_A + C_B + K_C^{rel} C_C} \tag{16.34c}$$

where C_i (i=A, B, C) are the reactant concentrations and K^{rel} the relative adsorption coefficients.

The hydrogenation of toluene was performed at ambient temperature and pressure in a semi-batch isothermal stirred reactor with commercial 5% Ru-act catalyst. Hydrogen was automatically added to the system at the same rate at which it was consumed. Particle size of the catalyst used and efficiency of stirring were sufficient for carrying out the reaction in the kinetic regime. Under the experimental conditions validity of Henry's law was assumed. The data are given below in Table 16.25.

You are asked to use the Gauss-Newton method and determine the parameters k_H, k_D, k_2, $K_{A\text{-rel}}$, and $K_{C\text{-rel}}$ as well as their 95% confidence intervals.

For comparison purposes, it is noted that Belohlav et al. (1997) reported the following parameter estimates: k_H=0.023 min^{-1}, k_D=0.005 min^{-1}, k_2=0.011 min^{-1}, $K_{A\text{-rel}}$=1.9, and $K_{C\text{-rel}}$=1.8.

Table 16.25 Data for the Hydrogenation of Toluene

τ (min)	C_A	C_B	C_C
0	1.000	0.000	0.000
15	0.695	0.312	0.001
30	0.492	0.430	0.080
45	0.276	0.575	0.151
60	0.225	0.570	0.195
75	0.163	0.575	0.224
90	0.134	0.533	0.330
120	0.064	0.462	0.471
180	0.056	0.362	0.580
240	0.041	0.211	0.747
320	0.031	0.146	0.822
360	0.022	0.080	0.898
380	0.021	0.070	0.909
400	0.019	0.073	0.908

Source: Belohlav et al. (1997).

16.4.2 Methylester Hydrogenation

Consider the following reaction scheme

$$A \xrightarrow{r_1} B \xrightarrow{r_2} C \xrightarrow{r_3} D$$

where A, B, C and D are the methyl esters of linolenic, linoleic, oleic and stearic acids and r_1, r_2 and r_3 are the hydrogenation rates. The proposed kinetic model describing the above system is given next (Belohlav et al. 1997).

$$\frac{dC_A}{dt} = -r_1 \quad ; \quad C_A(0) = 0.101 \tag{16.35a}$$

$$\frac{dC_B}{dt} = r_1 - r_2 \quad ; \quad C_B(0) = 0.221 \tag{16.35b}$$

$$\frac{dC_C}{dt} = r_2 - r_3 \quad ; \quad C_C(0) = 0.657 \tag{16.35c}$$

$$\frac{dC_D}{dt} = r_3 \quad ; \quad C_D(0) = 0.0208 \tag{16.35d}$$

The rate equations are as follows

$$r_1 = \frac{k_1 K_A^{rel} C_A}{C_A + K_B^{rel} C_B + K_C^{rel} C_C + K_D^{rel} C_D} \tag{16.36a}$$

$$r_2 = \frac{k_2 K_B^{rel} C_B}{C_A + K_B^{rel} C_B + K_C^{rel} C_C + K_D^{rel} C_D} \tag{16.36b}$$

$$r_3 = \frac{k_3 K_C^{rel} C_C}{C_A + K_B^{rel} C_B + K_C^{rel} C_C + K_D^{rel} C_D} \tag{16.36c}$$

Table 16.26 Data for the Hydrogenation of Methylesters

τ (min)	C_A	C_B	C_C	C_D
0	0.1012	0.2210	0.6570	0.0208
10	0.0150	0.1064	0.6941	0.1977
14	0.0044	0.0488	0.6386	0.3058
19	0.0028	0.0242	0.5361	0.4444
24	0.0029	0.0015	0.3956	0.6055
34	0.0017	0.0005	0.2188	0.7808
69	0.0003	0.0004	0.0299	0.9680
124	0.0001	0.0002	0.0001	0.9982

Source: Belohlav et al. (1997).

where C_i (i=A, B, C, D) are the reactant concentrations and K_{rel} the relative adsorption coefficients.

The experiments were performed in an autoclave at elevated pressure and temperature. The Ni catalyst DM2 was used. The data are given below in Table 16.26 (Belohlav et al., 1997).

The objective is to determine the parameters k_1, k_2, k_3, $K_{A\text{-}rel}$, $K_{B\text{-}rel}$, $K_{C\text{-}rel}$, $K_{D\text{-}rel}$ as well as their standard errors. It is noted that $K_{A\text{-}rel}=1$. Belohlav et al. (1997) reported the following parameter estimates: $k_1=1.44\ min^{-1}$, $k_2=0.03\ min^{-1}$, $k_3=0.09\ min^{-1}$, $K_{B\text{-}rel}=28.0$, $K_{C\text{-}rel}=1.8$ and $K_{D\text{-}rel}=2.9$.

16.4.3 Catalytic Hydrogenation of 3-Hydroxypropanal (HPA) to 1,3-Propanediol (PD) - Nonisothermal Data

Let us reconsider the hydrogenation of 3-hydroxypropanal (HPA) to 1,3-propanediol (PD) over $Ni/SiO_2/Al_2O_3$ catalyst powder that used as an example earlier. For the same mathematical model of the system you are asked to regress simultaneously the data provided in Table 16.23 as well as the additional data given here in Table 16.28 for experiments performed at 60°*C* (333 *K*) and 80°*C* (353 *K*). Obviously an Arrhenius type relationship must be used in this case. Zhu et al. (1997) reported parameters for the above conditions and they are shown in Table 16.28.

Table 16.27 HPA Hydrogenation: Estimated Parameter Values from the Data Collected at 333 K and 353 K

	Parameter Estimates ± Standard Error	
	T = 333 *K*	T = 353 *K*
k_1	19.54 ±0.102	52.84 ±0.401
k_2	1.537×10^{-3} $\pm0.27\times10^{-3}$	1.475 $\pm0.12\times10^{-2}$
k_3	3.058×10^{-4} $\pm0.75\times10^{-5}$	2.066×10^{-3} $\pm0.036\times10^{-3}$
k_{-3}	1.788×10^{-3} $\pm0.120\times10^{-3}$	5.315×10^{-3} $\pm0.025\times10^{-3}$
k_4	7.515×10^{-5} $\pm0.90\times10^{-5}$	2.624×10^{-3} $\pm0.023\times10^{-3}$
K_1	120.0 ±1.05	160.0 ±1.73
K_2	3.029 ±0.013	2.767 ±0.021

Source: Zhu et al. (1997).

Table 16.28 HPA Hydrogenation: Experimental Data Collected at 333 K and 353 K

Experimental Conditions	Time (*min*)	C_{HPA} (*mol/L*)	C_{PD} (*mol/L*)
5.15 *MPa* & 333 *K*	0.0	1.34953	0.0
	10	1.0854	0.196452
	20	0.663043	0.602365
	30	0.3165541	0.851117
	40	0.00982225	1.16938
	50	0.00515874	1.24704
	60	0.0020635	1.24836
5.15 *MPa* & 353 *K*	0.0	1.34953	0.0
	5	0.873513	0.388568
	10	0.44727	0.816032
	15	0.140925	0.967017
	20	0.0350076	1.05125
	25	0.0130859	1.08239
	30	0.00581597	1.12024
4.0 *MPa* & 333 *K*	0.0	1.34953	0.0
	10	1.13876	0.151380
	20	0.785521	0.558344
	30	0.448402	0.867148
	40	0.191058	1.12536
	50	0.0530074	1.29
	60	0.0199173	1.32

Source: Zhu et al. (1997).

17

Parameter Estimation in Biochemical Engineering Models

A number of examples from biochemical engineering are presented in this chapter. The mathematical models are either algebraic or differential and they cover a wide area of topics. These models are often employed in biochemical engineering for the development of bioreactor models for the production of bio-pharmaceuticals or in the environmental engineering field. In this chapter we have also included an example dealing with the determination of the average specific production rate from batch and continuous runs.

17.1 ALGEBRAIC EQUATION MODELS

17.1.1 Biological Oxygen Demand

Data on biological oxygen demand versus time are modeled by the following equation

$$y = k_1 [1 - exp(-k_2 t)] \tag{17.1}$$

where k_1 is the ultimate carbonaceous oxygen demand (*mg/L*) and k_2 is the BOD reaction rate constant (d^{-1}). A set of BOD data were obtained by 3rd year Environmental Engineering students at the Technical University of Crete and are given in Table 4.2.

Although this is a dynamic experiment where data are collected over time, it is considered as a simple algebraic equation model with two unknown parameters.

Using an initial guess of k_1=350 and k_2=1 the Gauss-Newton method converged in five iterations without the need for Marquardt's modification. The estimated parameters are k_1= 334.27±2.10% and k_2=0.38075±5.78%. The model-calculated values are compared with the experimental data in Table 17.1. As seen the agreement is very good in this case.

The quadratic convergence of the Gauss-Newton method is shown in Table 17.2 where the reduction of the LS objective function is shown for an initial guess of k_1=100 and k_2=0.1.

Table 17.1 BOD Data: Experimental Data and Model Calculated Values

Time (*d*)	BOD - Experimental Data (*mg/L*)	BOD - Model Calculations (*mg/L*)
1	110	105.8
2	180	178.2
3	230	227.6
4	260	261.4
5	280	284.5
6	290	300.2
7	310	311.0
8	330	318.4

Table 17.2 BOD Data: Reduction of the LS Objective Function

Iteration	Objective function	k_1	k_2
0	390384	100	0.1
1	380140	70.78	0.1897
2	19017.3	249.5	1.169
3	12160.3	271.9	0.4454
4	475.7	331.8	0.3686
5	289.0	334.3	0.3803
6	288.9	334.3	0.3807

17.1.2 Enzyme Kinetics

Let us consider the determination of two parameters, the maximum reaction rate (r_{max}) and the saturation constant (K_m) in an enzyme-catalyzed reaction following Michaelis-Menten kinetics. The Michaelis-Menten kinetic rate equation relates the reaction rate (r) to the substrate concentrations (S) by

$$r = \frac{r_{max} S}{K_m + S} \tag{17.2}$$

The parameters are usually obtained from a series of initial rate experiments performed at various substrate concentrations. Data for the hydrolysis of benzoyl-L-tyrosine ethyl ester (BTEE) by trypsin at 30°C and pH 7.5 are given in Table 17.3

Table 17.3 Enzyme Kinetics: Experimental Data for the *hydrolysis of benzoyl-L-tyrosine ethyl ester (BTEE) by trypsin at 30 °C and pH 7.5*

S *(μM)*	20	15	10	5.0	2.5
r *(μM/min)*	330	300	260	220	110

Source: Blanch and Clark (1996).

As we have discussed in Chapter 8, this is a typical transformably linear system. Using the well-known Lineweaver-Burk transformation, Equation 17.2 becomes

$$\left(\frac{1}{r}\right) = \frac{1}{r_{max}} + \frac{K_m}{r_{max}}\left(\frac{1}{S}\right) \tag{17.3}$$

The results from the simple linear regression of (1/r) versus (1/S) are shown graphically in Figure 17.1.

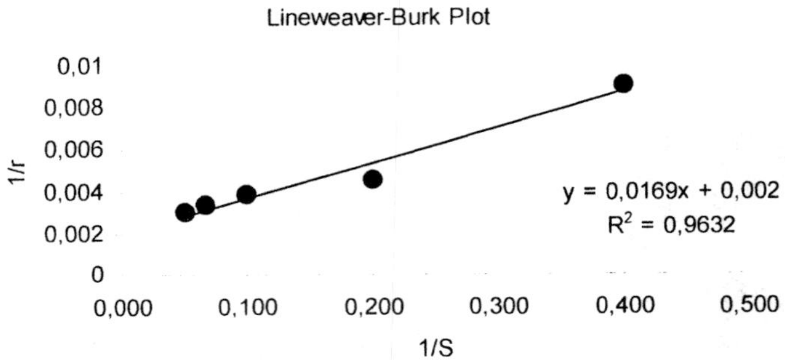

Figure 17.1 Enzyme Kinetics: Results from the Lineweaver-Burk transformation.

If instead we use the Eadie-Hofstee transformation, Equation 17.2 becomes

$$r = r_{max} - K_m\left(\frac{r}{S}\right) \tag{17.4}$$

The results from the simple linear regression of (r) versus (r/S) are shown graphically in Figure 17.2.

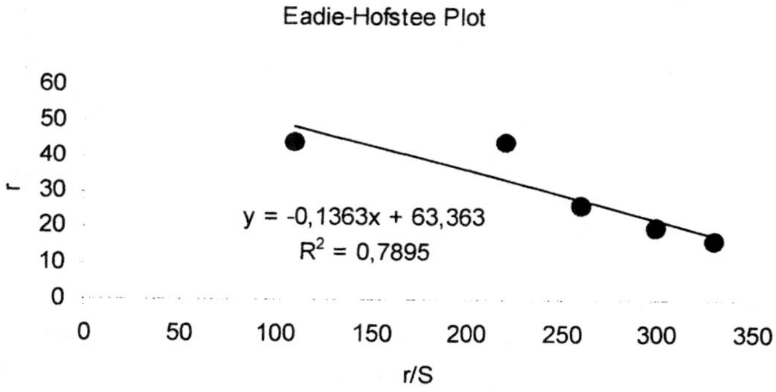

Figure 17.2 *Enzyme Kinetics: Results from the Eadie-Hofstee transformation.*

Similarly we can use the Hanes transformation whereby Equation 17.2 becomes

$$\left(\frac{S}{r}\right) = \frac{1}{r_{max}}S + \frac{K_m}{r_{max}} \tag{17.5}$$

The results from the simple linear regression of (S/r) versus (S) are shown graphically in Figure 17.3. The original parameters r_{max} and K_m can be obtained from the slopes and intercepts determined by the simple linear least squares for all three cases and are shown in Table 17.4. As seen there is significant variation in the estimates and in general they are not very reliable. Nonetheless, these estimates provide excellent initial guesses for the Gauss-Newton method.

Table 17.4 Enzyme Kinetics: Estimated Parameter Values

Estimation Method	k_1	k_2
Lineweaver-Burk plot	500	8.45
Eadie-Hofstee plot	63.36	0.136
Hanes plot	434.8	6.35
Nonlinear Regression	420.2	5.705

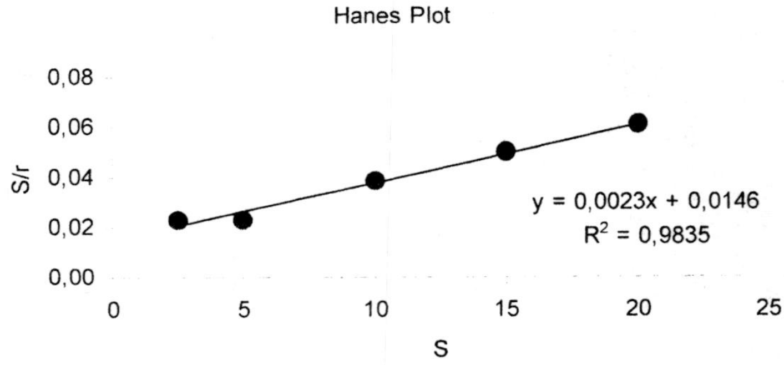

Figure 17.3 *Enzyme Kinetics: Results from the Hanes transformation.*

Indeed, using the Gauss-Newton method with an initial estimate of $\mathbf{k}^{(0)}$=(450, 7) convergence to the optimum was achieved in three iterations with no need to employ Marquardt's modification. The optimal parameter estimates are k_1= 420.2± 8.68% and k_2= 5.705±24.58%. It should be noted however that this type of a model can often lead to ill-conditioned estimation problems if the data have not been collected both at low and high values of the independent variable. The convergence to the optimum is shown in Table 17.5 starting with the initial guess $\mathbf{k}^{(0)}$=(1, 1).

Table 17.5 Enzyme Kinetics: Reduction of the LS Objective Function

Iteration	Objective Function	k_1	k_2
0	324816	1	1
1	312388	370.8	740.3
2	308393	36.52	43.38
3	295058	23.50	7.696
4	196487	71.92	1.693
5	39131.7	386.7	12.65
6	3026.12	359.1	4.356
7	981.260	417.8	5.685
8	974.582	420.2	5.706
9	974.582	420.2	5.705

17.1.3 Determination of Mass Transfer Coefficient (k_La) in a Municipal Wastewater Treatment Plant (with PULSAR aerators)

The PULSAR units are high efficiency static aerators that have been developed for municipal wastewater treatment plants and have successfully been used over extended periods of time without any operational problems such as unstable operation or plugging up during intermittent operation of the air pumps (Chourdakis, 1999). Data have been collected from a pilot plant unit at the Wastewater Treatment plant of the Industrial Park (Herakleion, Crete). A series of experiments were conducted for the determination of the mass transfer coefficient (k_La) and are shown in Figure 17.4. The data are also available in tabular form as part of the parameter estimation input files provided with the enclosed CD.

In a typical experiment by the dynamic gassing-in/gassing-out method during the normal operation of the plant, the air supply is shut off and the dissolved oxygen (DO) concentration is monitored as the DO is depleted. The dissolved oxygen dynamics during the gassing-out part of the experiment are described by

$$\frac{dC_{O2}}{dt} = -q_{O2}x_v \tag{17.6}$$

where C_{O2} is the dissolved oxygen concentration, x_v is the viable cell concentration and q_{O2} is the specific oxygen uptake rate by the cells. For the very short period of this experiment we can assume that x_v is constant. In addition, when the dissolved oxygen concentration is above a critical value (about 1.5 *mg/L*), the specific oxygen uptake rate is essentially constant and hence Equation 17.6 becomes,

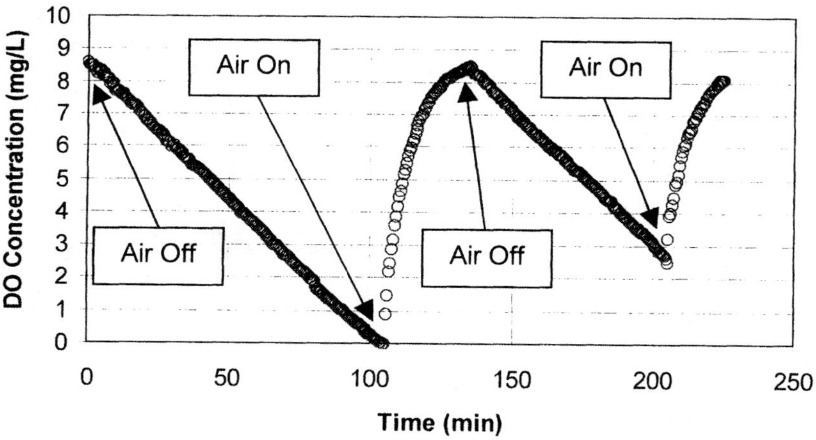

Figure 17.4: PULSAR: Measurements of Dissolved Oxygen (DO) Concentration During a Dynamic Gassing-in/Gassing-out Experiment.

$$C_{O2}(t) = C_{O2}(t_0) - q_{O2}x_v\,(t - t_0) \tag{17.7}$$

Equation 17.7 suggests that the oxygen uptake rate ($q_{O2}x_v$) can be obtained by simple linear regression of $C_{O2}(t)$ versus time. This is shown in Figure 17.5 where the oxygen uptake rate has been estimated to be 0.0813 *mg/L·min*.

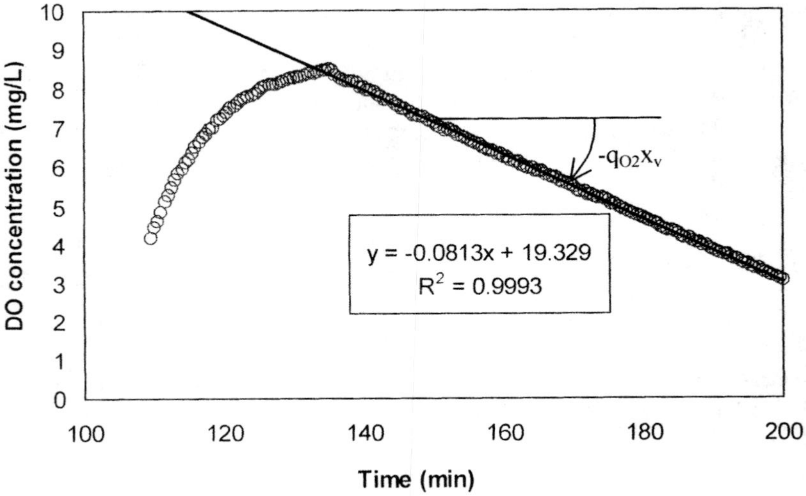

Figure 17.5: PULSAR: Determination of Oxygen Uptake Rate by Simple Linear Least Squares During the Gassing-out Period.

Subsequently, during the gassing-in part of the experiment, we can determine k_La. In this case, the dissolved oxygen dynamics are described by

$$\frac{dC_{O2}}{dt} = k_La[C^*_{O2} - C_{O2}] - q_{O2}x_v \tag{17.8}$$

where C^*_{O2} is the dissolved oxygen concentration in equilibrium with the oxygen in the air supply at the operating pressure and temperature of the system. This value can be obtained from the partial pressure of oxygen in the air supply and Henry's constant. However, Henry's constant values for wastewater are not readily available and hence, we shall consider the equilibrium concentration as an additional unknown parameter.

Analytical solution of Equation 17.8 yields

$$\frac{C_{O2}^{*} - C_{O2}(t) - \dfrac{q_{O2}x_v}{k_La}}{C_{O2}^{*} - C_{O2}(t_0) - \dfrac{q_{O2}x_v}{k_La}} = e^{-k_La[t-t_0]} \tag{17.9}$$

or equivalently,

$$C_{O2}(t) = C_{O2}^{*} - \frac{q_{O2}x_v}{k_La} - \left(C_{O2}^{*} - C_{O2}(t_0) - \frac{q_{O2}x_v}{k_La}\right)e^{-k_La[t-t_0]} \tag{17.10}$$

Equation 17.10 can now be used to obtain the two unknown parameters (k_La and C_{O2}^{*}) by fitting the data from the gassing-in period of the experiment. Indeed, using the Gauss-Newton method with an initial guess of (10, 10) convergence is achieved in 7 iterations as shown in Table 17.6. There was no need to employ Marquardt's modification. The FORTRAN program used for the above calculations is also provided in Appendix 2.

Table 17.6 PULSAR: Reduction of the LS Objective Function and Convergence to the Optimum (Two Parameters)

Iteration	Objective Function	k_La	C_{O2}^{*}
0	383.422	10	10
1	306.050	1.8771	9.8173
2	72.6579	0.7346	8.5173
3	9.01289	0.3058	8.0977
4	1.50009	0.1929	8.2668
5	0.678306	0.1540	8.5971
6	0.338945	0.1225	9.1203
7	0.0891261	0.1246	9.2033
Standard Deviation (%)		2.06	0.586

While using Equation 17.10 for the above computations, it has been assumed that $C_{O2}(t_0) = 4.25$ was known precisely from the measurement of the dissolved oxygen concentration taken at $t_0 = 207$ *min*. We can relax this assumption by treating $C_{O2}(t_0)$ as a third parameter. Indeed, by doing so a smaller value for the least squares objective function could be achieved as seen in Table 17.7 by a small adjustment in $C_{O2}(t_0)$ from 4.25 to 4.284.

Table 17.7 PULSAR: Reduction of the LS Objective Function and Convergence to the Optimum (Three Parameters)

Iteration	Objective Function	k_La	C^*_{O2}	$C_{O2}(t_0)$
0	383.422	10	10	4.25
1	194.529	0.6643	9.6291	4.250
2	9.17805	0.3087	7.6867	4.702
3	1.92088	0.1880	8.3528	4.428
4	0.944376	0.1529	8.8860	4.303
5	0.508239	0.1375	9.0923	4.285
6	0.246071	0.1299	9.1794	4.283
7	0.138029	0.1259	9.2173	4.283
8	0.101272	0.1239	9.2342	4.283
9	0.090090	0.1228	9.2420	4.283
10	0.086898	0.1223	9.2458	4.283
11	0.086018	0.1228	9.2420	4.283
12	0.085780	0.1219	9.2486	4.284
13	0.085716	0.1218	9.2490	4.284
14	0.085699	0.1218	9.2494	4.284
Standard Deviation (%)		1.45	0.637	0.667

In both cases the agreement between the experimental data and the model calculated values is very good.

17.1.4 Determination of Monoclonal Antibody Productivity in a Dialyzed Chemostat

Linardos et al. (1992) investigated the growth of an SP2/0 derived mouse-mouse hybridoma cell line and the production of an *anti-Lewis^b* IgM immunoglobulin in a dialyzed continuous suspension culture using an 1.5 *L* Celligen bioreactor. Growth medium supplemented with 1.5% serum was fed directly into the bioreactor at a dilution rate of 0.45 d^{-1}. Dialysis tubing with a molecular weight cut-off of 1000 was coiled inside the bioreactor. Fresh medium containing no serum or serum substitutes was passed through the dialysis tubing at flow rates of 2 to 5 *L/d*. The objective was to remove low molecular weight inhibitors such as lactic acid and ammonia while retaining high molecular weight components such as growth factors and antibody molecules. At the same time essential nutrients such as glucose, glutamine and other aminoacids are replenished by the same mechanism.

In the dialyzed batch start-up phase and the subsequent continuous operation a substantial increase in viable cell density and monoclonal antibody (MAb) titer was observed compared to a conventional suspension culture. The raw data, profiles of the viable cell density, viability and monoclonal antibody titer during the batch start-up and the continuous operation with a dialysis flow rate of 5 L/d are shown in Figures 17.6 and 17.7. The raw data are also available in tabular form in the corresponding input file for the FORTRAN program on data smoothing for short cut methods provided with the enclosed CD.

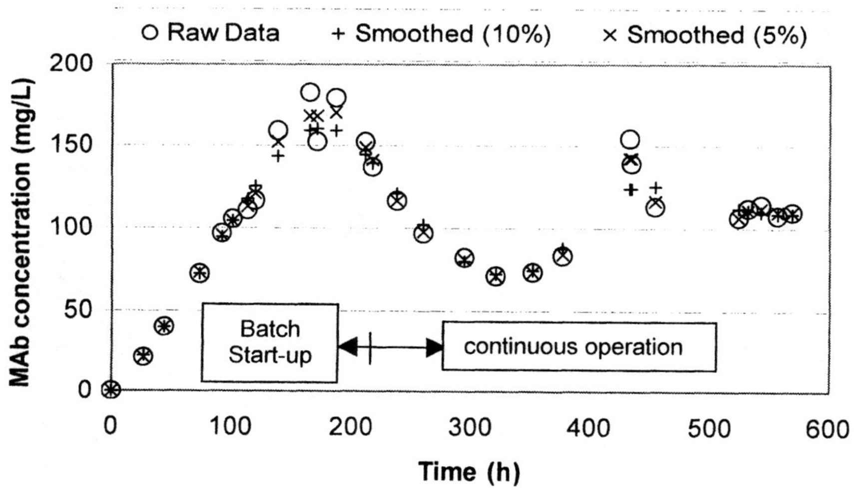

Figure 17.6: Dialyzed Chemostat: Monoclonal antibody concentration (raw and smoothed measurements) during initial batch start-up and subsequent dialyzed continuous operation with a dialysis flow rate of 5 L/d. [reprinted from the Journal of Biotechnology & Bioengineering with permission from J. Wiley].

The objective of this exercise is to use the techniques developed in Section 7.3 of this book to determine the specific monoclonal antibody production rate (q_M) during the batch start-up and the subsequent continuous operation.

<u>*Derivative Approach:*</u>

The operation of the bioreactor was in the batch mode up to time t=212 *h*. The dialysis flow rate was kept at 2 *L/d* up to time t=91.5 *h* when a sharp drop in the viability was observed. In order to increase further the viable cell density, the dialysis flow rate was increased to 4 *L/d* and at 180 *h* it was further increased to 5 *L/d* and kept at this value for the rest of the experiment.

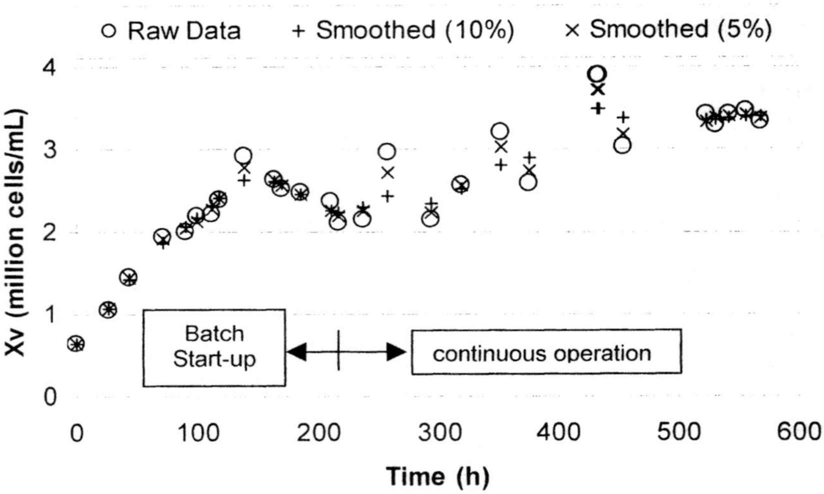

Figure 17.7: Dialyzed Chemostat: Viable cell density (raw and smoothed measurements) during initial batch start-up and subsequent dialyzed continuous operation with a dialysis flow rate of 5 L/d [reprinted from the Journal of Biotechnology & Bioengineering with permission from J. Wiley].

As described in Section 7.3.1 when the derivative method is used, the specific MAb production rate at any time t during the batch start-up period is determined by

$$q_M(t) = \frac{1}{X_v(t)} \frac{dM(t)}{dt} \tag{17.11}$$

where $X_v(t)$ and M(t) are the smoothed (filtered) values of viable cell density and monoclonal antibody titer at time t.

At time t=212 *h* the continuous feeding was initiated at 5 *L/d* corresponding to a dilution rate of 0.45 d^{-1}. Soon after continuous feeding started, a sharp increase in the viability was observed as a result of physically removing dead cells that had accumulated in the bioreactor. The viable cell density also increased as a result of the initiation of direct feeding. At time t≈550 *h* a steady state appeared to have been reached as judged by the stability of the viable cell density and viability for a period of at least 4 days. Linardos et al. (1992) used the steady state measurements to analyze the dialyzed chemostat. Our objective here is to use the techniques developed in Chapter 7 to determine the specific monoclonal antibody production rate in the period 212 to 570 h where an oscillatory behavior of the MAb titer is observed and examine whether it differs from the value computed during the start-up phase.

During the continuous operation of the bioreactor the specific MAb production rate at time t is determined by the derivative method as

$$q_M(t) = \frac{1}{X_v(t)}\left(\frac{dM(t)}{dt} + DM(t)\right) \tag{17.12}$$

In Figures 17.6 and 17.7 besides the raw data, the filtered values of MAb titer and viable cell density are also given. The smoothed values have been obtained using the IMSL routine CSSMH assuming either a 10% or a 5% standard error in the measurements and a value of 0.465 for the smoothing parameter s/N. This value corresponds to a value of -2 of the input parameter SLEVEL in the FORTRAN program provided in Appendix 2 for data smoothing. The computed derivatives from the smoothed MAb data are given in Figure 17.8. In Figure 17.9 the corresponding estimates of the specific MAb production rate (q_M) versus time are given.

Despite the differences between the estimated derivatives values, the computed profiles of the specific MAb production rate are quite similar. Upon inspection of the data, it is seen that during the batch period (up to t=212 *h*), q_M is decreasing almost monotonically. It has a mean value of about 0.5 $\mu g/(10^6\ cells \cdot h)$. Throughout the dialyzed continuous operation of the bioreactor, the average q_M is about 0.6 $\mu g/(10^6\ cells \cdot h)$ and it stays constant during the steady state around time t≈550 *h*.

In general, the computation of an instantaneous specific production rate is not particularly useful. Quite often the average production rate over specific periods of time is a more useful quantity to the experimentalist/analyst. Average rates are better determined by the integral approach which is illustrated next.

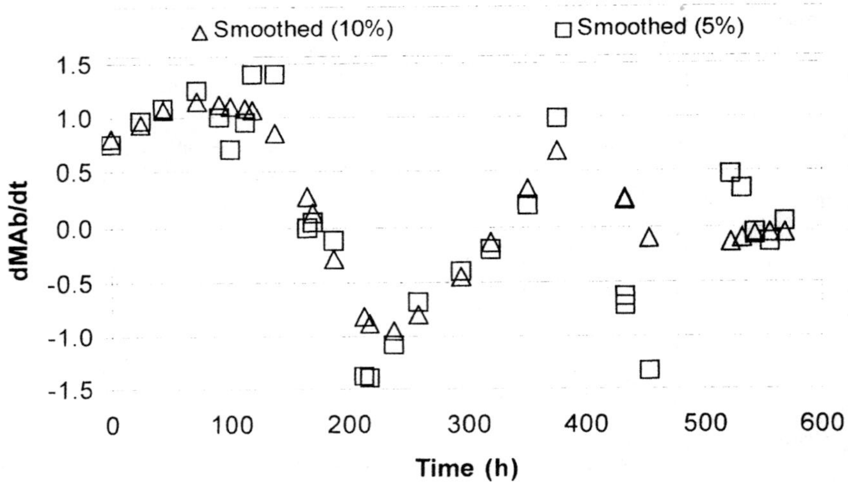

Figure 17.8: *Dialyzed Chemostat: Estimated values of dM/dt versus time during the initial batch start-up and subsequent dialyzed continuous operation with a dialysis flow rate of 5 L/d.*

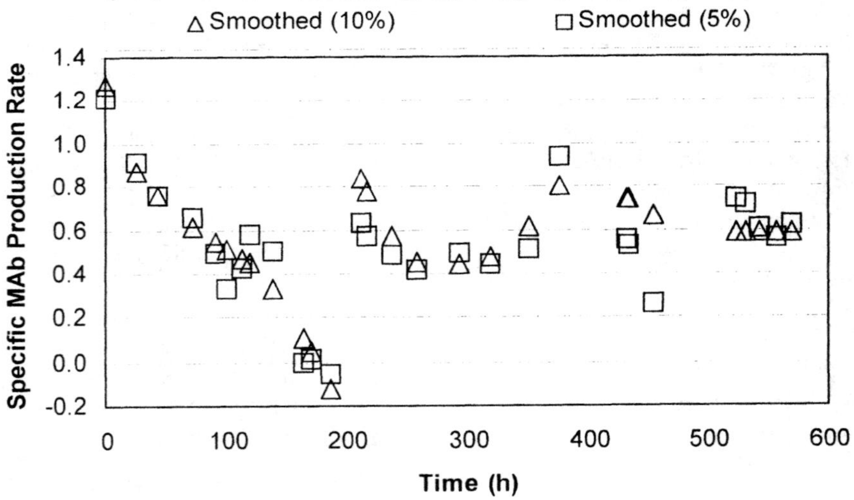

Figure 17.9: *Dialyzed Chemostat: Estimated values of specific MAb production rate versus time during the initial batch start-up and subsequent dialyzed continuous operation with a dialysis flow rate of 5 L/d.*

Integral Approach:
As described in Section 7.3.2 when the integral method is employed, the *average* specific MAb production rate during any time interval of the batch start-up period is estimated as the slope in a plot of $M(t_i)$ versus $\int_{t_0}^{t_i} X_v(t)dt$.

In order to implement the integral method, we must compute numerically the integrals $\int_{t_0}^{t_i} X_v(t)dt$, $t_i=1,\ldots,N$ where X_v is the smoothed value of the viable cell density. An efficient and robust way to perform the integration is through the use of the IMSL routine QDAGS. The necessary function calls to provide X_v at selected points in time are done by calling the IMSL routine CSVAL. Of course the latter two are used once the cubic splines coefficients and break points have been computed by CSSMH to smooth the raw data. The program that performs all the above calculations is also included in the enclosed CD. Two different values for the weighting factors (10% and 5%) have been used in the data smoothing.

In Figures 17.10 and 17.11 the plots of $M(t_i)$ versus $\int_{t_0}^{t_i} X_v(t)dt$ are shown for the batch start-up period.

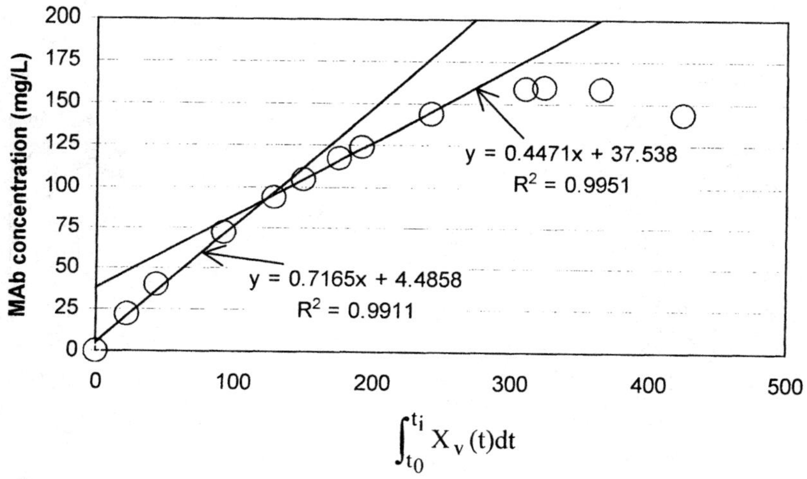

Figure 17.10: *Dialyzed Chemostat: Estimated values of specific MAb production rate versus time during the initial batch start-up period. A 10% standard error in raw data was assumed for data smoothing.*

The computation of an *average* specific production rate is particularly useful to the analyst if the instantaneous rate is approximately constant during the time segment of interest. By simple visual inspection of the plot of MAb titer versus the integral of X_v, one can readily identify segments of the data where the data points are essentially on a straight line and hence, an average rate describes the data satisfactorily. For example upon inspection of Figure 17.10 or 17.11, three periods can be identified. The first five data points (corresponding to the time period 0 to 91 *h*) yield an average q_M of 0.7165 $\mu g/(10^6\ cells\cdot h)$ when a 10% weighting in used in data smoothing or 0.7215 $\mu g/(10^6\ cells\cdot h)$ when a 5% weighting is used. The next period where a lower rate can be identified corresponds to the time period 91 to 140 *h* (i.e., 5th, 6th, ..., 9th data point). In this segment of the data the average specific MAb production rate is 0.4471 or 0.504 $\mu g/(10^6\ cells\cdot h)$ when a 10% or a 5% weighting is used respectively. The near unity values of the computed correlation coefficient (R^2>0.99) suggests that the selection of the data segment where the slope is determined by linear least squares estimation is most likely appropriate.

The third period corresponds to the last five data points where it is obvious that the assumption of a nearly constant q_M is not valid as the slope changes essentially from point to point. Such a segment can still be used and the integral method will provide an average q_M, however, it would not be representative of the behavior of the culture during this time interval. It would simply be a mathematical average of a time varying quantity.

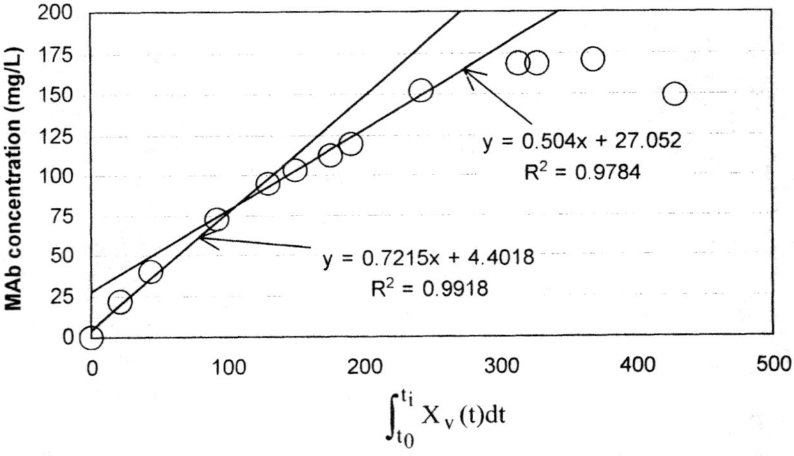

Figure 17.11: *Dialyzed Chemostat: Estimated values of specific MAb production rate versus time during the initial batch start-up period. A 5% standard error in raw data was assumed for data smoothing.*

Let us now turn our attention to the dialyzed continuous operation (212 to 570 *h*). By the integral method, the specific MAb production rate can be estimated as the slope in a plot of $\left\{M(t_i) + D\int_{t_0}^{t_i} M(t)dt\right\}$ versus $\int_{t_0}^{t_i} X_v(t)dt$.

In this case besides the integral of X_v, the integral of MAb titer must also be computed. Obviously the same FORTRAN program used for the integration of X_v can also be used for the computations (see Appendix 2). The results by the integral method are shown in Figures 17.12 and 17.13.

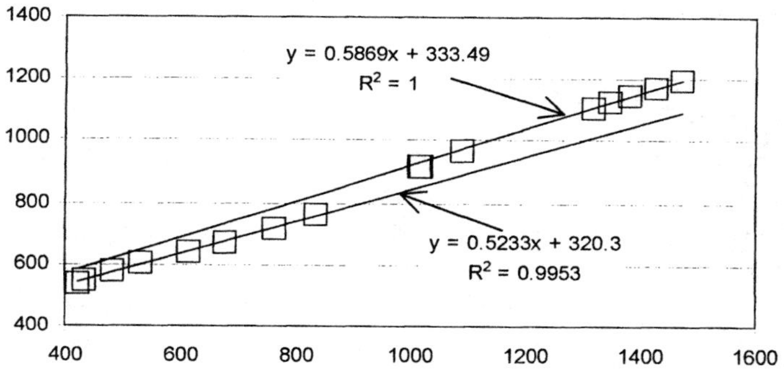

$\left\{\begin{matrix} M(t_i) + \\ D\int_{t_0}^{t_i} M(t)dt \end{matrix}\right\}$

$\int_{t_0}^{t_i} X_v(t)dt$

Figure 17.12: *Dialyzed Chemostat: Estimated values of specific MAb production rate versus time during the period of continuous operation. A 10% standard error in the raw data was assumed for data smoothing.*

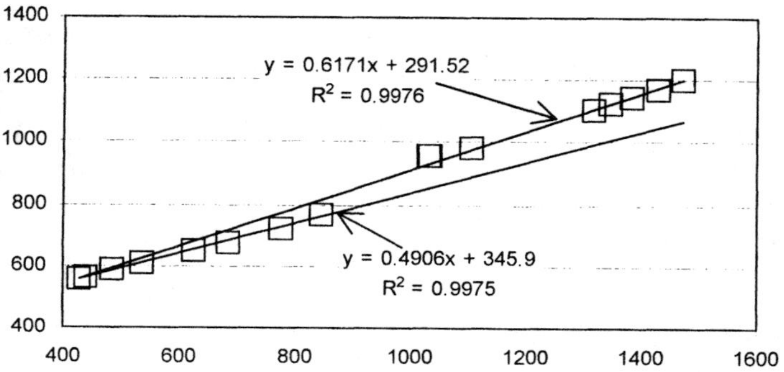

$\left\{\begin{matrix} M(t_i) + \\ D\int_{t_0}^{t_i} M(t)dt \end{matrix}\right\}$

$\int_{t_0}^{t_i} X_v(t)dt$

Figure 17.13: *Dialyzed Chemostat: Estimated values of specific MAb production rate versus time during the period of continuous operation. A 5% standard error in the raw data was assumed for data smoothing.*

In both cases, one can readily identify two segments of the data. The first one is comprised by the first 8 data points that yielded a q_M equal to 0.5233 or 0.4906 *μg/(10⁶ cells·h)* when the weighting factors for data smoothing are 10% or 5% respectively. The second segment is comprised of the last 5 data points corresponding to the steady state conditions used by Linardos et al. (1992) to estimate the specific MAb production rate. Using the integral method on this data segment q_M was estimated at 0.5869 or 0.6171 $\mu g/(10^6\ cells \cdot h)$ as seen in Figure 17.13.

This is the integral approach for the estimation of specific rates in biological systems. Generally speaking it is a simple and robust technique that allows a visual conformation of the computations by the analyst.

17.2 PROBLEMS WITH ALGEBRAIC EQUATION MODELS

17.2.1 Effect of Glucose to Glutamine Ratio on MAb Productivity in a Chemostat

At the Pharmaceutical Production Research Facility of the University of Calgary experimental data have been collected (Linardos, 1991) to investigate the effect of glucose to glutamine ratio on monoclonal antibody (*anti-Lewis*^{b} IgM) productivity in a chemostat and they are reproduced here in Tables 17.8, 17.9 and 17.10. Data are provided for a 5:1 (standard for cell culture media), 5:2 and 5:3 glucose to glutamine ratio in the feed. The dilution rate was kept constant at 0.45 d^{-1}.

For each data set you are asked to use the *integral approach* to estimate the following rates:

(a) Specific monoclonal antibody production rate
(b) Specific glucose uptake rate
(c) Specific glutamine uptake rate
(d) Specific lactate production rate, and
(e) Specific ammonia production rate

Does the analysis of data suggest that there is a significant effect of glucose to glutamine ratio on MAb productivity?

By computing the appropriate integrals with filtered data and generating the corresponding plots, you must determine first which section of the data is best suited for the estimation of the specific uptake and production rates.

In the next three tables, the following notation has been used:

t	Elapsed Time (h)
x_v	Viable Cell Density ($10^6\ cells/mL$)
v_b	Viability *(%)*
MAb	Monoclonal Antibody Concentration (*mg/L*)
Lac	Lactate (*mmol/L*)
Gls	Glucose (*mmol/L*)
Amm	Ammonia (*mmol/L*)

Glm	Glutamine (*mmol/L*)
Glt	Glutamate (*mmol/L*)
x_d	Nonviable Cell Density (10^6 *cells/mL*)

Table 17.8: Glucose/Glutamine Ratio: Experimental Data from a Chemostat Run with a Glucose to Glutamine Ratio in the Feed of 5:1

t	x_v	v_b	*MAb*	*Lac*	*Gls*	*Amm*	*Glm*	*Glt*	x_d
0.0	1.91	0.84	58.46	24.38	7.00	1.25	0.081	0.141	0.364
17.0	1.99	0.69	61.25	33.21	5.85	1 531	0.041	0.131	0.894
73.0	1.42	0.87	60.3	32.19	5.62	1.45	0.131	0.141	0.212
91.0	2.13	0.89	60	32.04	6.22	1.64	0.18	0.141	0.263
118.0	1.16	0.89	63.86	31.17	6.02	1 551	0.21	0.141	0.143
160.0	1.57	0.93	52.8	35.58	5.67	1 521	0.161	0.161	0.118
191.3	1.6	0.92	54.3	36.05	5.50	1.63	0.161	0.15	0.139
210.4	1.66	0.9	54	34.94	5.13	1 281	0.141	0.131	0.184
305.9	2.03	0.83	66.6	34.54	4.75	1 121	0.06	0.131	0.416
311.0	2.21	0.8	58.71	30.61	3.61	1 071	0.041	0.111	0.553
330.0	2.03	0.85	61.5	34.50	3.96	0.98	0.03	0.101	0.358
339.0	2.5	0.85	73.21	38.51	5.50	1.14	0.071	0.12	0.441
365.0	2.4	0.82	67.6	34.71	5.92	1 051	0.051	0.131	0.527

Source: Linardos (1991).

Table 17.9: Glucose/Glutamine Ratio: Experimental Data from a Chemostat Run with a Glucose to Glutamine Ratio in the Feed of 5:2

t	x_v	v_b	*MAb*	*Lac*	*Gls*	*Amm*	*Glm*	*Glt*	x_d
189.8	1.49	0.65	55.50	37.12	4.92	1.64	1.90	0.09	0.802
203.0	2.05	0.76	53.65	37.53	5.34	2.50	1.86	0.09	0.647
212.0	1.92	0.75	55.90	41.01	5.49	2.32	2.03	0.09	0.640
231.8	1.70	0.71	53.65	37.57	4.77	2.53	1.45	0.11	0.694
251.8	1.90	0.77	53.65	36.58	5.17	1.65	2.05	0.08	0.568
298.0	1.74	0.80	38.71	36.22	5.99	1.84	2.17	0.09	0.435
309.3	2.09	0.78	15.81	35.97	4.00	1.44	1.56	0.06	0.589
322.8	1.99	0.80	35.71	40.35	6.67	1.31	2.01	0.09	0.498
334.0	1.60	0.81	23.00	40.07	5.29	1.87	1.73	0.08	0.375
346.8	2.15	0.84	31.00	42.93	6.45	1.43	1.65	0.09	0.410

Source: Linardos (1991).

Table 17.10: Glucose/Glutamine Ratio: Experimental Data from a Chemostat Run with a Glucose to Glutamine Ratio in the Feed of 5:3.

t	x_v	v_b	MAb	Lac	Gls	Amm	Glm	Glt	x_d
23.9	1.95	0.83	32.71	42.61	5.62	3.11	3.59	0.03	0.399
33.5	2.05	0.82	19.61	41.46	4.05	2.90	3.40	0.09	0.450
39.4	1.85	0.80	29.90	42.25	5.09	2.61	4.60	0.03	0.463
49.7	2.05	0.80	30.00	43.29	4.51	2.80	4.67	0.55	0.513
58.2	1.99	0.81	16.20	38.51	3.25	2.80	3.01	0.52	0.467
72.9	2.21	0.78	32.00	26.18	3.78	3.90	6.26	0.57	0.623
97.1	2.44	0.81	25.50	24.06	2.90	3.90	4.96	0.56	0.572
119.8	2.07	0.78	28.31	24.99	2.90	3.59	6.00	0.59	0.584
153.9	2.09	0.71	24.81	22.99	2.53	4.63	6.38	0.60	0.854
181.6	1.94	0.70	37.05	23.09	2.09	6.88	7.39	0.68	0.831
192.3	2.25	0.69	23.81	24.27	1.56	6.66	6.26	0.66	1 011

Source: Linardos (1991).

17.2.2 Enzyme Inhibition Kinetics

Blanch and Clark (1996) reported the following data on an enzyme catalyzed reaction in the presence of inhibitors A or B. The data are shown in Table 17.11.

Table 17.11 Inhibition Kinetics: Initial Reaction Rates in the Presence of Inhibitors A or B at Different Substrate Concentrations

Substrate Concentration (mM)	Reaction Rate (nM/min)		
	No Inhibitor	A at 5 μM	B at 25 μM
0.2	8.34	3.15	5.32
0.33	12.48	5.06	6.26
0.5	16.67	7.12	7.07
1.0	25.0	13.3	8.56
2.5	36.2	26.2	9.45
4.0	40.0	28.9	9.60
5.0	42.6	31.8	9.75

Source: Blanch and Clark (1996).

In this problem you are asked to do the following:

(a) Use the Michaelis-Menten kinetic model and the data from the inhibitor free-experiments to estimate the two unknown parameters (r_{max} and K_m)

$$r = \frac{r_{max} S}{K_m + S} \tag{17.13}$$

where r is the measured reaction rate and S is the substrate concentration.

(b) For each data set with inhibitor A or B, estimate the inhibition parameter K_i for each alternative inhibition model given below:

Competitive Inhibition

$$r = \frac{r_{max} S}{K_m \left(1 + \frac{I}{K_i}\right) + S} \tag{17.14}$$

Uncompetitive Inhibition

$$r = \frac{r_{max} S}{K_m + S\left(1 + \frac{I}{K_i}\right)} \tag{17.15}$$

Nonncompetitive Inhibition

$$r = \frac{r_{max} S}{\left(1 + \frac{I}{K_i}\right)(K_m + S)} \tag{17.16}$$

where I is the concentration of inhibitor (A or B) and K_i is the inhibition kinetic constant.

(c) Use the appropriate model adequacy tests to determine which one of the inhibition models is the best.

17.2.3 Determination of k_La in Bubble-free Bioreactors

Kalogerakis and Behie (1997) have reported experimental data from dynamic gassing-in experiments in three identical 1000 L bioreactors designed for the cultivation of anchorage dependent animal cells. The bioreactors have an internal conical aeration area while the remaining bubble-free region contains the microcarriers. A compartmental mathematical model was developed by Kalogerakis and Behie (1997) that relates the apparent k_La to operational and bioreactor design parameters. The raw data from a typical run are shown in Table 17.12.

In the absence of viable cells in the bioreactor, an *effective* mass transfer coefficient can be obtained from

$$\frac{dC_{O2}}{dt} = k_L a[C^*_{O2} - C_{O2}] \tag{17.17}$$

where C_{O2} is the dissolved oxygen concentration in the bubble-free region and C^*_{O2} is the dissolved oxygen concentration in equilibrium with the oxygen in the air supply at the operating pressure and temperature of the system. Since the dissolved oxygen concentration is measured as percent of saturation, DO (%), the above equation can be rewritten as

$$\frac{dDO}{dt} = k_L a[DO_{100\%} - DO] \tag{17.18}$$

where the following linear calibration curve has been assumed for the probe

$$\frac{C_{O2}}{C^*_{O2}} = \frac{DO - DO_{0\%}}{DO_{100\%} - DO_{0\%}} \tag{17.19}$$

The constants $DO_{100\%}$ and $DO_{0\%}$ are the DO values measured by the probe at 100% or 0% saturation respectively. With this formulation one can readily analyze the operation with oxygen enriched air.

Upon integration of Equation 17.18 we obtain

$$DO(t) = DO_{100\%} - [DO_{100\%} - DO(t_0)]e^{-k_L at} \tag{17.20}$$

Using the data shown in Table 17.12 you are asked to determine the effective $k_L a$ of the bioreactor employing different assumptions:

(a) Plot the data (DO(t) versus time) and determine DO100%, i.e., the steady state value of the DO transient. Generate the differences $[DO_{100\%}$-DO(t)] and plot them in a semi-log scale with respect to time. In this case $k_L a$ can be simply obtained as the slope the line formed by the transformed data points.

(b) Besides $k_L a$ consider $DO_{100\%}$ as an additional parameter and estimate simultaneous both parameters using nonlinear least squares estimation. In this case assume that DO(t0) is known precisely and its value is given in the first data entry of Table 17.12.

(c) Redo part (b) above, however, consider that $DO(t_0)$ is also an unknown parameter.

Finally, based on your findings discuss the reliability of each solution approach.

Table 17.12 Bubble-free Oxygenation: Experimental Data From a Typical Gassing-in Experiment[†]

Time (*min*)	DO (%)	Time (*min*)	DO (%)	Time (*min*)	DO (%)
1	-0.41	28	15.31	55	20.30
2	0.41	29	15.61	56	20.40
3	1.21	30	15.90	57	20.50
4	2.10	31	16.11	58	20.50
5	3.00	32	16.40	59	20.61
6	3.71	33	16.71	60	20.71
7	4.60	34	16.90	61	20.80
8	5.31	35	17.21	62	20.90
9	6.10	36	17.40	63	20.90
10	6.81	37	17.61	64	21.00
11	7.50	38	17.80	65	21.11
12	8.20	39	18.00	66	21.11
13	8.81	40	18.11	67	21.21
14	9.40	41	18.40	68	21.21
15	9.90	42	18.50	69	21.21
16	10.40	43	18.71	70	21.21
17	11.00	44	18.90	71	21.30
18	11.40	45	19.00	72	21.40
19	11.90	46	19.21	73	21.40
20	12.40	47	19.30	74	21.50
21	12.81	48	19.50	75	21.61
22	13.20	49	19.61	76	21.61
23	13.61	50	19.71	77	21.61
24	14.00	51	19.80	78	21.61
25	14.31	52	20.00	79	21.71
26	14.61	53	20.00	80	21.71
27	15.00	54	20.11		

[†] Bioreactor No. 1, Working volume 500 *L*, 40 *RPM*, Air flow 0.08 *vvm*.
Source: Kalogerakis and Behie (1997).

17.3 ORDINARY DIFFERENTIAL EQUATION MODELS

17.3.1 Contact Inhibition in Microcarrier Cultures of MRC-5 Cells

Contact inhibition is a characteristic of the growth of anchorage dependent cells grown on microcarriers as a monolayer. Hawboldt et al. (1994) reported data on MRC5 cells grown on Cytodex II microcarriers and they are reproduced here in Table 17.13.

Growth inhibition on microcarriers cultures can be best quantified by cellular automata (Hawboldt et al., 1994; Zygourakis et al., 1991) but simpler models have also been proposed. For example, Frame and Hu (1988) proposed the following model

$$\frac{dx}{dt} = \mu_{max}\left(1 - exp\left(-C\frac{x_\infty - x}{x_\infty}\right)\right)x \qquad x(0)=x_0 \qquad (17.21)$$

where x is the average cell density in the culture and μ_{max}, and C are adjustable parameters. The constant x_∞ represents the maximum cell density that can be achieved at confluence. For microcarrier cultures this constant depends on the inoculation level since different inocula result in different percentages of beads with no cells attached. This portion of the beads does not reach confluence since at a low bead loading there is no cell transfer from bead to bead (Forestell et al., 1992; Hawboldt et al., 1994). The maximum specific growth rate, μ_{max} and the constant C are expected to be the same for the two experiments presented here as the only difference is the inoculation level. The other parameter, x_∞, depends on the inoculation level (Forestell et al., 1992) and hence it cannot be considered the same. As initial condition we shall consider the first measurement of each data set.

A rather good estimate of the maximum growth rate, μ_{max}, can be readily obtained from the first few data points in each experiment as the slope of *ln*(x) versus time. During the early measurements the contact inhibition effects are rather negligible and hence μ_{max} is estimated satisfactorily. In particular for the first experiment inoculated with 2.26 *cells/bead*, μ_{max} was found to be 0.026 (h^{-1}) whereas for the second experiment inoculated with 4.76 *cells/bead* μ_{max} was estimated to be 0.0253 (h^{-1}). Estimates of the unknown parameter, x_∞ can be obtained directly from the data as the maximum cell density achieved in each experiment; namely, 1.8 and 2.1 (10^6 *cells/mL*). To aid convergence and simplify programming we have used $1/x_\infty$ rather than x_∞ as the parameter to be estimated.

Let us consider first the regression of the data from the first experiment (inoculation level=2.26). Using as an initial guess $[1/x_\infty, C, \mu_{max}]^{(0)} = [0.55, 5, 0.026]$, the Gauss-Newton method converged to the optimum within 13 iterations. The optimal parameter values were [0.6142 ± 2.8%, 2.86 ± 41.9%, 0.0280 ± 12.1%] corresponding to a LS objective function value of 0.091621. The raw data together with the model calculated values are shown in Figure 17.14.

Table 17.13: Growth of MRC5 Cells: Cell Density versus Time for MRC5 Cells Grown on Cytodex I Microcarriers in 250 mL Spinner Flasks Inoculated at 2.26 and 4.76 cells/bead

Inoculation level: 2.26 *(cells/bead)*		Inoculation level: 4.76 *(cells/bead)*	
Time *(h)*	Cell Density *(10^6 cells/mL)*	Time *(h)*	Cell Density *(10^6 cells/mL)*
21.8	0.06	23.0	0.16
24.2	0.06	48.4	0.37
30.2	0.07	58.1	0.47
41.7	0.08	73.8	0.58
48.4	0.13	94.4	0.99
66.5	0.09	101.6	1.33
73.8	0.15	133.1	1.88
91.9	0.34	150.0	1.78
99.2	0.39	166.9	1.89
111.3	0.65	181.5	1.78
118.5	0.67	196.0	2.09
134.3	0.92		
144.0	1.24		
158.5	1.47		
166.9	1.36		
182.7	1.56		
205.6	1.52		
215.3	1.61		
239.5	1.78		

Source: Hawboldt et al. (1994).

Next using the converged parameter values as an initial guess for the second set of data (inoculation level=4.76) we encounter convergence problems as the parameter estimation problem is severly ill-conditioned. Convergence was only achieved with a nonzero value for Marquardt's parameter (10^{-5}). Two of the parameters, x_∞ and μ_{max}, were estimated quite accurately; however, parameter C was very poorly estimated. The best parameter values were [0.5308 ± 2.5%, 119.7 ± 4.1×10^8%, 0.0265 ± 3.1%] corresponding to a LS objective function value of 0.079807. The raw data together with the model calculated values are shown in Figure 17.15. As seen, the overall match is quite acceptable.

Finally, we consider the simulatneous regression of both data sets. Using the converged parameter values from the first data set as an initial guess convergence was obtained in 11 iterations. In this case parameters C and μ_{max} were common to both data sets; however, two different values for x_∞ were used - one for each data set. The best parameter values were $1/x_\infty = 0.6310 \pm 2.9\%$ for the first data set and $1/x_\infty = 0.5233 \pm 2.6\%$ for the second one. Parameters C and μ_{max} were found to have values equal to $4.23 \pm 40.3^8\%$ and $0.0265 \pm 6.5\%$ respectively. The value of the LS objective function was found to be equal to 0.21890. The model calculated values using the parameters from the simultaneous regression of both data sets are also shown in Figures 17.14 and 17.15.

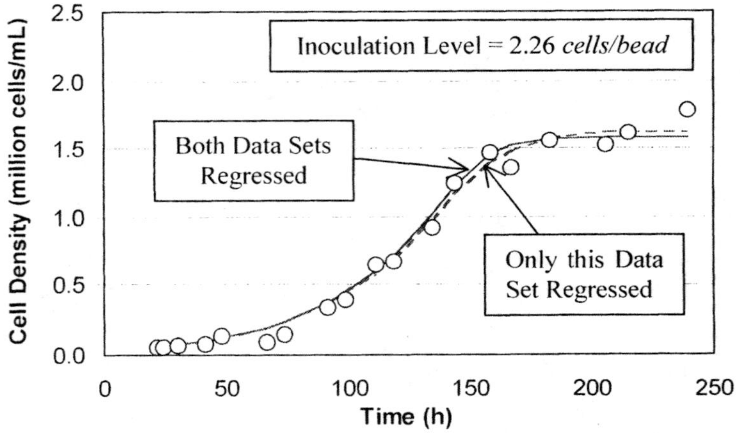

Figure 17.14: Growth of MRC5 Cells: Measurements and model calculated values of cell density versus time for MRC5 cells grown on Cytodex II microcarriers. Inoculation level was 2.26 cells/bead.

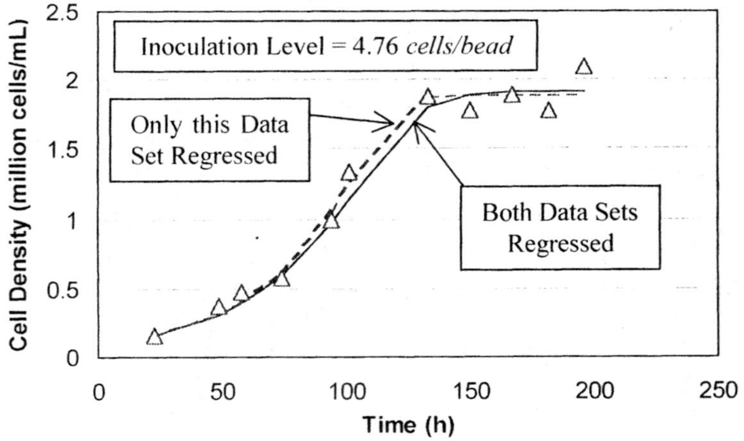

Figure 17.15: Growth of MRC5 Cells: Measurements and model calculated values of cell density versus time for MRC5 cells grown on Cytodex II microcarriers. Inoculation level was 4.76 cells/bead.

17.4 PROBLEMS WITH ODE MODELS

17.4.1 Vero Cells Grown on Microcarriers (Contact Inhibition)

Hawboldt et al. (1994) have also reported data on anchorage dependent Vero cells grown on Cytodex I microcarriers in 250 *mL* spinner flasks and they are reproduced here in Table 17.14.

You are asked to determine the adjustable parameters in the growth model proposed by Frame and Hu (1988)

$$\frac{dx}{dt} = \mu_{max}\left(1 - exp\left(-C\frac{x_{\infty} - x}{x_{\infty}}\right)\right)x \qquad x(0)=x_0 \tag{17.22}$$

Table 17.14: Growth of Vero Cells: Cell Density versus Time for Vero Cells Grown on Cytodex I Microcarriers in 250 mL Spinner Flasks at Two Inoculation Levels (2.00 and 9.72 cells/bead).

Time *(h)*	Cell Density *(10^6 cells/mL)*	
	Inoculation level: 2.00 *(cells/bead)*	Inoculation level: 9.72 *(cells/bead)*
0.0	0.07	0.33
9.9	0.12	0.37
24.1	0.18	0.52
36.8	0.21	0.56
51.0	0.24	0.65
76.5	0.46	1.15
85.0	0.54	1.49
99.2	0.74	1.84
108.4	0.99	2.07
119.0	1.24	2.60
144.5	1.98	3.64
155.9	2.19	3.53
168.6	2.64	3.64
212.6	3.68	4.36
223.9	3.43	4.42
236.6	3.80	4.48
260.7	3.97	4.77
286.9	3.99	4.63

Source: Hawboldt et al. (1994)

The parameter estimation should be performed for the following two cases:

(1) x(0) is known precisely and it is equal to x_0 (i.e., the value given below at time t=0).

(2) x(0) is not known and hence, x_0 is to be considered just as an additional measurement.

Furthermore, you are asked to determine:

(3) What is the effect of assuming that x_0 is unknown on the standard error of the model parameters?

(4) Is parameter C independent of the inoculation level?

17.4.2 Effect of Temperature on Insect Cell Growth Kinetics

Andersen et al. (1996) and Andersen (1995) have studied the effect of temperature on the recombinant protein production using a baulovirus/insect cell expression system. In Tables 17.15, 17.16, 17.17, 17.18 and 17.19 we reproduce the growth data obtained in spinner flasks (batch cultures) using *Bombyx mori (Bm5)* cells adapted to serum-free media (Ex-Cell 400). The working volume was 125 *mL* and samples were taken twice daily. The cultures were carried out at six different incubation temperatures (22, 26, 28, 30 and 32 ℃).

Table 17.15: Growth of Bm5 Cells: Growth Data taken from a Batch Culture of Bm5 Cells Incubated at 22 ℃

t	x_v	x_d	*Gls*	*Lac*
0.0	2.09	2.18	2.267	0.040
24.0	2.43	2.66	2.170	0.020
48.0	2.70	2.98	2.130	0.040
69.3	3.32	3.74	2.044	0.040
89.6	3.66	4.02	1.967	0.028
113.6	4.16	4.55	1.753	0.020
142.9	7.41	8.03	1.400	0.027
165.4	10.95	11.44	1.240	0.040
188.9	17.29	18.43	0.927	0.020
217.0	18.50	19.50	0.487	0.040
241.8	19.04	21.00	0.267	0.027
264.8	20.15	23.00	0.100	0.027
290.0	19.30	22.01		
338.0	18.40	22.00		

Source: Andersen et al. (1996) & Andersen (1995).

Table 17.16: Growth of Bm5 Cells: Growth Data taken from a Batch Culture of Bm5 Cells Incubated at 26℃

t	x_v	x_d	*Gls*	*Lac*
0.0	2.65	2.85	2.23	0.027
24.0	2.73	3.07	2.17	0.020
48.0	4.04	4.53	2.05	0.030
70.5	6.21	7.08	1.85	0.020
90.5	8.90	9.30	1.55	0.020
115.0	13.18	13.61	1.14	0.020
137.8	17.10	17.55	0.67	0.033
161.8	21.69	22.13	0.20	0.087
190.8	28.80	29.80	0.06	0.040
217.8	29.41	30.17	0.08	0.040
234.0	28.72	30.13	0.10	0.033
257.0	29.00	30.40	0.14	0.040
283.0	25.20	29.05	0.04	0.020

Source: Andersen et al. (1996) & Andersen (1995)

Table 17.17: Growth of Bm5 Cells: Growth Data taken from a Batch Culture of Bm5 Cells Incubated at 28℃

t	x_v	x_d	*Gls*	*Lac*
0.0	2.142	2.292	2.320	0.020
24.0	2.981	3.300	2.130	0.040
48.0	4.369	4.813	1.990	0.020
70.5	7.583	8.066	1.667	0.027
90.5	10.513	10.725	1.300	0.030
114.0	16.416	17.020	0.700	0.020
137.0	20.677	21.468	0.220	0.080
161.0	27.458	28.040	0.060	0.053
190.0	29.685	30.438	0.087	0.020
216.7	30.452	32.030	0.080	0.027
237.0	25.890	33.000	0.080	0.040
259.0	18.900	31.500		

Source: Andersen et al. (1996) & Andersen (1995).

Table 17.18: Growth of Bm5 Cells: Growth Data taken from a Batch Culture of Bm5 Cells Incubated at 30°C

t	x_v	x_d	*Gls*	*Lac*
0.0	2.800	2.950	2.50	0.030
24.0	3.500	4.066	2.40	0.020
48.0	4.800	5.296	1.88	0.040
70.5	6.775	7.700	1.42	0.020
91.5	11.925	12.499	0.98	0.047
114.0	16.713	17.500	0.44	0.040
137.0	20.958	22.208	0.07	0.040
161.0	24.500	26.354	0.12	0.040
190.5	24.416	25.833	0.06	0.027
216.0	24.290	26.708	0.08	0.053
262.0	15.000	26.600	0.11	0.073

Source: Andersen et al. (1996) & Andersen (1995).

Table 17.19: Growth of Bm5 Cells: Growth Data taken from a Batch Culture of Bm5 Cells Incubated at 32°C

t	x_v	x_d	*Gls*	*Lac*
0.0	1.98	2.18	2.260	0.020
24.0	3.12	3.45	2.100	0.020
40.0	3.64	4.11	1.950	0.020
67.0	7.19	7.58	1.610	0.020
91.0	11.57	11.95	1.060	0.020
113.0	15.55	16.10	0.527	0.040
137.0	20.50	21.09	0.087	0.073
166.8	23.63	24.92	0.093	0.020
192.3	24.06	26.03	0.090	0.040
212.3	24.31	26.83	0.140	0.020
235.0	22.75	28.13	0.140	0.040
261.0	12.00	26.63	0.126	0.220

Source: Andersen et al. (1996) & Andersen (1995).

In the above tables the following notation has been used:

t Elapsed Time (h)
x_v Viable Cell Density (10^6 *cells/mL*)
x_d Nonviable Cell Density (10^6 *cells/mL*)

Gls	Glucose *(g/L)*
Lac	Lactate *(g/ L)*

At each temperature the simple Monod kinetic model can be used that can be combined with material balances to arrive at the following unstructured model

$$\frac{dx_v}{dt} = (\mu - k_d) \cdot x_v \tag{17.23a}$$

$$\frac{dx_d}{dt} = k_d \cdot x_v \tag{17.23b}$$

$$\frac{dS}{dt} = -\frac{\mu}{Y} \cdot x_v \tag{17.23c}$$

where

$$\mu = \frac{\mu_{max} S}{K_s + S} \tag{17.23d}$$

The limiting substrate (glucose) concentration is denoted by S. There are four parameters: μ_{max} is the maximum specific growth rate, K_s is the saturation constant for S, k_d is the specific death rate and Y is the average yield coefficient (assumed constant).

In this problem you are asked to:

(1) Estimate the parameters (μ_{max}, K_s, k_d and Y) for each operating temperature. Use the portion of the data where glucose is above the threshold value of 0.1 g/L that corresponds approximately to the exponential growth period of the batch cultures.

(2) Examine whether any of the estimated parameters follow an Arrhenius-type relationship. If they do, re-estimate these parameters simultaneously. A better way to numerically evaluate Arrhenius type constants is through the use of a reference value. For example, if we consider the death rate, k_d as a function of temperature we have

$$k_d = A \cdot e^{-\frac{E}{RT}} \tag{17.24}$$

At the reference temperature, T_0 (usually taken close to the mean operating temperature), the previous equation becomes

$$k_{d0} = A \cdot e^{-\frac{E}{RT_0}} \qquad (17.25)$$

By dividing the two expressions, we arrive at

$$k_d = k_{d0} \cdot e^{\frac{E}{R}\left(\frac{1}{T_0}-\frac{1}{T}\right)} \qquad (17.26)$$

Equation 17.26 behaves much better numerically than the standard Arrhenius equation and it is particularly suited for parameter estimation and/or simulation purposes. In this case instead of A and E/R we estimate k_{d0} and E/R. In this example you may choose $T_0 = 28\,°C$.

18

Parameter Estimation in Petroleum Engineering

Parameter estimation is routinely used in many areas of petroleum engineering. In this chapter, we present several such applications. First, we demonstrate how multiple linear regression is employed to estimate parameters for a drilling penetration rate model. Second, we use kinetic data to estimate parameters from simple models that are used to describe the complex kinetics of bitumen low temperature oxidation and high temperature cracking reactions of Alberta oil sands. Finally, we describe an application of the Gauss-Newton method to PDE systems. In particular we present the development of an efficient automatic history matching simulator for reservoir engineering analysis.

18.1 MODELING OF DRILLING RATE USING CANADIAN OFFSHORE WELL DATA

Offshore drilling costs may exceed similar land operations by 30 to 40 times and hence it is important to be able to minimize the overall drilling time. This is accomplished through mathematical modeling of the drilling penetration rate and operation (Wee and Kalogerakis, 1989).

The processes involved in rotary drilling are complex and our current understanding is far from complete. Nonetheless, a basic understanding has come from field and laboratory experience over the years. The most comprehensive model is the one developed by Bourgoyne and Young (1986) that relates the penetration rate (dD/dt) to eight process variables. The model is transformably linear with respect to its eight adjustable parameters. The model for the penetration rate is given by the exponential relationship,

$$\frac{dD}{dt} = exp\left(a_1 + \sum_{j=2}^{8} a_j x_j \right) \tag{18.1}$$

which can be transformed to a linear model by taking natural logarithms from both sides to yield

$$ln\left(\frac{dD}{dt} \right) = a_1 + \sum_{j=2}^{8} a_j x_j \tag{18.2}$$

The explanation of each drilling parameter (a_j) related to the corresponding drilling variable (x_j) is given below:

a_1	formation strength constant
a_2	normal compaction trend constant
a_3	undercompaction constant
a_4	pressure differential constant
a_5	bit weight constant
a_6	rotary speed constant
a_7	tooth wear constant
a_8	hydraulics constant

The above model is referred to as the *Bourgoyne-Young model.* After careful manipulation of a set of raw drilling data for a given formation type, a set of penetration rate data is usually obtained. The drilling variables are also measured and the measurements become part of the raw data set. The objective of the regression is then to estimate the parameters a_j by matching the model to the drilling penetration data.

Parameter a_1 in Equation 18.1 accounts for the lumped effects of factors other than those described by the drilling variables x_2, x_3,...,x_8. Hence, its value is expected to be different from well to well whereas the parameter values for a_2,a_3,..,a_8 are expected to be similar. Thus, when data from two wells (well A and well B) are simultaneously analyzed, the model takes the form

$$\frac{dD}{dt} = exp\left(\sum_{j=2}^{8} a_j x_j + a_9 x_9 + a_{10} x_{10} \right) \tag{18.3}$$

where the following additional definitions apply.

a_9	formation strength constant (parameter a_1) for well A
a_{10}	formation strength constant (parameter a_1) for well B
x_9	=1 for data from well A and 0 for all other data
x_{10}	=1 for data from well B and 0 for all other data

The above model is referred to as the *Extended Bourgoyne-Young model.* Similarly, analysis of combined data from more than two wells is straightforward and it can be done by adding a new variable and parameter for each additional well. A large amount of field data from the drilling process are needed to reliably estimate all model parameters.

Wee and Kalogerakis (1989) have also considered the simple *three-parameter model* given next

$$\frac{dD}{dt} = exp(a_1 + a_5x_5 + a_6x_6) \tag{18.4}$$

18.1.1 Application to Canadian Offshore Well Data

Wee and Kalogerakis (1989) tested the above models using Canadian offshore well penetration data (offshore drill operated by Husky Oil). Considerable effort was required to convert the raw data into a set of data suitable for regression. The complete dataset is given in the above reference.

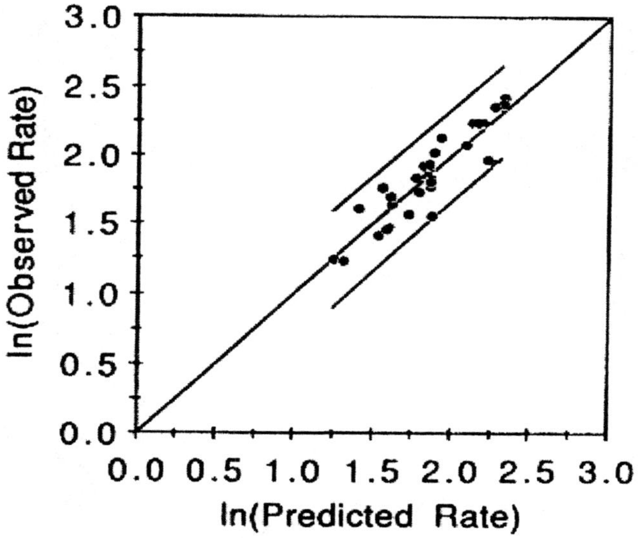

Figure 18.1 Observed versus calculated penetration rate and 95% confidence intervals for well A using the Bourgoyne-Young model [reprinted from the Journal of Canadian Petroleum Technology with permission].

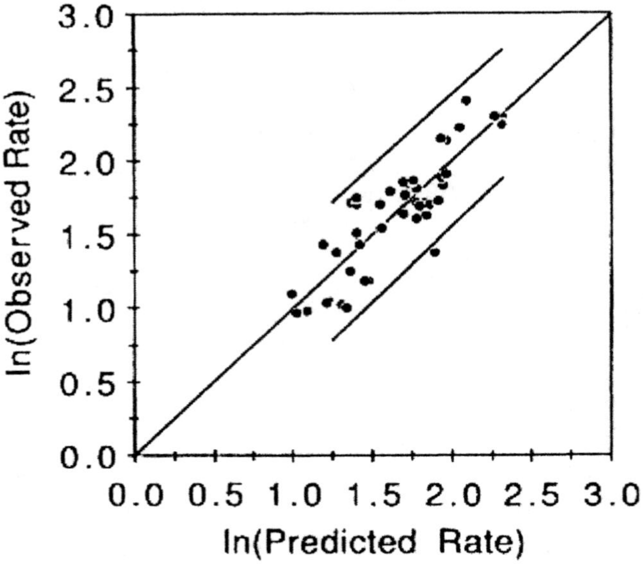

Figure 18.2 *Observed versus calculated penetration rate and 95% confidence intervals for well B using the Bourgoyne-Young model [reprinted from the Journal of Canadian Petroleum Technology with permission].*

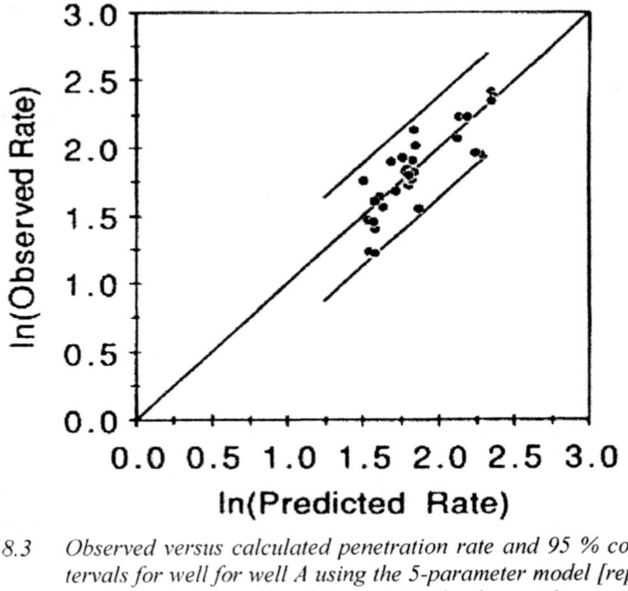

Figure 18.3 *Observed versus calculated penetration rate and 95 % confidence intervals for well for well A using the 5-parameter model [reprinted from the Journal of Canadian Petroleum Technology with permission].*

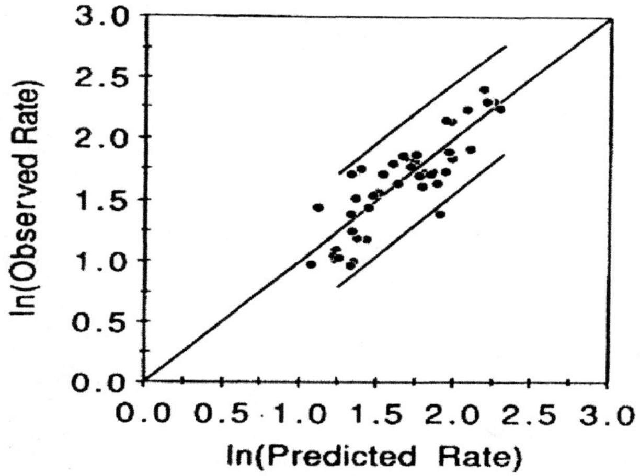

Figure 18.4 *Observed versus calculated penetration rate and 95 % confidence intervals for well for well B using the 5-parameter model [reprinted from the Journal of Canadian Petroleum Technology with permission].*

As expected, the authors found that the eight-parameter model was sufficient to model the data; however, they questioned the need to have eight parameters. Figure 18.1 shows a plot of the logarithm of the observed penetration rate versus the logarithm of the calculated rate using the Bourgoyne-Young model. The 95 % confidence intervals are also shown. The results for well B are shown in Figure 18.2.

Results with the three parameter model showed that the fit was poorer than that obtained by the Bourgoyne-Young model. In addition the dispersion about the 45 degree line was more significant. The authors concluded that even though the Bourgoyne-Young model gave good results it was worthwhile to eliminate possible redundant parameters. This would reduce the data requirements. Indeed, by using appropriate statistical procedures it was demonstrated that for the data examined five out of the eight parameters were adequate to calculate the penetration rate and match the data sufficiently well. The five parameters were a_1, a_2, a_4, a_6 and a_7 and the corresponding *five-parameter model* is given by

$$\frac{dD}{dt} = exp\left(a_1 + a_2x_2 + a_4x_4 + a_6x_6 + a_6x_6\right) \qquad (18.5)$$

Figures 18.3 and 18.4 show the observed versus calculated penetration rates for wells A and B using the five-parameter model. As seen the results have not changed significantly by eliminating three of the parameters (a_3, a_5 and a_8). The elimination of each parameter was done sequentially through hypothesis testing. Obviously, the fact that only the above five variables affect significantly the penetration rate means that the three-parameter model is actually inadequate even though it might be able to fit the data to some extent.

18.2 MODELING OF BITUMEN OXIDATION AND CRACKING KINETICS USING DATA FROM ALBERTA OIL SANDS

In the laboratory of Professor R.G. Moore at the University of Calgary, kinetic data were obtained using bitumen samples of the North Bodo and Athabasca oil sands of northern Alberta. Low temperature oxidation data were taken at 50, 75, 100, 125 and 150^oC whereas the high temperature thermal cracking data at 360, 397 and 420^oC.

Preliminary work showed that first order reaction models are adequate for the description of these phenomena even though the actual reaction mechanisms are extremely complex and hence difficult to determine. This simplification is a desired feature of the models since such simple models are to be used in numerical simulators of in situ combustion processes. The bitumen is divided into five major pseudo-components: coke (COK), asphaltene (ASP), heavy oil (HO), light oil (LO) and gas (GAS). These pseudo-components were lumped together as needed to produce two, three and four component models. Two, three and four-component models were considered to describe these complicated reactions (Hanson and Kalogerakis, 1984).

18.2.1 Two-Component Models

In this class of models, the five bitumen pseudo-components are lumped into two in an effort to describe the following reaction

$$R \xrightarrow{k} P \tag{18.6}$$

where R (reactant) and P (product) are the two lumped pseudo-components. Of all possible combinations, the four mathematical models that are of interest here are shown in Table 18.1. An Arrhenius type temperature dependence of the unknown rate constants is always assumed.

Table 18.1 Bitumen Oxidation and Cracking: Formulation of Two-Component Models

Model	*Reactant*	*Product*	*Model ODE*
A	R=HO+LO	P=COK+ASP	$\frac{dC_P}{dt} = (1 - C_R)k_A\, exp\left(\frac{-E_A}{RT}\right)$
B	R=HO+LO	P=COK+ASP	$\frac{dC_P}{dt} = (1 - C_R)^{N_B} k_B\, exp\left(\frac{-E_B}{RT}\right)$
D	R=HO+LO+ASP	P=COK	$\frac{dC_P}{dt} = (1 - C_R)k_D\, exp\left(\frac{-E_D}{RT}\right)$
E	R=HO+LO+ASP	P=COK	$\frac{dC_P}{dt} = (1 - C_R)^{N_E} k_B\, exp\left(\frac{-E_E}{RT}\right)$

In Table 18.1 the following variables are used in the model equations:

C_P Product concentration (*weight %*)
C_R Reactant concentration (*weight %*)
k_j Reaction rate constant in model j=A,B,D,E
E_j Energy of activation in model j=A,B,D,E.
N_B Exponent used in model B
N_E Exponent used in model E

18.2.2 Three-Component Models

By lumping pseudo-components, we can formulate five three-component models of interest. Pseudo-components shown together in a circle are treated as one pseudo-component for the corresponding kinetic model.

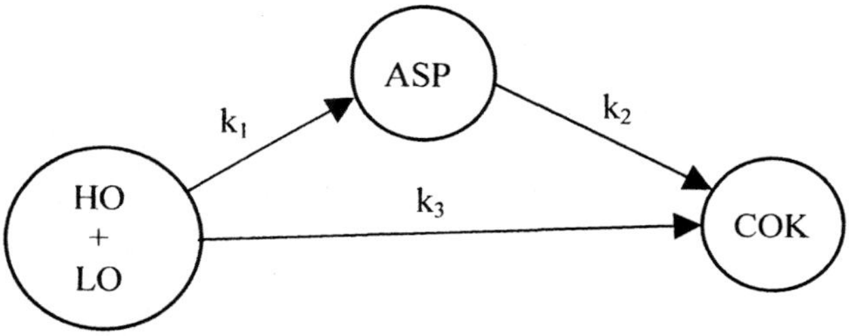

Figure 18.5 Schematic of reaction network for model F.

Model F is depicted schematically in Figure 18.5 and the corresponding mathematical model is given by the following two ODEs:

$$\frac{dC_{ASP}}{dt} = k_1(1 - C_{ASP} - C_{COK}) - k_2 C_{ASP} \tag{18.7a}$$

$$\frac{dC_{COK}}{dt} = k_3(1 - C_{ASP} - C_{COK}) + k_2 C_{ASP} \tag{18.7b}$$

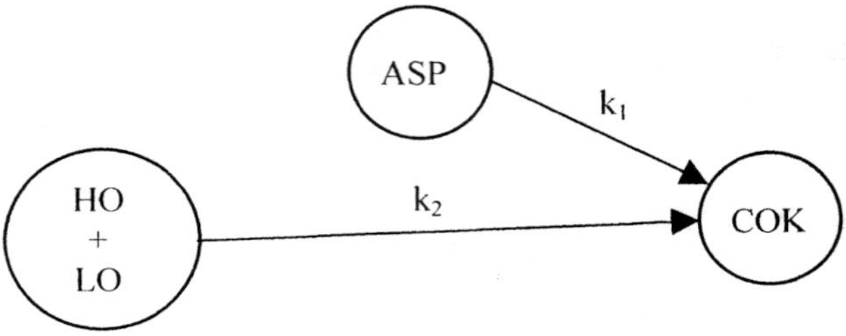

Figure 18.6 Schematic of reaction network for model G.

Model G is depicted schematically in Figure 18.6 and the corresponding mathematical model is given by the following two ODEs:

$$\frac{dC_{ASP}}{dt} = -k_1 C_{ASP} \tag{18.8a}$$

$$\frac{dC_{COK}}{dt} = k_2(1 - C_{ASP} - C_{COK}) + k_1 C_{ASP} \tag{18.8b}$$

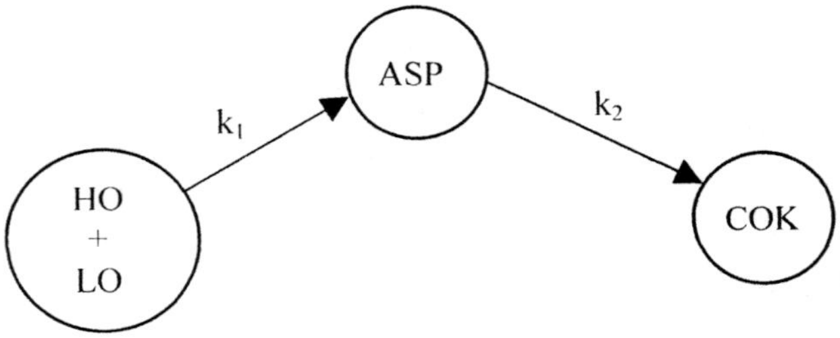

Figure 18.7 Schematic of reaction network for model I.

Model I is depicted schematically in Figure 18.7 and the corresponding mathematical model is given by the following two ODEs:

$$\frac{dC_{ASP}}{dt} = k_1(1 - C_{ASP} - C_{COK}) - k_2 C_{ASP} \tag{18.9a}$$

$$\frac{dC_{COK}}{dt} = k_2 C_{ASP} \tag{18.9b}$$

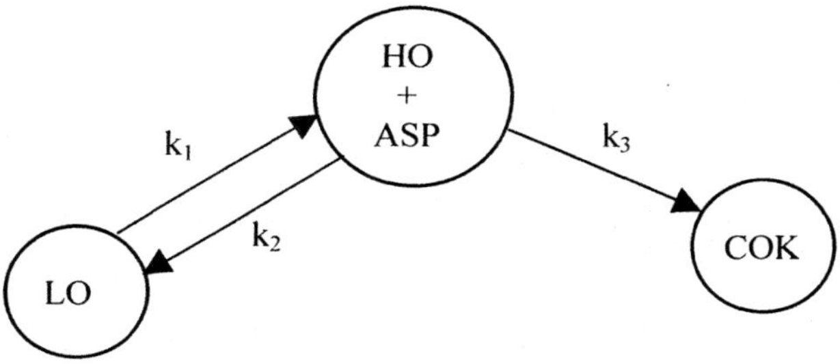

Figure 18.8 Schematic of reaction network for model K.

Model K is depicted schematically in Figure 18.8 and the corresponding mathematical model is given by the following two ODEs:

$$\frac{dC_{ASP+HO}}{dt} = k_1(1 - C_{ASP+HO} - C_{COK}) - k_2 C_{ASP+HO} - k_3 C_{ASP+HO} \tag{18.10a}$$

$$\frac{dC_{COK}}{dt} = k_3 C_{ASP+HO} \tag{18.10b}$$

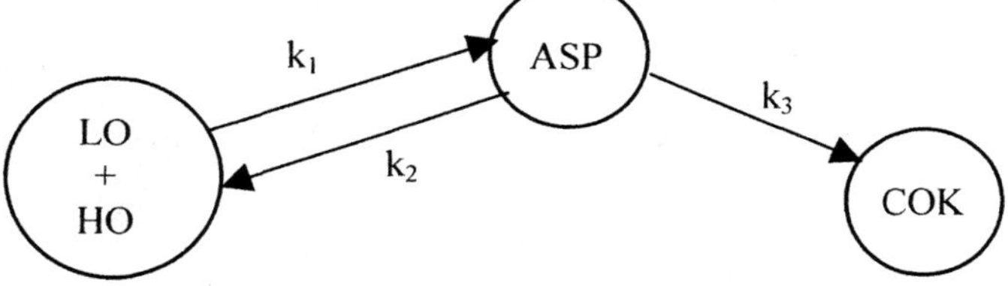

Figure 18.9 Schematic of reaction network for model J.

Model J is depicted schematically in Figure 18.9 and the corresponding mathematical model is given by the following two ODEs:

$$\frac{dC_{ASP}}{dt} = k_1(1 - C_{ASP} - C_{COK}) - k_2C_{ASP} - k_3C_{ASP} \tag{18.11a}$$

$$\frac{dC_{COK}}{dt} = k_3C_{ASP} \tag{18.11b}$$

In all the above three-component models as well as in the four-component models presented next, an Arrhenius-type temperature dependence is assumed for all the kinetic parameters. Namely each parameter k_i is of the form $A_i exp(-E_i/RT)$.

18.2.3 Four-Component Models

We consider the following four-component models. Model N is depicted schematically in Figure 18.10 and the corresponding mathematical model is given by the following three ODEs:

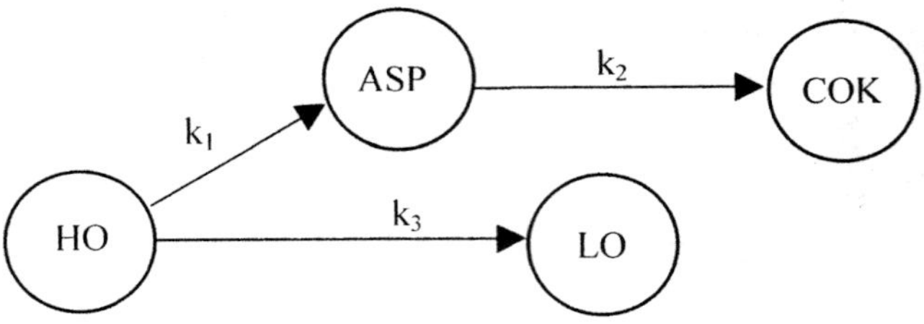

Figure 18.10 Schematic of reaction network for model N.

$$\frac{dC_{HO}}{dt} = -k_3C_{HO} - k_1C_{HO} \tag{18.12a}$$

$$\frac{dC_{ASP}}{dt} = k_1C_{HO} - k_2C_{ASP} \tag{18.12b}$$

$$\frac{dC_{COK}}{dt} = k_2C_{ASP} \tag{18.12c}$$

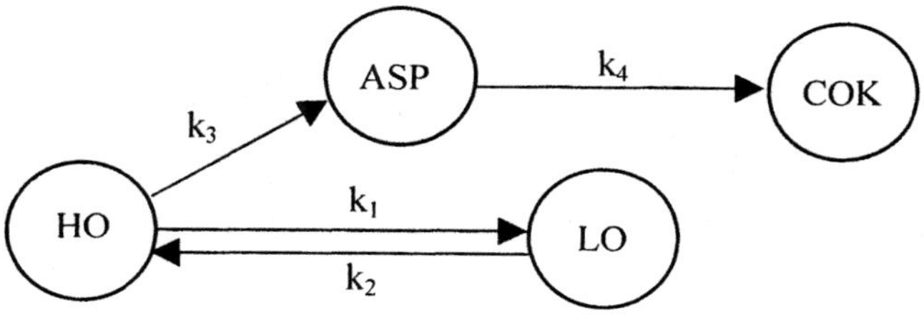

Figure 18.11 Schematic of reaction network for model M.

Model M is depicted schematically in Figure 18.11 and the corresponding mathematical model is given by the following three ODEs:

$$\frac{dC_{HO}}{dt} = k_1(1 - C_{HO} - C_{ASP} - C_{COK}) - k_2 C_{HO} - k_3 C_{HO} \tag{18.13a}$$

$$\frac{dC_{ASP}}{dt} = k_3 C_{HO} - k_4 C_{ASP} \tag{18.13b}$$

$$\frac{dC_{COK}}{dt} = k_4 C_{ASP} \tag{18.13c}$$

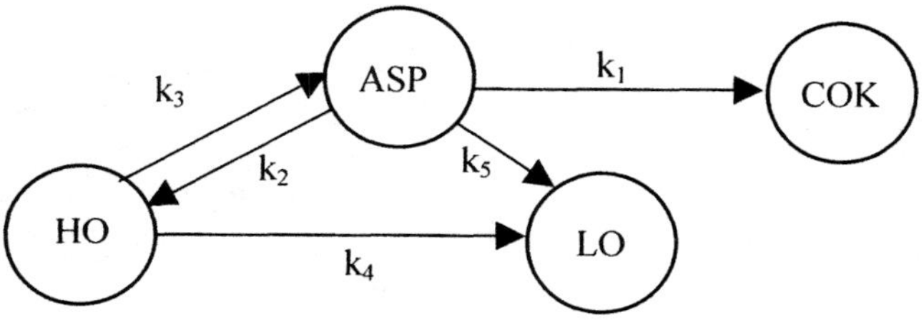

Figure 18.12 Schematic of reaction network for model O.

Model O is depicted schematically in Figure 18.12 and the corresponding mathematical model is given by the following three ODEs:

$$\frac{dC_{ASP}}{dt} = -k_1 C_{ASP} - k_2 C_{ASP} + k_3 C_{HO} - k_5 C_{ASP} \tag{18.14a}$$

$$\frac{dC_{HO}}{dt} = k_2 C_{ASP} - k_3 C_{HO} - k_4 C_{HO} \tag{18.14b}$$

$$\frac{dC_{LO}}{dt} = k_5 C_{ASP} + k_4 C_{HO} \tag{18.14c}$$

18.2.4 Results and Discussion

The two-component models are "too simple" to be able to describe the complex reactions taking place. Only model D was found to describe early coke (COK) production adequately. For Low Temperature Oxidation (LTO) conditions the model was adequate only up to 45 *h* and for cracking conditions up to 25 *h*.

The three-component models were found to fit the experimental data better than the two-component ones. Model I was found to be able to fit both LTO and cracking data very well. This model was considered *the best of all models even though it is unable to calculate the HO/LO split* (Hanson and Kalogerakis, 1984).

Four component models were found very difficult or impossible to converge. Models K, M and O are more complicated and have more reaction paths compared to models I or N. Whenever the parameter with the highest variance was eliminated in any of these three models, it would revert back to the simpler ones: Model I or N. Model N was the only four pseudo-component model that converged. This model also provides an estimate of the HO/LO split. This model together with model I were recommended for use in situ combustion simulators (Hanson and Kalogerakis, 1984). Typical results are presented next for model I.

Figures 18.13, through 18.17 show the experimental data and the calculations based on model I for the low temperature oxidation at 50, 75, 100, 125 and 150 ℃ of a North Bodo oil sands bitumen with a 5% oxygen gas. As seen, there is generally good agreement between the experimental data and the results obtained by the simple three pseudo-component model at all temperatures except the run at 125 ℃. The only drawback of the model is that it cannot calculate the HO/LO split. The estimated parameter values for model I and N are shown in Table 18.2. The observed large standard deviations in the parameter estimates is rather typical for Arrhenius type expressions.

In Figures 18.18, 18.19 and 18.20 the experimental data and the calculations based on model I are shown for the high temperature cracking at 360, 397 and 420 ℃ of an Athabasca oil sands bitumen (Drum 20). Similar results are seen in Figures 18.21, 18.22 and 18.23 for another Athabasca oil sands bitumen (Drum 433). The estimated parameter values for model I are shown in Table 18.3 for Drums 20 and 433.

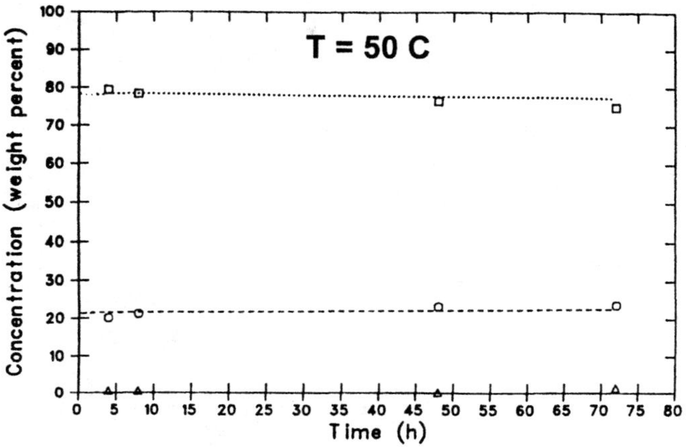

Figure 18.13 *Experimental and calculated concentrations of Coke (COK) "Δ", Asphaltene (ASP) "o" and Heavy Oil + Light Oil (HO+LO) "□" at 50 °C for the low temperature oxidation of North Bodo oil sands bitumen using model I.*

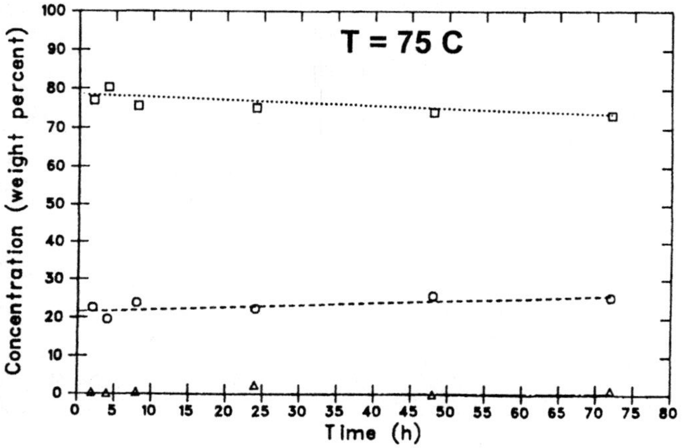

Figure 18.14 *Experimental and calculated concentrations of Coke (COK) "Δ", Asphaltene (ASP) "o" and Heavy Oil + Light Oil (HO+LO) "□" at 75 °C for the low temperature oxidation of North Bodo oil sands bitumen using model I.*

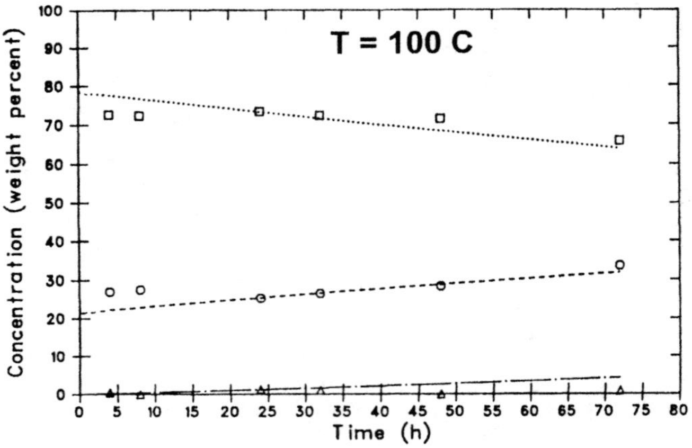

Figure 18.15 *Experimental and calculated concentrations of Coke (COK) "Δ", Asphaltene (ASP) "○" and Heavy Oil + Light Oil (HO+LO) "□" at 100 °C for the low temperature oxidation of North Bodo oil sands bitumen using model I.*

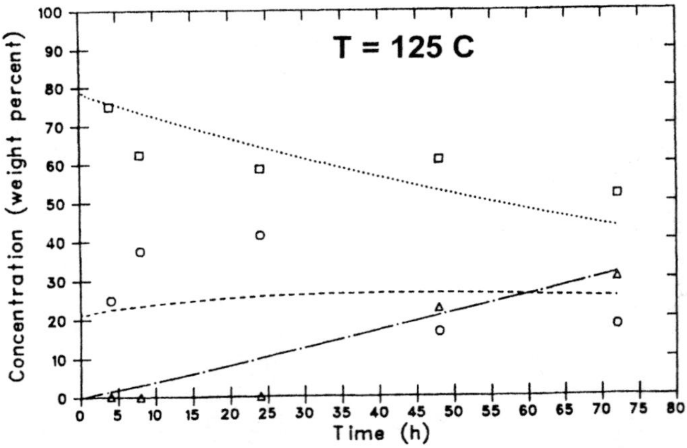

Figure 18.16 *Experimental and calculated concentrations of Coke (COK) "Δ", Asphaltene (ASP) "○" and Heavy Oil + Light Oil (HO+LO) "□" at 125 °C for the low temperature oxidation of North Bodo oil sands bitumen using model I.*

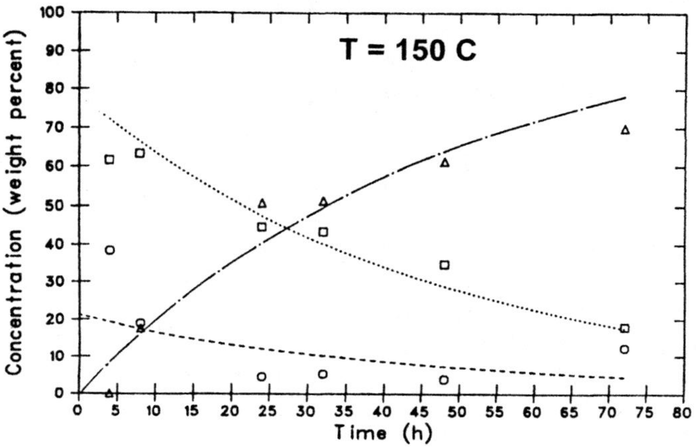

Figure 18.17 *Experimental and calculated concentrations of Coke (COK) "Δ", Asphaltene (ASP) "o" and Heavy Oil + Light Oil (HO+LO) "□" at 150 °C for the low temperature oxidation of North Bodo oil sands bitumen using model I.*

Table 18.2 *Estimated Parameter Values for Models I and N for the Low Temperature Oxidation of North Bodo Oil Sands Bitumen*

Model	Parameter	Parameter Values	Standard Deviation (%)
I	A_1	5.499×10^4	97
	E_1	6255	6.6
	A_2	3.075×10^{11}	272
	E_2	12150	9.2
N	A_1	1.885×10^6	117
	E_1	7866	6.4
	A_2	4.934×10^{12}	435
	E_2	13298	13.5
	A_3	3.450×10^{14}	3218
	E_3	17706	90

Source: Hanson and Kalogerakis (1984).

Table 18.3 Estimated Parameter Values for Model I for High Temperature Cracking of Athabasca Oil Sands Bitumen

Bitumen Code	Parameter	Parameter Values	Standard Deviation (%)
Drum 20	A_1	3.54×10^{4}	8770
	E_1	59320	102
	A_2	8.76×10^{11}	255
	E_2	19485	8.8
Drum 433	A_1	1.23×10^{31}	4297
	E_1	52340	57
	A_2	9.51×10^{15}	235
	E_2	26232	6.1

Source: Hanson and Kalogerakis (1984).

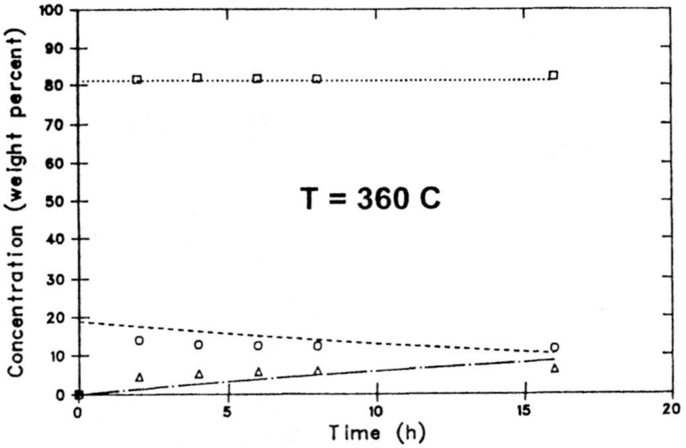

Figure 18.18 Experimental and calculated concentrations of Coke (COK) "Δ", Asphaltene (ASP) "○" and Heavy Oil + Light Oil (HO+LO) "□" at 360 °C for the high temperature cracking of Athabasca oil sands bitumen (Drum 20) using model I.

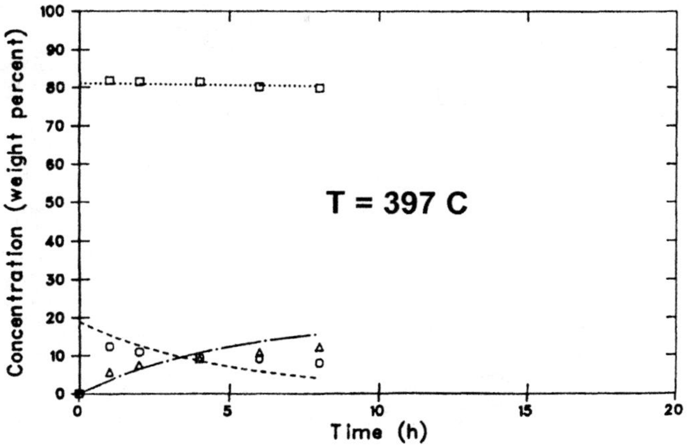

Figure 18.19 *Experimental and calculated concentrations of Coke (COK) "Δ", Asphaltene (ASP) "○" and Heavy Oil + Light Oil (HO+LO) "□" at 397 °C for the high temperature cracking of Athabasca oil sands bitumen (Drum 20) using model I.*

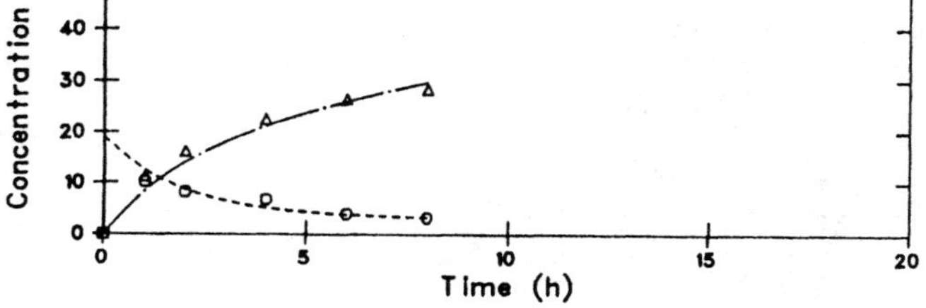

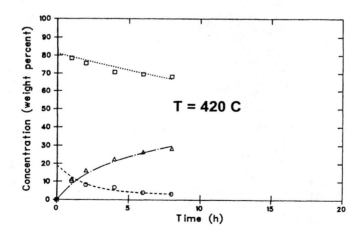

Figure 18.20 *Experimental and calculated concentrations of Coke (COK) "Δ", Asphaltene (ASP) "○" and Heavy Oil + Light Oil (HO+LO) "□" at 420 °C for the high temperature cracking of Athabasca oil sands bitumen (Drum 20) using model I.*

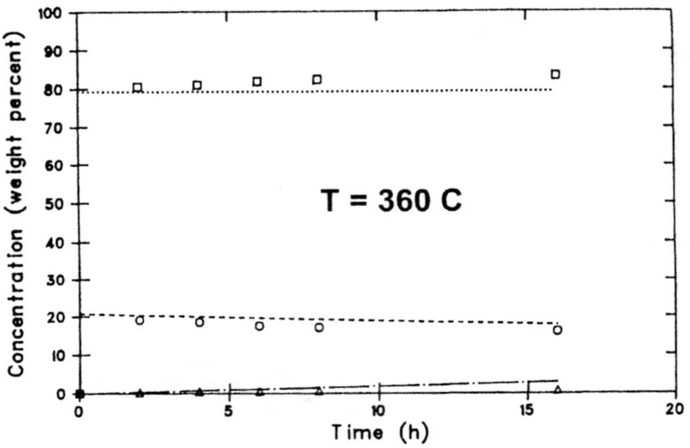

Figure 18.21 *Experimental and calculated concentrations of Coke (COK) "Δ", Asphaltene (ASP) "○" and Heavy Oil + Light Oil (HO+LO) "□" at 360 °C for the high temperature cracking of Athabasca oil sands bitumen (Drum 433) using model I.*

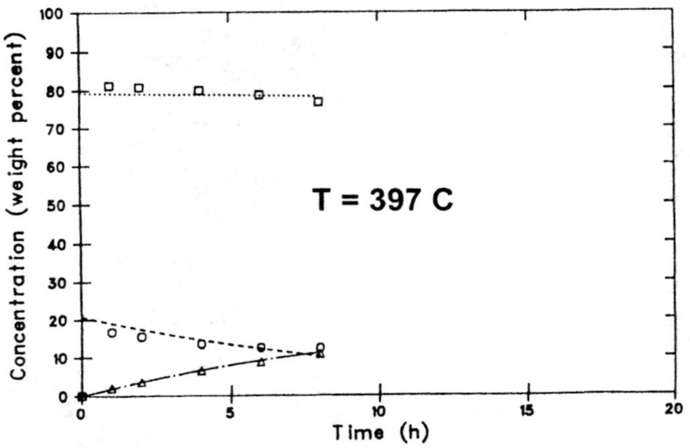

Figure 18.22 *Experimental and calculated concentrations of Coke (COK) "Δ", Asphaltene (ASP) "○" and Heavy Oil + Light Oil (HO+LO) "□" at 397 °C for the high temperature cracking of Athabasca oil sands bitumen (Drum 433) using model I.*

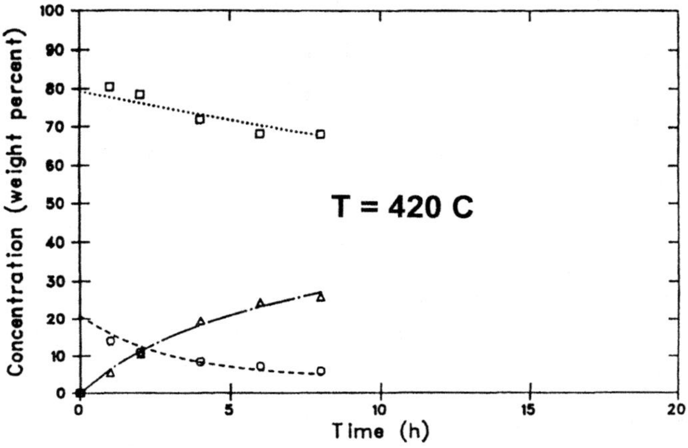

Figure 18.23 Experimental and calculated concentrations of Coke (COK) "Δ", Asphaltene (ASP) "o" and Heavy Oil + Light Oil (HO+LO) "□" at 420 °C for the high temperature cracking of Athabasca oil sands bitumen (Drum 433) using model I.

18.3 AUTOMATIC HISTORY MATCHING IN RESERVOIR ENGINEERING

History matching in reservoir engineering refers to the process of estimating hydrocarbon reservoir parameters (like porosity and permeability distributions) so that the reservoir simulator matches the observed field data in some optimal fashion. The intention is to use the history matched-model to forecast future behavior of the reservoir under different depletion plans and thus optimize production.

18.3.1 A Fully Implicit, Three Dimensional, Three-Phase Simulator with Automatic History-Matching Capability

The mathematical model for a hydrocarbon reservoir consists of a number of partial differential equations (PDEs) as well as algebraic equations. The number of equations depends on the scope/capabilities of the model. The set of PDEs is often reduced to a set of ODES by grid discretization. *The estimation of the reservoir parameters of each grid cell is the essence of history matching.*

The discretized reservoir model can be written in the general form presented in Section 10.3. The state variables are the pressure and the oil, water and gas saturations at each grid cell in a "black oil" reservoir model. The vector of calculated variables **y** is related to the state variables through a nonlinear relationship. For example, the computed water to oil ratio from a production oil well is a nonlinear function of pressure, saturation levels and fluid composition of the grid cells the well is completed in. In some cases we may also observe a very limited number of the state variables. Normally measurements are available from production or observation wells placed throughout the reservoir. In general, function evaluations are time consuming and consequently the number of function evaluations required by the iterative algorithm should be minimized (Tan and Kalogerakis, 1991).

The parameters are estimated by minimizing the usual LS objective function where the weighting matrix is often chosen as a diagonal matrix that normalizes the data and makes all measurements be of the same order of magnitude. Statistically, this is the correct approach if the error in the measurements is proportional to the magnitude of the measured variable.

Tan and Kalogerakis (1991) modified a three-dimensional, three-phase (oil, water, gas) multi-component (N_c components) industrially available simulator. The reservoir simulation model consisted of a set of ODEs that described the component material balances subject to constraints for the saturations, S, and the mole fractions, x_{ip}, i=1,...,Nc, in each of the three phases (oil, gas, water). Each component could be found in any one of the three phases. The state variables were pressure, saturations (S_{gas}, S_{water}, S_{oil}) and the master mole fractions at each grid cell of the reservoir. The master mole fractions are related to the actual mole fractions through the equilibrium distribution ratios also known as "K-values." It is known from phase equilibrium thermodynamics that the "K-values" are a convenient way to describe phase behavior. The parameters to be estimated were porosity (φ) and horizontal and vertical permeability (k_h and k_v) in each grid cell. The observed variables that could be matched were the pressures of the grid cells, water-oil ratios and gas-oil ratios of the individual production wells, and flowing bottom hole pressures. The Gauss-Newton algorithm was implemented with an optimal step size policy to avoid overstepping. An automatic timestep selector adjusted the timesteps in order to ensure that the simulator performs the calculations of the pressures and the producing ratios at the corresponding observation times.

Furthermore, the implementation of the Gauss-Newton method also incorporated the use of the pseudo-inverse method to avoid instabilities caused by the ill-conditioning of matrix **A** as discussed in Chapter 8. In reservoir simulation this may occur for example when a parameter zone is outside the drainage radius of a well and is therefore not observable from the well data. Most importantly, in order to realize substantial savings in computation time, the sequential computation of the sensitivity coefficients discussed in detail in Section 10.3.1 was implemented. Finally, the numerical integration procedure that was used was a fully implicit one to ensure stability and convergence over a wide range of parameter estimates.

18.3.2 Application to a Radial Coning Problem (Second SPE Comparative Solution Problem)

For illustration purposes, Tan and Kalogerakis (1991) applied their history matching approach to the Second SPE Comparative Solution Problem. The problem is described in detail by Chappelear and Nolen (1986). The reservoir consists of ten concentric rings and fifteen layers. A gas cap and oil zone and a water zone are all present. Starting with an arbitrary initial guess of the reservoir porosity and permeability distribution it was attempted to match: (a) the observed reservoir pressure; (b) the water-oil ratio; (c) the gas-oil ratio; (d) the bottom hole pressure; and (e) combinations of observed data. It is noted that the simulator with its built in parameter estimation capability can match reservoir pressures, water-oil ratios, gas-oil ratios and flowing bottom hole pressures either individually or all of them simultaneously. The reservoir model was used to generate artificial observations using the original values of the reservoir parameters and by adding a random noise term. Subsequently, starting with an arbitrary initial guess of the reservoir parameters, it was attempted to recover the original reservoir parameters by matching the "observed" data.

The various simulation runs revealed that the Gauss-Newton implementation by Tan and Kalogerakis (1991) was extremely efficient compared to other reservoir history matching methods reported earlier in the literature.

18.3.2.1 Matching Reservoir Pressure

The fifteen layers of constant permeability and porosity were taken as the reservoir zones for which these parameters would be estimated. The reservoir pressure is a state variable and hence in this case the relationship between the output vector (observed variables) and the state variables is of the form $\mathbf{y}(t_i)=\mathbf{C}\mathbf{x}(t_i)$.

Tan and Kalogerakis (1991) performed four different runs. In the 1st run, layers 3 to 12 were selected as the zones whose horizontal permeabilities were the unknown parameters. The initial guess for all the permeabilities was 200 md. It was found that the original permeability values were obtained in nine iterations of the Gauss-Newton method by matching the pressure profile. In Figures 18.24a and 18.24b the parameter values and the LS objective function are shown during the course of the iterations. The cpu time required for the nine iterations was equivalent to 19.51 model-runs by the simulator. Therefore, the average cpu time for one iteration was 2.17 model-runs. This is the time required for one model-run (computation of all state variables) and for the computation of the sensitivity coefficients of the ten parameters which was obviously 1.17 model-runs. This compares exceptionally well to 10 model-runs that would have normally been required by the standard formulation of the Gauss-Netwon method. These savings in computation time are solely due to the efficient computation of the sensitivity coefficients as described in Section 10.3.1. This result represents an order of magnitude reduction in computational requirements and makes automatic history matching through nonlinear regression practical and economically feasible for the first time in reservoir simulation (Tan and Kalogerakis, 1991).

Next, in the 2nd run the horizontal permeabilities of all 15 layers were estimated by using the value of 200 *md* as initial guess. It required 12 iterations to converge to the optimal permeability values.

In the 3rd run the porosity of the ten zones was estimated by using an initial guess of 0.1. Finally, in the 4th run the porosity of all fifteen zones was estimated by using the same initial guess (0.1) as above. In this case, matrix **A** was found to be extremely ill-conditioned and the pseudo-inverse option had to be used.

Unlike the permeability runs, the results showed that, the observed data were not sufficient to distinguish all fifteen values of the porosity. The ill-conditioning of the problem was mainly due to the limited observability and it could be overcome by supplying more information such as additional data or by a re-parameterization of the reservoir model itself (rezoning the reservoir).

18.3.2.2 Matching Water-Oil Ratio, Gas-Oil Ratio or Bottom Hole Pressure

The water-oil ratio is a complex time-dependent function of the state variables since a well can produce oil from several grid cells at the same time. In this case the relationship of the output vector and the state variables is nonlinear of the form $\mathbf{y}(t_i)=\mathbf{h}(\mathbf{x}(t_i))$.

The centrally located well is completed in layers 7 and 8. The set of observations consisted of water-oil ratios at 16 timesteps that were obtained from a base run. These data were then used in the next three runs to be matched with calculated values. In the 1st run the horizontal permeabilities of layers 7 and 8 were estimated using an initial guess of 200 *md*. The optimum was reached within five iterations of the Gauss-Newton method. In the 2nd run the objective was to estimate the unknown permeabilities in layers 6 to 9 (four zones) using an initial guess of 200 *md*. In spite of the fact that the calculated water-oil ratios agreed with the observed values very well, the calculated permeabilities were not found to be close to the reservoir values. Finally, the porosity of layers 6 to 9 was estimated in the 3rd run in five Gauss-Newton iterations and the estimated values were found to be in good agreement with the reservoir porosity values.

Similar findings were observed for the gas-oil ratio or the bottom hole pressure of each well which is also a state variable when the well production rate is capacity restricted (Tan and Kalogerakis, 1991).

18.3.2.3 Matching All Observed Data

In this case the observed data consisted of the water-oil ratios, gas-oil ratios, flowing bottom hole pressure measurements and the reservoir pressures at two locations of the well (layers 7 and 8). In the first run, the horizontal permeabilities of layers 6 to 9 were estimated by using the value of 200 *md* as the initial guess.

As expected, the estimated values were found to be closer to the correct ones compared with the estimated values when the water-oil ratios are only matched. In the 2nd run, the horizontal permeabilities of layers 5 to 10 (6 zones) were estimated using the value of 200 *md* as initial guess. It was found necessary to use the pseudo-inverse option in this case to ensure convergence of the computations. The initial and converged profiles generated by the model are compared to the observed data in Figures 18.25a and 18.25b.

It was also attempted to estimate permeability values for eight zones but it was not successful. It was concluded that in order to extent the reservoir that can be identified from measurements one needs observation data over a longer history. Finally, in another run, it was shown that the porosities of layers 5 to 10 could be readily estimated within 10 iterations. However, it was not possible to estimate the porosity values for eight layers due to the same reason as the permeabilities.

Having performed all the above computer runs, a simple linear correlation was found relating the required cpu time as model equivalent runs (MER) with the number of unknown parameters regressed. The correlation was

$$MER = 0.319 + 0.07 \times p \tag{18.15}$$

This indicates that after an initial overhead of 0.319 model runs to set up the algorithm, an additional 0.07 of a model-run was required for the computation of the sensitivity coefficients for each additional parameter. This is about 14 times less compared to the one additional model-run required by the standard implementation of the Gauss-Newton method. Obviously these numbers serve only as a guideline however, the computational savings realized through the efficient integration of the sensitivity ODEs are expected to be very significant whenever an implicit or semi-implicit reservoir simulator is involved.

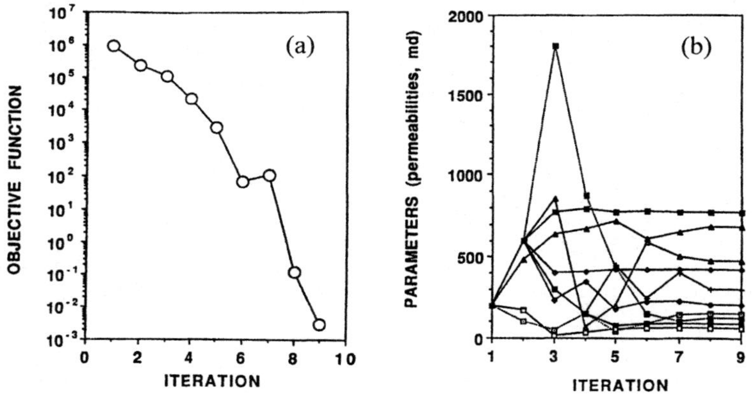

Figure 18.24 *2nd SPE problem with 15 permeability zones and using all measurements: (a) Reduction of LS objective function and (b) Parameter values during the course of the iterations [reprinted with permission from the Society of Petroleum Engineers].*

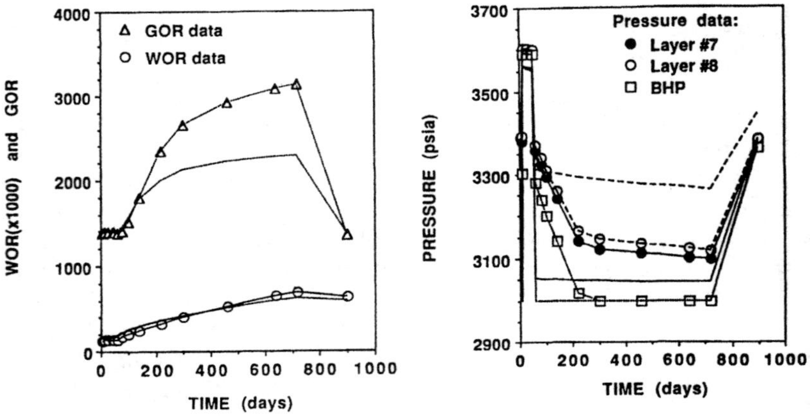

Figure 18.25 Observed data and model calculations for initial and converged parameter values for the 2nd SPE problem. (a) Match of gas-oil ratio and water-oil ratio, (b) Match of bottom-hole pressure and reservoir pressures at layers 7 and 8 [reprinted with permission from the Society of Petroleum Engineers].

18.3.3 A Three-Dimensional, Three-Phase Automatic History-Matching Model: Reliability of Parameter Estimates

An important benefit of using a three-dimensional three-phase automatic history matching simulator is that besides the estimation of the unknown parameters, it is possible to analyze the quality of the fit that has been achieved between the model and the field data. In the history matching phase of a reservoir simulation study, the practicing engineer would like to know how accurately the parameters have been estimated, whether a particular parameter has a significant influence on the history match and to what degree. The traditional procedure of history matching by varying different reservoir parameters based on engineering judgment may provide an adequate match; however, the engineer has no measure of the reliability of the fitted parameters.

As already discussed in Chapter 11, matrix **A** calculated during each iteration of the Gauss-Newton method can be used to determine the covariance matrix of the estimated parameters, which in turn provides a measure of the accuracy of the parameter estimates (Tan and Kalogerakis, 1992).

The covariance matrix, $COV(\mathbf{k}^*)$, of the parameters is given by Equation 11.1, i.e., $COV(\mathbf{k}^*)=[\mathbf{A}^*]^{-1}S_{LS}(\mathbf{k}^*)/(d.f.)$ and the variances of the estimated parameters are simply obtained as the diagonal elements of $COV(\mathbf{k}^*)$. It is reminded that the square root of the variances are the standard deviations of the parameter estimates (also known as the standard estimation error). The magnitude of the standard deviation of a particular parameter estimate indicates the degree of confidence that should be placed on that parameter value.

The covariances between the parameters are the off-diagonal elements of the covariance matrix. The covariance indicates how closely two parameters are correlated. A large value for the covariance between two parameter estimates indicates a very close correlation. Practically, this means that these two parameters may not be possible to be estimated separately. This is shown better through the correlation matrix. The correlation matrix, **R**, is obtained by transforming the covariance matrix as follows

$$\mathbf{R} = \mathbf{D}^{-1}\left[COV(\mathbf{k})\right]\mathbf{D}^{-1} \tag{18.16}$$

where **D** is a diagonal matrix with elements the square root of the diagonal elements of $COV(\mathbf{k}^*)$. The diagonal elements of the correlation matrix **R** will all be equal to one whereas the off-diagonal ones will have values between −1 and 1. If an off-diagonal element has an absolute value very close to one then the corresponding parameters are highly correlated. Hence, the off-diagonal elements of matrix **R** provide a direct indication of two-parameter correlation.

Correlation between three or more parameters is very difficult to detect unless an eigenvalue decomposition of matrix **A*** is performed. As already discussed in Chapter 8, matrix **A*** is symmetric and hence an *eigenvalue decomposition* is also an orthogonal decomposition

$$\mathbf{A}^* = \mathbf{V}\,\mathbf{\Lambda}\,\mathbf{V}^T \tag{18.17}$$

where $\mathbf{\Lambda}$ is a diagonal matrix with positive elements that are the eigenvalues of matrix **A***, **V** is an orthogonal matrix whose columns are the normalized eigenvectors of matrix **A*** and hence, $\mathbf{V}^T\mathbf{V}=\mathbf{I}$. Furthermore, as shown in Chapter 11, the *(1-α)100% joint confidence region* for the parameter vector **k** is described by the ellipsoid

$$[\mathbf{k}-\mathbf{k}^*]^T\,[\mathbf{A}^*]^{-1}\,[\mathbf{k}-\mathbf{k}^*] = \frac{pS(\mathbf{k}^*)}{(d.f.)}\,F^{\alpha}_{p,Nm-p} \tag{18.18}$$

If this ellipsoid is highly elongated in one direction then the uncertainty in the parameter values in that direction is significant. The p eigenvectors of **A** form the direction of the p principal axes of the ellipsoid where each eigenvalue is equal to the inverse of the length of the corresponding principal axis. The longest axis, corresponds to the smallest eigenvalue whereas the shortest axis to the largest eigenvalue. The ratio of the largest to the smallest eigenvalue is the condition number of matrix **A**. As discussed in Chapter 8, the condition number can be calculated during the minimization steps to provide an indication of the ill-conditioning of matrix **A**. If some parameters are correlated then the value of the condition number will be very large. One should be careful, however, to note that the eigenvalues are not in a one-to-one correspondence with the parameters. It requires inspection of the eigenvectors corresponding to the smallest eigenvalues to know which parameters are highly correlated.

The elements of each eigenvector are the cosines of the angles the eigenvector makes with the axes corresponding to the *p* parameters. If some of the parameters are highly correlated then their axes will lie in the same direction and the angle between them will be very small. Hence, the eigenvector corresponding to the smallest eigenvalue will have significant contributions from the correlated parameters and the *cosines* of the angles will tend to 1 since the angles tend to 0. Hence, the larger elements of the eigenvector will enable identification of the parameters which are correlated.

18.3.3.1 Implementation and Numerical Results

Tan and Kalogerakis (1992) illustrated the above points using two simple reservoir problems representing a uniform areal 3×3 model with a producing well at grid block (1,1). In this section their results from the SPE second comparative solution problem are only reported. This is the well-known three-phase radial coning problem described by Chappelear and Nolen (1986). The reservoir has 15 layers each of a constant horizontal permeability. The well is completed at layers 7 and 8. The objective here is to estimate the permeability of each layer using the measurements made at the wells. The measurements include water-oil ratio, gas-oil ratio, flowing bottom-hole pressure and reservoir pressures of the well locations.

Observations were generated running the model with the original description and by adding noise to the model calculated values as follows

$$\hat{y}_j(t_i) = (1 + z_{ij}\sigma_\varepsilon)y_j(t_i) \tag{18.19}$$

where $\hat{y}_j(t_i)$ are the noisy observations of variable j at time t_i, σ_ε is the normalized standard measurement error and z_{ij} is a random variable distributed normally with zero mean and standard deviation one. A value of 0.001 was used for σ_ε.

In the first attempt to characterize the reservoir, all 15 layers were treated as separate parameter zones. An initial guess of 300 *md* was used for the permeability of each zone. A diagonal weighting matrix $\mathbf{Q}_i$ with elements $\hat{y}_j(t_i)^{-2}$ was used. After 10 iterations of the Gauss-Newton method the LS objective function was reduced to 0.0147. The estimation problem as defined, is severely ill-conditioned. Although the algorithm did not converged, the estimation part of the program provided estimates of the standard deviation in the parameter values obtained thus far.

Table 18.4 2nd SPE Problem: Permeability Estimates of Original Zonation

Layer	True Permeability	Estimated Permeability	Standard Deviation (%)	Eigenvector of Smallest Eigenvalue
1	35	237.5	867.7	0.1280
2	47.5	329.2	1416	-0.2501
3	148	22.4	725.2	0.1037
4	202	289.2	40.7	-0.0041
5	90	240.8	69.1	0.0108
6	418.5	324.6	40.3	-0.0056
7	775	676.2	5.33	0.0003
8	60	52.6	20.9	0.0001
9	682	636	116.8	-0.0194
10	472	411.4	121.0	-0.0074
11	125	821.1	261.4	-0.0428
12	300	323.7	1284	-0.2848
13	137.5	159.9	3392	0.8613
14	191	246.1	1241	0.1693
15	350	256.4	1256	-0.2373

Source: Tan and Kalogerakis (1992).

The estimated permeabilities together with the true permeabilities of each layer are shown in Table 18.4. In Table 18.4, the eigenvector corresponding to the smallest eigenvalue of matrix **A** is also shown. The largest elements of the eigenvector correspond to zones 1,2,3,12,13,14 and 15 suggesting that these zones are correlated. Based on the above and the fact that the eigenvector of the second smallest eigenvalue suggests that layers 9,10,11 are also correlated, it was decided to rezone the reservoir in the following manner: Layers 1,2 and 3 are lumped into one zone, layers 4,5,6,7 and 8 remain as distinct zones and layers 9,10,11,12,13,14 and 15 are lumped into one zone. With the revised zonation there are only 7 unknown parameters.

With the revised zonation the LS objective function was reduced to 0.00392. The final match is shown in Figures 18.26 and 18.27. The estimated parameters and their standard deviation are shown in Table 18.5. As seen, the estimated permeability for layer 7 has the lowest standard error. This is simply due to the fact that the well is completed in layers 7 and 8. It is interesting to note that contrary to expectation, the estimated permeability corresponding to layer 5 has a higher uncertainty compared to zones even further away from it. The true permeability of layer 5 is low and most likely its effects have been missed in the timing of the 16 observations taken.

Table 18.5 2nd SPE Problem: Permeability Estimates of Rezoned Reservoir

Layer	True Permeability	Zone	Estimated Permeability	Standard Deviation (%)
1	35	1	246.2	14.4
2	47.5			
3	148			
4	202	2	248.0	21.4
5	90	3	128.1	38.1
6	418.5	4	478.6	10.1
7	775	5	696.8	1.9
8	60	6	67.5	8.7
9	682	7	480.5	12.1
10	472			
11	125			
12	300			
13	137.5			
14	191			
15	350			

Source: Tan and Kalogerakis (1992).

18.3.4 Improved Reservoir Characterization Through Automatic History Matching

In the history matching phase of a reservoir simulation study, the reservoir engineer is phased with two problems: First, a grid cell model that represents the geological structure of the underground reservoir must be developed and second the porosity and permeability distributions across the reservoir must be determined so that the reservoir simulation model matches satisfactorily the field data. In an actual field study, the postulated grid cell model may not accurately represent the reservoir description since the geological interpretation of seismic data can easily be in error. Furthermore, the variation in rock properties could be such that the postulated grid cell model may not have enough detail. Unfortunately, even with current computer hardware one cannot use very fine grids with hundred of thousands of cells to model reservoir heterogeneity in porosity or permeability distribution. The simplest way to address the reservoir description problem is to use a *zonation approach* whereby the reservoir is divided into a relatively small number of zones of constant porosity and permeability (each zone may have many grid cells dictated by accuracy and stability consideration). Chances are that the size and shape of these zones in the postulated model will be quite different than the true distribution in the reservoir. In this section it is shown how automatic history matching can be used to guide the practicing engineer towards a reservoir characterization that is more representative of the actual underground reservoir (Tan and Kalogerakis, 1993).

In practice when reservoir parameters such as porosities and permeabilities are estimated by matching reservoir model calculated values to field data, one has some prior information about the parameter values. For example, porosity and permeability values may be available from core data analysis and well test analysis. In addiction, the parameter values are known to be within certain bounds for a particular area. All this information can be incorporated in the estimation method of the simulator by introducing prior parameter distributions and by imposing constraints on the parameters (Tan and Kalogerakis, 1993).

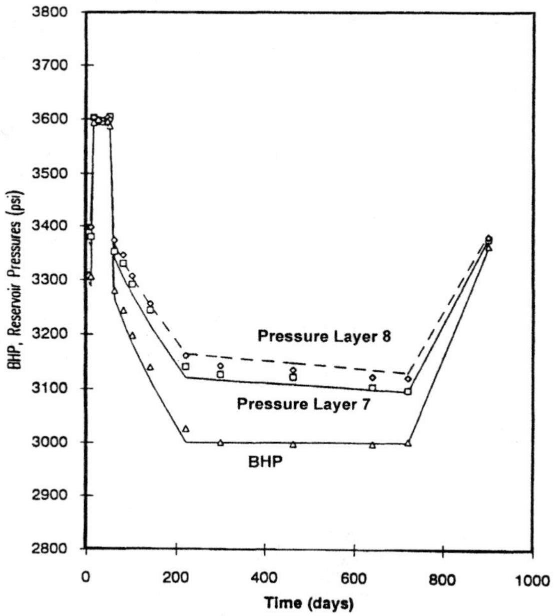

Figure 18.26 *Observed and calculated bottom-hole pressure and reservoir pressures at layers 7 and 8 for the 2nd SPE problem using 7 permeability zones [reprinted from the Journal of the Canadian Petroleum Technology with permission].*

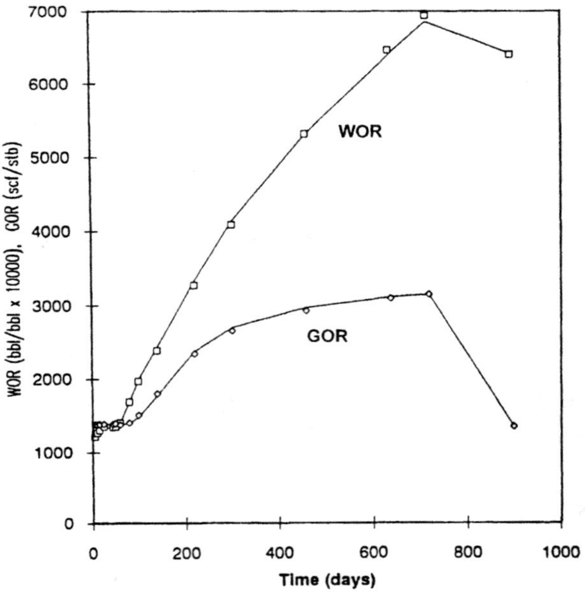

Figure 18.27 *Observed and calculated water-oil ratio and gas-oil ratio for the 2nd SPE problem using 7 permeability zones [reprinted from the Journal of the Canadian Petroleum Technology with permission].*

18.3.4.1 Incorporation of Prior Information and Constraints on the Parameters

It is reasonable to assume that the most probable values of the parameters have normal distributions with means equal to the values that were obtained from well test and core data analyses. These are the prior estimates. Each one of these most probable parameter values (k_{Bj}, j=1,...,p) also has a corresponding standard deviation σ_{kj} which is a measure of the uncertainty of the prior parameter estimate. As already discussed in Chapter 8 (Section 8.5) using maximum likelihood arguments the prior information is introduced by augmenting the LS objective function to include

$$S_{prior} = [\mathbf{k} - \mathbf{k}_B]^T \mathbf{V}_B^{-1} [\mathbf{k} - \mathbf{k}_B] \tag{18.20}$$

The prior covariance matrix of the parameters ($\mathbf{V}_B$) is a diagonal matrix and since in the solution of the problem only the inverse of $\mathbf{V}_B$ is used, it is preferable to use as input to the program the inverse itself. Namely, the information is entered into the program as $\mathbf{V}_B^{-1} = diag(\sigma_{k1}^{-2}, \sigma_{k2}^{-2},..., \sigma_{kp}^{-2})$.

Several authors have suggested the use of a weighting factor for S_{prior} in the overall objective function. As there is really no rigorous way to assign an optimized weighting factor, it is better to be left up to the user (Tan and Kalogerakis, 1993). If there is a significant amount of historical observation data available, the objective function will be dominated by the least squares term whereas the significance of the prior term increases as the amount of such data decreases. If the estimated parameter values differ significantly from the prior estimates then the user should either revise the prior information or assign it a greater significance by increasing the value of the weighting factor or equivalently decrease the value of the prior estimates σ_{kj}. As pointed out in Section 8.5, the crucial role of the prior term should be emphasized when the estimation problem is severely ill-conditioned. In such cases, the parameters that are not affected by the data maintain essentially their prior estimates (k_j and σ_{kj}) and at the same time, the condition number of matrix **A** does not become prohibitively large.

While prior information may be used to influence the parameter estimates towards realistic values, there is no guarantee that the final estimates will not reach extreme values particularly when the postulated grid cell model is incorrect and there is a large amount of data available. A simple way to impose inequality constraints on the parameters is through the incorporation of a penalty function as already discussed in Chapter 9 (Section 9.2.1.2). By this approach extra terms are added in the objective function that tend to explode when the parameters approach near the boundary and become negligible when the parameters are far. One can easily construct such penalty functions. For example a simple and yet very effective penalty function that keeps the parameters in the interval ($k_{min,i}$, $k_{max,i}$) is

$$\Theta_i(k_i) = \frac{k_{max,i} - k_{min,i}}{k_i - k_{min,i}} + \frac{k_{max,i} - k_{min,i}}{k_{max,i} - k_i} \quad ; \; i = 1,...,p \tag{18.21}$$

These functions are also multiplied by a user supplied weighting constant, ω (≥ 0) which should have a large value during the early iterations of the Gauss-Newton method when the parameters are away from their optimal values. In general, ω should be reduced as the parameters approach the optimum so that the contribution of the penalty function is essentially negligible (so that no bias is introduced in the parameter estimates). If p penalty functions are incorporated then the overall objective function becomes

$$S_{total}(\mathbf{k}) = S_{LS} + S_{prior} + S_{penalty} \tag{18.22}$$

where

$$S_{penalty} = \omega \sum_{i=1}^{p} \Theta_i(k_i) \tag{18.23}$$

As seen above, the objective function is modified in a way that it increases quite significantly as the solution approaches the constraints but remains practically unchanged inside the feasible region. Inside the feasible region, $\Theta_i(k_i)$ is small and results in a small contribution to the main diagonal element of matrix **A**. Consequently, $\Delta\mathbf{k}^{(j+1)}$ will not be affected significantly. If on the other hand, the parameter values are near the boundary $\Theta_i(k_i)$ becomes large and dominates the diagonal element of the parameter to which the constraint is applied. As a result, the value of $\Delta\mathbf{k}^{(j+1)}$ is very small and k_i is not allowed to cross that boundary.

18.3.4.2 Reservoir Characterization Using Automatic History Matching

As it was pointed out earlier, there are two major problems associated with the history matching of a reservoir. The first is the correct representation of the reservoir with a grid cell model with a limited number of zones, and the second is the regression analysis to obtain the optimal parameter values. While the second problem has been addressed in this chapter, there is no rigorous method available that someone may follow to address the first problem. Engineers rely on information from other sources such as seismic mapping and geological data. However, Tan and Kalogerakis (1993) have shown that automatic history matching can also be useful in determining the existence of impermeable boundaries, assessing the possibility of reservoir extensions, providing estimates of the volumes of oil, gas and water in place while at the same time limiting the parameter estimates to stay within realistic limits.

They performed an extensive case study to demonstrate the use of automatic history matching to reservoir characterization. For example, if the estimated permeability of a particular zone is unrealistically small compared to geological information, there is a good chance that an impermeable barrier is present. Similarly if the estimated porosity of a zone approaches unrealistically high values, chances are the zone of the reservoir should be expanded beyond its current boundary.

Based on the analysis of the case study, Tan and Kalogerakis (1993) recommended the following practical guidelines.

Step 1. Construct a postulated grid cell model of the reservoir by using all the available information.

Step 2. Subdivide the model into zones by following any geological zonation as closely as possible. The zones for porosity and permeability need not be identical and the cells allocated to a zone need not be contiguous.

Step 3. Provide estimates of the most probable values of the parameters and their standard deviation (k_j, σ_{kj}, j=1,...,p).

Step 4. Provide the boundary (minimum and maximum values) that each parameter should be constrained with.

Step 5. Run the automatic history matching model to obtain estimates of the unknown parameters.

Step 6. When a converged solution has been achieved or at the minimum of the objective function check the variances of the parameters and the eigenvectors corresponding to the smallest eigenvalues to identify any highly correlated zones. Combine any adjacent zones of high variance.

Step 7. In order to modify and improve the postulated grid cell representation of the reservoir, analyze any zones with values close to these constraints.

Step 8. Go to Step 3 if you have made any changes to the model. Otherwise proceed with the study to predict future reservoir performance, etc.

It should be noted that the nature of the problem is such that it is practically impossible to obtain a postulated model which is able to *uniquely* represent the reservoir. As a result, it is required to continuously update the match when additional information becomes available and possibly also change the grid cell description of the reservoir.

By using automatic history matching, the reservoir engineer is not faced with the usual dilemma whether to reject a particular grid cell model because it is not a good approximation to the reservoir or to proceed with the parameter search because the best set of parameters has not been determined yet.

18.3.5 Reliability of Predicted Well Performance Through Automatic History Matching

It is of interest in a reservoir simulation study to compute future production levels of the history matched reservoir under alternative depletion plans. In addition, the sensitivity of the anticipated performance to different reservoir descriptions is also evaluated. Such studies contribute towards assessing the risk associated with a particular depletion plan.

Kalogerakis and Tomos (1995) presented an efficient procedure for the determination of the 95 percent confidence intervals of forecasted horizontal well production rates. Such computations are usually the final task of a reservoir simulation study. The procedure is based on the Gauss-Newton method for the estimation of the reservoir parameters as was described earlier in this chapter. The estimation of the 95 % confidence intervals of well production rates was done as follows.

The production rates for oil (Q_{oj}), water (Q_{wj}) and gas (Q_{gj}) from a vertical or horizontal well completed in layer j are given by the following equations

$$Q_{oj} = W_{pi} M_o (P_{bh} + H - P_{grid}) / B_o \tag{18.24a}$$

$$Q_{wj} = W_{pi} M_w (P_{bh} + H - P_{grid}) / B_w \tag{18.24b}$$

$$Q_{gj} = W_{pi} M_g (P_{bh} + H - P_{grid}) E_g + R_s Q_o \tag{18.24c}$$

where W_{pi} is the well productivity index given by

$$W_{pi} = \frac{2\pi h \sqrt{k_h k_v}}{ln\left(\frac{r_o}{r_w}\right) + F_{skin}} \tag{18.25}$$

and M is defined for each phase as the ratio of the relative permeability to the viscosity of the oil, water or gas, P_{bh} is the flowing bottom hole pressure, H is the hydrostatic pressure (assumed zero for the top layer), B_o and B_w are the formation volume factors for oil and water, R_s is the solution gas-oil ratio, E_g is the gas expansion factor, k_h and k_v are the horizontal and vertical permeabilities and h is the block thickness. The determination of r_o, the effective well radius and F_{skin}, the skin factor, for horizontal wells has been the subject of considerable discussion in the literature.

The total oil production rate is obtained by the sum of all producing layers of the well, i.e.,

$$Q_o \equiv Q_o(\mathbf{x}, \mathbf{k}) = \sum_{j=1}^{N_L} Q_{oj} \tag{18.26}$$

Similar expressions are obtained for the total gas (Q_g) and water (Q_w) production rates by the sum of the Q_{gj} and Q_{wj} terms over all N_L layers.

In order to derive estimates of the uncertainty of the computed well production rates, the expression for the well production rate Q_c (c=o, w, g) is linearized around the optimal reservoir parameters to arrive at an expression for the dependence of Q_c on the parameters (**k**) at any point in time t, namely,

$$Q_c(\mathbf{k},t) = Q_c(\mathbf{k}^*,t) + \left(\frac{dQ_c}{d\mathbf{k}}\right)^T (\mathbf{k}-\mathbf{k}^*) \tag{18.27}$$

where **k*** represents the converged values of the parameter vector **k**.

The *overall sensitivity of the well production rate* can now be estimated through implicit differentiation from the sensitivity coefficients already computed as part of the Gauss-Newton method computations as follows

$$\frac{dQ_c}{d\mathbf{k}} = \mathbf{G}(t)^T \left(\frac{\partial Q_c}{\partial \mathbf{x}}\right) + \left(\frac{\partial Q_c}{\partial \mathbf{k}}\right) \tag{18.28}$$

where **G**(t) is the parameter sensitivity matrix described in Chapter 6. The evaluation of the partial derivatives in the above equation is performed at **k**=**k*** and at time t.

Once, ($dQ_c/d\mathbf{k}$) has been computed, the variance $\sigma^2_{Q_c}$ of the well production rate can be readily obtained by taking variances from both sides of Equation 18.27 to yield,

$$\sigma^2_{Q_c} = \left(\frac{dQ_c}{d\mathbf{k}}\right)^T COV(\mathbf{k}) \left(\frac{dQ_c}{d\mathbf{k}}\right) \tag{18.29}$$

Having estimated $\sigma^2_{Q_c}$, the (1-α)% confidence interval for Q_c at any point in time is given by

$$Q_c(\mathbf{k}^*,t) - t^{\nu}_{\alpha/2}\sigma_{Q_c} \le Q_c(\mathbf{k},t) \le Q_c(\mathbf{k}^*,t) + t^{\nu}_{\alpha/2}\sigma_{Q_c} \tag{18.30}$$

where $t^{\nu}_{\alpha/2}$ is obtained from the statistical tables for the *t-distribution* for ν degrees of freedom which are equal to the total number of observations minus the number of unknown parameters.

If there are N_w production wells then the (1-α)% confidence interval for the *total production* from the reservoir is given by

$$Q_{c|tot}(\mathbf{k}^*,t) - t^{\nu}_{\alpha/2}\sigma_{Q_{c|tot}} \le Q_{c|tot}(\mathbf{k},t) \le Q_{c|tot}(\mathbf{k}^*,t) + t^{\nu}_{\alpha/2}\sigma_{Q_{c|tot}} \tag{18.31}$$

where

$$Q_{c|tot}(\mathbf{k}^*, t) = \sum_{i=1}^{N_w} Q_{c,i}(\mathbf{k}^*, t) \tag{18.32}$$

and $\sigma_{Q_c|tot}$ is the standard error of estimate of the total reservoir production rate which is given by

$$\sigma^2_{Q_c|tot} = \left[\sum_{i=1}^{N_w}\left(\frac{dQ_{c,i}}{d\mathbf{k}}\right)^T\right] COV(\mathbf{k}) \left[\sum_{i=1}^{N_w}\left(\frac{dQ_{c,i}}{d\mathbf{k}}\right)\right] \tag{18.33}$$

It is noted that the partial derivatives in the above variance expressions depend on time t and therefore the variances should be computed simultaneously with the state variables and sensitivity coefficients. Finally, the confidence intervals of the *cumulative production* of each well and of the total reservoir are calculated by integration with respect to time (Kalogerakis and Tomos, 1995).

18.3.5.1 Quantification of Risk

Once the standard error of estimate of the mean forecasted response has been estimated, i.e., the uncertainty in the total production rate, one can compute the probability level, α, for which the minimum total production rate is below some pre-determined value based on a previously conducted economic analysis. Such calculations can be performed as part of the post-processing calculations.

18.3.5.2 Multiple Reservoir Descriptions

With the help of automatic history matching, the reservoir engineer can arrive at several plausible history matched descriptions of the reservoir. These descriptions may differ in the grid block representation of the reservoir, existence of sealing and non-sealing faults, or simply different zonations of constant porosity or permeability (Kalogerakis and Tomos, 1995). In addition, we may assign to each one of these reservoir models a probability of being the correct one. This probability can be based on additional geological information about the reservoir as well as the plausibility of the values of the estimated reservoir parameters.

Thus, for the rth description of the reservoir, based on the material presented earlier, one can compute expected total oil production rate as well as the minimum and maximum production rates corresponding to a desired confidence level (1-α).

The minimum and maximum total oil production rates for the rth reservoir description are given by

$$Q^{(r)}_{o|tot,min}(\mathbf{k},t) = Q^{(r)}_{o|tot}(\mathbf{k}^*,t) - t^{\nu}_{\alpha/2}\sigma^{(r)}_{Q_{o|tot}} \qquad (18.34)$$

and

$$Q^{(r)}_{o|tot,max}(\mathbf{k},t) = Q^{(r)}_{o|tot}(\mathbf{k}^*,t) + t^{\nu}_{\alpha/2}\sigma^{(r)}_{Q_{o|tot}} \qquad (18.35)$$

Next, the probability, $P_b(r)$ that the rth model is the correct one, can be used to compute the *expected overall field production rate* based on the production data from N_m different reservoir models, namely

$$E\left\{Q_{o|tot}(t)\right\} = \sum_{r=1}^{N_m} P_b(r) Q^{(r)}_{o|tot}(t) \qquad (18.36)$$

Similarly the minimum and maximum bounds at $(1-\alpha)100\%$ confidence level are computed as

$$E\left\{Q_{o|tot,min}(t)\right\} = \sum_{r=1}^{N_m} P_b(r)\, Q^{(r)}_{o|tot,min}(t) \qquad (18.37)$$

and

$$E\left\{Q_{o|tot,max}(t)\right\} = \sum_{r=1}^{N_m} P_b(r)\, Q^{(r)}_{o|tot,max}(t) \qquad (18.38)$$

Again, the above limits can be used to compute the risk level to meet a desired production rate from the reservoir.

18.3.5.3 Case Study – Reliability of a Horizontal Well Performance

Kalogerakis and Tomos (1995) demonstrated the above methodology for a horizontal well by adapting the example used by Collins et al. (1992). They assumed a period of 1,250 days for history matching purposes and postulated three different descriptions of the reservoir. All three models matched the history of the reservoir practically with the same success. However, the forecasted performance from 1,250 to 3,000 days was quite different. This is shown in Figures 18.28a and 18.28b where the predictions by Models B and C are compared to the actual reservoir performance (given by Model A). In addition, the 95% confidence intervals for Model A are shown in Figures 18.29a and 18.29b. Form these plots it appears that the uncertainty is not very high; however, it should be kept in mind that these computations were made under the null hypothesis that all other parameters in the reservoir model (i.e., relative permeabilities) and the grid cell description of the are all known precisely.

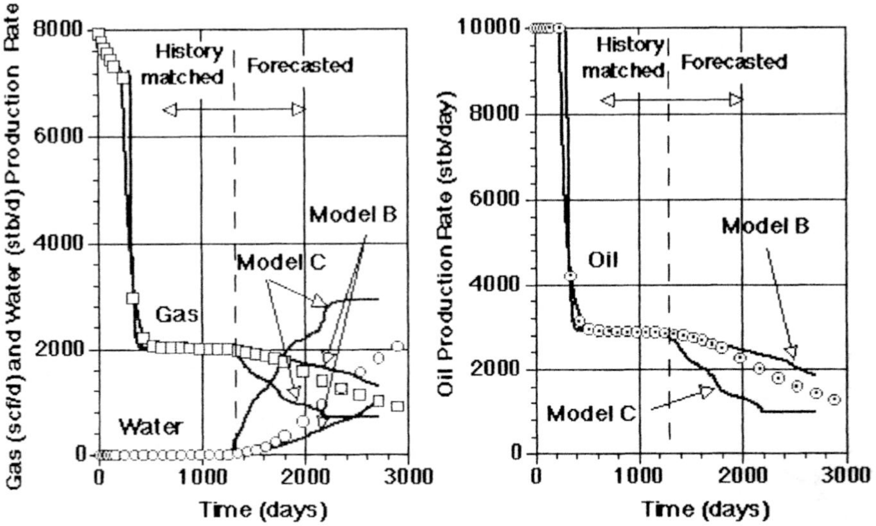

Figure 18.28 *History matched and forecasted production by Models B and C and comparison with actual : (a) Water and Gas production, (b) Oil production [reprinted from the Journal of the Canadian Petroleum Technology with permission].*

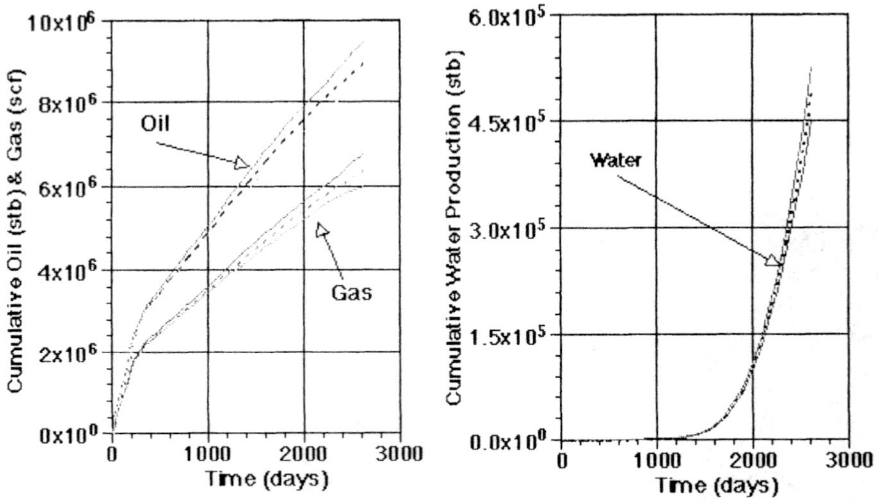

Figure 18.29 *95% confidence intervals of history matched and forecasted production by Model A: (a) Oil and Gas production, (b) Water production [reprinted from the Journal of the Canadian Petroleum Technology with permission].*

References

Andersen, J.N., "Temperature Effect on Recombinant Protein Production Using a Baculovirus/Insect Cell Expression System", *Diploma Thesis*, University of Calgary and Technical University of Denmark, 1995.

Andersen, J.N., P.G. Sriram, N. Kalogerakis and L.A. Behie, "Effect of Temperature on Recombinant protein Production Using the Bm5/Bm5.NPV Expression System", *Can. J. Chem. Eng.*, 74, 511-517 (1996).

Anderson T.F, D.S. Adams and E.A. Greens II, "Evaluation of Parameters for Nonlinear Thermodynamic Models", *AIChE J.,* 24: 20-29 (1978).

Anderson, VL. and R.A. McLean, *Design of Experiments*, M. Dekker, NewYork, NY, 1974.

Appleyard, J.R. and L.M. Chesire, "Nested Factorization" paper SPE 12264 presented at the 1983 SPE Symposium on Reservoir Simulation, San Francisco, CA (1983).

Aris, R., *Mathematical Modelling Techniques*, Dover Publications, Inc., New York, NY, 1994.

Ayen, R.J., and M.S. Peters, "Catalytic Reduction of Nitric Oxide", *Ind Eng Chem Proc Des Dev*, 1, 204-207 (1962).

Bard, Y., "Comparison of Gradient Methods for the Solution of Nonlinear Parameter Estimation Problems", *SIAM J. Numer. Anal.*, 7, 157-186 (1970).

Bard, Y., *Nonlinear Parameter Estimation*, Academic Press, New York, NY, 1974.

Barnett, V., T. Lewis, and V. Rothamsted, *Outliers in Statistical Data*, 3rd edition, Wiley, New York, NY, 1994.

Basmadjian, D., *The Art of Modeling in Science and Engineering*, Chapman & Hall, CRC, Boca Raton, FL, 1999.

Bates, D.M., and D.G. Watts, *Nonlinear Regression Analysis and its Applications*, J. Wiley, New York, NY, 1988.

Bates, D.M., and D.G. Watts, *Nonlinear Regression Analysis and its Applications*, J. Wiley, New York, NY, 1988.

Beck, J.V., and K.J. Arnold, *Parameter Estimation in Engineering and Science*, J. Wiley, New York, NY, 1977.

Bellman, R., and R. Kalaba, *Quasilinearization and Nonlinear Boundary Value Problems*, American Elsevier, New York, NY, 1965.

Bellman, R., J. Jacquez, R. Kalaba, and S. Schwimmer, "Quasilinearization and the Estimation of Chemical rate Constants from Raw Kinetic Data", *Math. Biosc. 1*, 71-76 (1967).

Belohlav, Z., P. Zamostny, P. Kluson, and J. Volf, "Application of Random-Search Algorithm for Regression Analysis of Catalytic Hydrogenations", *Can J. Chem. Eng.*, 75, 735-742 (1997).

Bevington, P.R., and D.K. Robinson, *Data Reduction and Error Analysis for the Physical Sciences*, 2nd ed., McGraw Hill, New York, NY, 1992.

Bishop, C.M., *Neural Networks for Pattern Recognition*, Oxford Univ. Press, UK, 1995.

Blanch, H.W. and D.S. Clark, *Biochemical Engineering*, Marcel Dekker, New York, NY, 1996.

Blanco, B., S. Beltran, J.L. Cabezas, and J. Coca, "Vapor-Liquid Equilibria of Coal-Derived Liquids. 3. Binary Systems with Tetralin at 200 mmHg", *J. Chem. Eng. Data*, 39, 23-26 (1994).

Bourgoyne, A.T., Jr., K. K. Millheim, M. E. Chenevert, and F.S. Young, Jr., *Applied Drilling Engineering*, SPE Textbook Series Vol. 2, 1986.

Box, G.E.P. and W.J. Hill, "Discrimination Among Mechanistic Models", *Technometrics*, 9, 57-71 (1967)

Box, G.E.P., G.M. Jenkins, G.C. Reinsel and G. Jenkins, *Time Series Analysis: Forecasting & Control*, 3rd Ed., Prentice Hall, Eglewood-Cliffs, NJ, 1994.

Box, M.J., "A Comparison of Several Current Optimization Methods, and the Use of Transformations in Constrained Problems", *Computer J.*, 9, 67-77 (1966).

Box, M.J., "Improved Parameter Estimation", *Technometrics*, 12, 219-229 (1970).

Brelland, E. and P. Englezos, "Equilibrium Hydrate Formation Data for carbon Dioxide in Aqueous Glycerol Solutions", *J. Chem. Eng. Data*, 41, 11-13 (1996).

Britt, H.I., and R.H. Luecke, "The Estimation of Parameters in Nonlinear, Implicit Models", *Technometrics*, 15, 233-247 (1973).

Buzzi-Ferraris, G. and P. Forzatti, "A New Sequential Experimental Design Procedure for Discriminating Among Rival Models", *Chem. Eng. Sci.*, 38, 225-232 (1983).

Buzzi-Ferraris, G., P. Forzatti, G. Emig and H. Hofmann, "Sequential Experimental Design for Model Discrimination in the Case of Multiple Responses", *Chem. Eng. Sci.*, 39(1), 81-85 (1984).

Bygrave, G., "Thermodynamics of Metal Partitioning in Kraft Pulps", *M.A.Sc. Thesis*, Dept of Chemical and Bio-Resource Engineering, Univ. of British Columbia, BC, Canada, 1997.

Campbell S.W., R.A. Wilsak and G. Thodos, "Isothermal Vapor-Liquid Equilibrium Measurements for the n-Pentane-Acetone System at 327.7, 397.7 and 422.6 K", *J. Chem. Eng. Data.* 31, 424-430 (1986).

Carroll J.J. and A.E. Mather, "Phase Equilibrium in the System Water-Hydrogen Sulfide: Experimental Determination of the LLV Locus", *Can. J. Chem. Eng.*, *67*, 468-470 (1989a).

Carroll J.J. and A.E. Mather, "Phase Equilibrium in the System Water-Hydrogen Sulfide: Modelling the Phase Behavior with an Equation of State", *Can. J. Chem. Eng.*, *67*, 999-1003 (1989b).

Černy, V., "Thermodynamic Approach to the Travelling Salesman Problem: An Efficient Simulation Algorithm", *J. Opt. Theory Applic.,* 45, 41-51 (1985).

Chappelear, J.E., and Nolen, J.S., "Second Comparative Project: A Three Phase Coning Study", *Journal of Petroleum Technology*, 345-353, March (1986).

Chourdakis, A., "PULSAR Oxygenation System", Operation Manual, Herakleion, Crete (1999).

Collins, D., L. Nghiem, R.. Sharma, Y.K. Li, and K.Jha, "Field-scale Simulation of Horizontal Wells"; *J. Can. Petr. Technology*, 31(1), 14-21 (1992).

Copp J. L., and D.H. Everett, "Thermodynamics of Binary Mixtures Containing Amines", *Disc. Faraday Soc.,* 15, 174-181 (1953).

Donnely, J.K. and D. Quon, "Identification of Parameters in Systems of Ordinary Differential Equations Using Quasilinearization and Data Pertrubation", *Can. J. Chem. Eng.,* 48, 114 (1970).

Draper, N. and H. Smith, *Applied Regression Analysis*, 3rd ed., Wiley, New York, NY, 1998.

Dueck, G. and T. Scheuer, "Threshold Accepting: A General Purpose Optimization Algorithm Appearing Superior to Simulated Annealing", *J. Computational Physics*, 90, 61 (1990).

Duever, T.A., S. E. Keeler, P.M. Reilly, J.H. Vera, and P.A. Williams, "An Application of the Error-In-Variables Model-Parameter Estimation from van Ness-type Vapour-Liqud Equilibrium Experiments", *Chem. Eng. Sci.*, 42, 403-412 (1987).

Dumez, F.J. and G.F. Froment, "Dehydrogenation of 1-Butene into Butadiene. Kinetics, Catalyst Coking, and Reactor Design", *Ind. Eng. Chem. Proc. Des. Devt.,* 15, 291-301 (1976).

Dumez, F.J., L.H. Hosten and G.F. Froment, "The Use of Sequential Discrimination in the Kinetic Study of 1-Butene Dehydrogenetion", *Ind. Eng. Chem. Fundam.,* 16, 298-301 (1977).

Dunker, A.M., "The Decoupled Direct Method for Calculating Sensitivity Coefficients in Chemical Kinetics", *J. Chem. Phys.*, 81, 2385-2393 (1984).

Edgar, T.F. and D.M. Himmelblau, *Optimization of Chemical Processes*, McGraw-Hill, New York, NY, 1988.

Edsberg, L., "Numerical Methods for Mass Action Kinetics" in *Numerical Methods for Differential Systems*, Lapidus, L. and Schlesser W.E., Eds., Academic Press, New York, NY, 1976.

Englezos, P. and N. Kalogerakis, "Constrained Least Squares Estimation of Binary Interaction Parameters in Equations of State", *Computers Chem. Eng.,* 17, 117-121 (1993).

Englezos, P. and S. Hull, "Phase Equilibrium Data on Carbon Dioxide Hydrate in the Presence of Electrolytes, Water Soluble Polymers and Montmorillonite", *Can J. Chem. Eng.*, 72, 887-893 (1994).

Englezos, P., G. Bygrave, and N. Kalogerakis, "Interaction Parameter Estimation in Cubic Equations of State Using Binary Phase Equilibrium & Critical Point Data", *Ind. Eng. Chem. Res.*37(5), 1613-1618 (1998).

Englezos, P., N. Kalogerakis and P.R. Bishnoi, "A Systematic Approach for the Efficient Estimation of Interaction Parameters in Equations of State Using Binary VLE Data", *Can. J. Chem. Eng.*, 71, 322-326 (1993).

Englezos, P., N. Kalogerakis and P.R. Bishnoi, "Estimation of Binary Interaction Parameters for Equations of State Subject to Liquid Phase Stability Requirements", *Fluid Phase Equilibria*, 53, 81-88, (1989).

Englezos, P., N. Kalogerakis and P.R. Bishnoi, "Simultaneous Regression of Binary VLE and VLLE Data", *Fluid Phase Equilibria*, 61, 1-15 (1990b).

Englezos, P., N. Kalogerakis, M.A. Trebble and P.R. Bishnoi, "Estimation of Multiple Binary Interaction Parameters in Equations of State Using VLE Data: Application to the Trebble-Bishnoi EOS", *Fluid Phase Equilibria*, 58, 117-132 (1990a).

Englezos, P., N. Kalogerakis, P.D. Dholabhai and P.R. Bishnoi, "Kinetics of Formation of Methane and Ethane Gas Hydrates", *Chem. Eng. Sci.*, 42, 2647-2658 (1987).

Ferguson, E.S., *Engineering and the Mind's Eye*, MIT Press, Boston, MA, 1992.

Forestell, S.P., N. Kalogerakis, L.A. Behie and D.F. Gerson, "Development of the Optimal Inoculation Conditions for Microcarrier Cultures", *Biotechnol. Bioeng.,* 39, 305-313 (1992).

Frame, K.K. and W.S. Hu, "A Model for Density-Dependent Growth of Anchorage Dependent Mammalian Cells", *Biotechnol. Bioeng.,* 32, 1061-1066 (1988).

Franckaerts, J. and G.F. Froment, "Kinetic Study of the Dehydrogenation of Ethanol.", *Chem. Eng. Sci.,* 19, 807-818 (1964).

Freund, R.J. and P.D. Minton, *Regression Methods*, Marcel Dekker, New York, NY, 1979.

Froment, G.F., "Model Discrimination and Parameter Estimation in Heterogeneous Catalysis", *AIChE J.,* 21 1041 (1975).

Froment, G.F., and K. B. Bischoff, *Chemical Reactor Analysis and Design,* 2nd ed., J. Wiley, New York, NY, 1990.

Galivel - Solastiuk F., S. Laugier and D. Richon, "Vapor-Liquid Equilibrium Data for the Propane-Methanol-CO_2 System", *Fluid Phase Equilibria,* 28, 73-85 (1986).

Gallot, J.E., M.P. Kapoor and S. Kaliaguine, "Kinetics of 2-hexanol and 3-Hexanol Oxidation Reaction over TS-1 Catalysts", *AIChE Journal*, 44 (6), 1438-1454 (1998).

Gans, P., *Data Fitting in the Chemical Sciences by the Method of Least Squares*, Wiley, New York, NY, (1992).

Gill, P.E. and W. Murray, "Newton-type Methods for Unconstrained and Linearly Constrained Optimization", *Mathematical Programming*, 7, 311-350 (1974).

Gill, P.E. and W. Murray, "Nonlinear Least Squares and Nonlinearly Constrained Optimization", *Lecture Notes in Mathematics No 506*, G.A. Watson, Ed, Springer-Verlag, Berlin and Heidelberg, pp 134-147 (1975).

Gill, P.E. and W. Murray, "Quasi-Newton Methods for Unconstrained Optimization", *J. Inst. Maths* Applics, 9, 91-108 (1972).

Gill, P.E., W. Murray, and M. H. Wright, *Practical Optimization*, Academic Press, London, UK, 1981.

Glover, LK., "A New Simulated Annealing Algorithm for Standard Cell Placement", *proc. IEEE Int. Conf. On Computer-Aided Design*, Santa Clara, 378-380 (1986).

Goldfard, D., "Factorized Variable Metric Methods for Unconstrained Optimization", *Mathematics of Computation,* 30 (136) 796-811 (1976).

Hanson, K. and N. Kalogerakis, "Kinetic Reaction Models for Low Temperature Oxidation and High Temperature Cracking of Athabasca and North Bodo Oil Sands Bitumen", *NSERC Report*, University of Calgary, AB, Canada, 1984.

Hartley, H.O., "The Modified Gauss-Newton Method for the Fitting of Non-Linear Regression Functions by Least Squares", *Technometrics*, 3(2), 269-280 (1961).

Hawboldt, K.A., N. Kalogerakis and L.A. Behie, "A Cellular Automaton Model for Microcarrier Cultures", *Biotechnol. Bioeng.,* 43, 90-100 (1994).

Hissong D.W.; Kay, W.B. "The Calculation of the Critical Locus Curve of Binary Hydrocarbon Systems", *AIChE J.*, 16, 580 (1970).

Hocking, R.R., *Methods and Applications of Linear Models*, J. Wiley, New York, NY, 1996.

Holland, J.H., *Adaptation in Natural and Artificial Systems*, Univ. of Michigan Press, Ann Arbor, MI, 1975.

Hong, J.H. and Kobayashi, R., "vapor-Liquid Equilibrium Studies for the carbon Dioxide-Methanol System", *Fluid Phase Equilibria*, 41, 269-276 (1988).

Hong, J.H., Malone, P.V., Jett, M.D., and R. Kobayashi, "The measurement and Interpretation of the Fluid Phase Equilibria of a Normal Fluid in a Hydrogen Bonding Solvent: The Methane-Methanol System", *Fluid Phase Equilibria*, 38, 83-86 (1987).

Hosten, L.H. and G. Emig, "Sequential Experimental Design Procedures for Precise Parameter Estimation in Ordinary Differential Equations", *Chem. Eng. Sci.,* 30, 1357 (1975)

Hougen, O., and K.M. Watson, *Chemical Process Principles*, Vol. 3, J. Wiley, New York, NY, 1948.

Hunter, W.G. and A.M. Reimer, "Designs for Discriminating Between Two Rival Models", *Technometrics*, 7, 307-323 (1965).

Kalogerakis, N. and L.A. Behie, "Oxygenation Capabilities of New Basket-Type Bioreactors for Microcarrier Cultures of Anchorage Dependent Cells", *Bioprocess Eng.,* 17, 151-156 (1997).

Kalogerakis, N. and R. Luus, "Increasing the Size of the Region of Convergence in Parameter Estimation", *proc. American Control Conference*, Arlington, Virginia, 1, 358-364 (1982).

Kalogerakis, N., "Parameter Estimation of Systems Described by Ordinary Differential Equations", *Ph.D. thesis,* Dept. of Chemical Engineering and Applied Chemistry, University of Toronto, ON, Canada, 1983.

Kalogerakis, N., "Sequential Experimental Design for Model Discrimination in Dynamic Systems", *proc. 34th Can. Chem. Eng. Conference*, Quebec City, Oct. 3-6 (1984).

Kalogerakis, N., and C. Tomos, "Reliability of Horizontal Well Performance on a Field Scale Through Automatic History Matching", *J. Can. Petr. Technology*, 34(9), 4755 (1995).

Kalogerakis, N., and R. Luus, "Effect of Data-Length on the Region of Convergence in Parameter Estimation Using Quasilinearization", *AIChE J.*, 26, 670-673 (1980).

Kalogerakis, N., and R. Luus, "Simplification of Quasilinearization Method for Parameter Estimation", *AIChE J.*, 29, 858-864 (1983a).

Kalogerakis, N., and R. Luus, "Improvement of Gauss-Newton Method for Parameter Estimation through the Use of Information Index", *Ind. Eng. Chem. Fundam.,* 22, 436-445 (1983b).

Kalogerakis, N., and R. Luus, "Sequential Experimental Design of Dynamic Systems Through the Use of Information Index", *Can. J. Chem. Eng.*, 62, 730-737 (1984).

Kirkpatrick S., C.D. Gelatt Jr. and M.P. Vecchi, "Optimization by Simulated Annnealing", *Science*, 220, 671-680 (1983).

Kittrell, J.R., R. Mezaki, and C.C. Watson, "Estimation of Parameters for Nonlinear Least Squares Analysis", *Ind. Eng. Chem.*, 57(12), 18-27 (1965b).

Kittrell, J.R., W.G. Hunter, and C.C. Watson, "Nonlinear Least Squares Analysis of Catalytic Rate Models", *AIChE J.*, 11(6), 1051-1057 (1965a).

Koch, K.R., *Parameter Estimation and Hypothesis Testing in Linear Models*, Springer-Verlag, New York, NY, 1987.

Kowalik, J., and M.R. Osborne, *Methods of Unconstrained Optimization Problems*, Elsevier, New York, NY, 1968.

Kurniawan, P., "A Study of In-Situ Brightening of Mechanical Pulp via the Electro-Oxidation of Sodium Carbonate", *M.A.Sc. Thesis*, Dept. of Chemical and Bio-Resource Engineering, The University of British Columbia, 1998.

Lawson, C.L., and R.J. Hanson, *Solving Least Squares Problems*, Prentice Hall, Eglewood-Cliffs, NJ, 1974.

Leis, J.R., and Kramer, M.A., "The Simultaneous Solution and Sensitivity Analysis of Systems Described by Ordinary Differential Equations", *ACM transactions on Mathematical Software*, 14, 45-60 (1988).

Leu, A-D., and D.B. Robinson, "Equilibrium Properties of the Methanol-Isobutane Binary System", *J. Chem. Eng. Data*, 37, 10-13 (1992).

Levenberg, K., "A Method for the Solution of Certain Non-linear Problems in Least Squares", *Quart. Appl. Math.*, II(2), 164-168 (1944).

Li Y-H., K.H. Dillard and R.L. Robinson, "Vapor-Liquid Phase Equilibrium for the CO2-n-Hexane at 40, 80, 120°C", *J. Chem. Eng. Data*, 26, 53-58 (1981).

Lim, H.C., B.J. Chen and C.C. Creagan, "Analysis of Extended and Exponentially Fed Batch Cultures", *Biotechnol. Bioeng.*, 19, 425-435 (1977).

Lin Y-N.; Chen, R.J.J., Chappelear, P.S.; Kobayashi, R. "Vapor-Liquid Equilibrium of the Methane-n-Hexane System at Low Temperature", *J. Chem. Eng. Data*, 22, 402-408 (1977).

Linardos, T., "Kinetics of Monoclonal Antibody Production in Chemostat Hybridoma Cultures", *Ph.D. thesis*, Dept. of Chemical & Petroleum Engineering, University of Calgary, AB, Canada, 1991.

Linardos, T.I., N. Kalogerakis, L.A. Behie and L.R. Lamontagne, "Monoclonal Antibody Production in Dialyzed Continuous Suspension Culture", *Biotechnol. Bioeng.*, 39, 504-510 (1992).

Ljung, L and Söderström, T., *Theory and Practice of Recursive Identification*, MIT press, Boston, MA (1983).

Luus, R., "Determination of the Region Sizes for LJ Optimization", *Hung. J. Ind. Chem.*, 26, 281-286 (1998).

Luus, R., "Optimization in Model Reduction", *Int. J. Control*, 32, 741-747 (1980).

Luus, R., and T.H.I. Jaakola, "Optimization by Direct Search and Systematic Reduction of the Search Region", *AIChE J*, 19, 760 (1973).

Luus, R., *Iterative Dynamic Programming*, Chapman & Hall, CRC, London, UK, 2000.

Marquardt, D.W., "An Algorithm for Least-Squares Estimation of Nonlinear Parameters", *J. Soc. Indust. Appl. Math.*, 11(2), 431-441 (1963).

Metropolis, N., A.W. Rosenbluthh, M.N. Rosenbluth, A.H. Teller and E. Teller, "Equations of State Calculations by Fast Computing Machines", *J. Chem. Physics*, 21, 1087-1092 (1953).

Michelsen, M.L. Phase Equilibrium Calculations. What is Easy and What is Difficult? *Computers Chem. Eng.*, 17, 431-439 (1993).

Modell, M., and R.C. Reid. *Thermodynamics and its Applications*; 2nd ed., Prentice-Hall: Englewood Cliffs, NJ, 1983.

Montgomery, D.C,. *Design and Analysis of Experiments*, 4th ed., J Wiley, New York, NY, 1997.

Montgomery, D.C. and E.A. Peck, *Introduction to Linear Regression Analysis*, J. Wiley, New York, NY, 1992.

Monton, J.B., and F. J. Llopis, "Isobaric Vapor-Liquid Equilibria of Ethylbenzene + m-Xylene and Ethylbenzene + o-Xylene Systems at 6.66 and 26.66 kPa", *J. Chem. Eng. Data*, 39, 50-52 (1994).

Murray, L.E. and E.K. Reiff Jr., "Design of Transient Experiments for Identification of Fixed Bed Thermal Transport Properties", *Can. J. Chem. Eng.,* 62, 55-61 (1984)

Nelder, J.A., and R. Mead, "A Simplex Method for Function Minimization", *Comp. J.*, 7, 308-313 (1965).

Nieman, R.E. and D.G. Fisher, "Parameter Estimation Using Linear programming and Quasilinearization", *Can. J. Chem. Eng.,* 50, 802 (1972).

Ohe, S. *Vapor-Liquid Equilibrium Data at High Pressure*; Elsevier Science Publishers: Amsterdam, NL, 1990.

Otten R.H.J.M. and L.P.P.P. Van Ginneken, *The annealing algorithm*, Kluwer International Series in Engineering and Computer Science, Kluwer Academic Publishers, NL, 1989.

Park, H., "Thermodynamics of Sodium Aluminosilicate Formation in Aqueous Alkaline Solutions Relevant to Closed-Cycle Kraft Pulp Mills", *PhD Thesis*, Dept. of Chemical and Bio-Resource Engineering, University of British Columbia, Vancouver, BC, Canada, 1999.

Park, H., and P. Englezos, "Osmotic Coefficient Data for Na_2SiO_3 and Na_2SiO_3-NaOH by an Isopiestic Method and Modelling Using Pitzer's Model", *Fluid Phase Equilibria*, 153, 87-104 (1998).

Park, H., and P. Englezos, "Osmotic Coefficient Data for the NaOH-NaCl-$NaAl(OH)_4$ System Measured by an Isopiestic Method and Modelling Using Pitzer's Model at 298.15 K", *Fluid Phase Equilibria*, 155, 251-260 (1999).

Patino-Leal, H., and P.M. Reilly, "Statistical Estimation of Parameters in Vapor-Liquid Equilibrium", *AIChE J.*, 28(4), 580-587 (1982).

Peneloux, A.; Neau, E.; Gramajo, A. Variance Analysis Fifteen Years Ago and Now. *Fluid Phase Equilibria,* 56, 1-16 (1990).

Peng, D-Y., and D.B. Robinson, "A New Two-Constant Equation of State", *Ind. Eng. Chem. Fundam.*, 15, 59-64 (1976).

Peressini, A.L., F.E. Sullivan, J.J. Uhl, Jr., *The Mathematics of Nonlinear programming*, Springer-Verlag, New York, NY, 1988.

Pitzer, K.S., *Activity Coefficients in Electrolyte Solutions*, 2nd Ed.; CRC Press: Boca Raton, FL, 1991.

Plackett, R.L., "Studies in the History of Probability and Statistic. XXIX The Discovery of the method of least squares", *Biometrika*, 59 (2), 239-251 (1972).

Poston, R.S.; McKetta, J. "Vapor-Liquid Equilibrium in the Methane-n-Hexane System", *J. Chem. Eng. Data*, 11, 362-363 (1966).

Powell, M.J. D., "A Method of Minimizing a Sum of Squares of Non-linear Functions without Calculating Derivatives", *Computer J.*, 7, 303-307 (1965).

Prausnitz, J.M., R.N. Lichtenthaler and E.G. de Azevedo, *Molecular Thermodynamics of Fluid Phase Equilibria*, 2nd ed., Prentice Hall, Englewood Cliffs, NJ, 1986.

Press, W.H., S.A. Teukolsky, W.T. Vetterling, and B. P. Flannery, *Numerical Recipes in Fortran: The Art of Scientific Computing,* Cambridge University Press, Cambridge, UK, 1992.

Ramaker, B.L., C.L. Smith and P.W. Murrill, "Determination of Dynamic Model Parameters Using Quasilinearization", *Ind. Eng. Chem. Fundam.,* 9, 28 (1970).

Rard, J.A., "Isopiestic Investigation of Water Activities of Aqueous $NiCl_2$ and $CuCl_2$ Solutions and the Thermodynamic Solubility Product of $NiCl_2$-$6H_2O$ at 298.15 K", *J. Chem. Eng. Data*, 37, 433-442 (1992).

Ratkowsky, D.A., *Nonlinear Regression Modelling*, Marcel Dekker, New York, NY, 1983.

Reamer, H.H., B.H. Sage and W.N. Lacey, "Phase Equibria in Hydrocarbon Systems", *Ind. Eng. Chem.,* 42, 534 (1950).

Reilly, P.M. and H. Patino-Leal, "A Bayessian Study of the Error in Variables Model", *Technometrics*, 23, 221-231 (1981).

Reklaitis, G.V., A. Ravindran, and K.M. Ragsdell, *Engineering Optimization: Methods and Aplications*, J. Wiley, New York, NY, 1983.

Rippin, D.W.T., L.M. Rose and C. Schifferli, "Nonlinear Experimental Design with Approximate Models in Reactor Studies for Process Development", *Chem. Eng. Sci.,* 35, 356 (1980).

Sachs, W.H., "Implicit Multifunctional Nonlinear Regression Analysis", Technometrics, 18, 161-173 (1976).

Salazar-Sotelo D., A. Boireaut and H. Renon, "Computer Calculations of the Optimal Parameters of a Model for the Simultaneous Representation of Experimental, Binary and Ternary Data", *Fluid Phase Equilibria*, 27, 383-403 (1986).

Sargent, R.W.H. "A Review of Optimization Methods for Nonlinear Problems", in *Computer Applications in Chemical Engineering*, R.G. Squires and G.V. Reklaitis, Eds., ACS Symposium Series 124, 37-52, 1980.

Scales, L.E., *Introduction to Non-linear Optimization*, Springer-Verlag, New York, NY, 1985.

Schwartzentruber J., F. Galivel-Solastiuk and H. Renon, "Representation of the Vapor-Liquid Equilibrium of the Ternary System Carbon Dioxide-Propane-Methanol and its Binaries with a Cubic Equation of State. A new Mixing Rule", *Fluid Phase Equilibria*, 38, 217-226 (1987).

Schwetlick, H. and V. Tiller, "Numerical Methods for Estimating Parameters in Nonlinear Models with Errors in the Variables", *Technometrics*, 27, 17-24 (1985).

Seber, G.A.F. and C.J. Wild, *Nonlinear Regression*, J. Wiley, New York, NY, 1989.

Seber, G.A.F., *Linear Regression Analysis*, J. Wiley, New York, NY, 1977.

Seinfeld, J.H., and G.R. Gavalas, "Analysis of Kinetic Parameters from Batch and Integral Reaction Experiments", *AIChE J.,* 16, 644-647 (1970).

Seinfeld, J.S., and L. Lapidus, *Mathematical Models in Chemical Engineering,* Vol. 3, Prentice-Hall, Inc., Englewood Cliffs, NJ, 1974.

Selleck, F.T., L.T. Carmichael and B.H. Sage, "Phase Behavior in the Hydrogen Sulfide - Water System", *Ind. Eng. Chem.*, 44, 2219-2226 (1952).

Shah, M.J., "Kinetic Models for Consecutive Heterogeneous Reactions", *Industrial and Engineering Chemistry*, 57, 18-23 (1965).

Shanmugan, K.S.and A. M. Breipohl, *Random Signals: Detection, Estimation and Data Analysis,* J. Wiley, New York, NY, 1988.

Shibata, S.K., and S.I. Sandler, "High-Pressure Vapor-Liquid Equilibria of Mixtures of Nitrogen, Carbon Dioxide, and Cyclohexane", *J. Chem. Eng. Data*, 34, 419-424 (1989).

Silva, A.M., and L.A. Weber, "Ebulliometric Measurement of the Vapor Pressure of 1-Chloro-1,1-Difluoroethane and 1,1-Difluoro Ethane, *J. Chem. Eng. Data*, 38, 644-646 (1993).

Smith, J.M., *Chemical Engineering Kinetics*, 3rd ed., McGraw-Hill, New York, NY, 1981.

Söderström, T., Ljung, L, and Gustavsson, I. (1978) "A theoretical analysis of recursive identification methods, *Automatica*, 14, 231-244.

Sorenson, H.W., *Parameter Estimation*, Marcel Dekker, New York, NY, 1980.

Spiegel, M.R., "*Theory and Problems of Advanced mathematics for Engineers and Scientists*", McGraw Hill, New York, NY, 1983.

Srinivasan, R., and A.A. Levi, "Kinetics of the Thermal Isomerization of Bicyclo [2.1.1] hexane", *J. Amer. Chem. Soc.*, 85, 3363-3365 (1963).

Srivastava, M.S. and Carter, E.M., *Introduction to Applied Multivariate Statistics*, North-Holland, Elsevier Science Publ., NewYork., N.Y. (1983)

Stone, H.L., "Iterative Solution of Implicit Approximations of Multi-Dimensional Partial Differential Equations", SIAM J. Numerical Analysis, 5, 530-558 (1968).

Sutton, T.L., and J.F. MacGregor, "The Analysis and Design of Binary Vapour-Liquid Equilibrium Experiments Part I: Parameter Estimation and Consistency Tests", *Can. J. Chem. Eng.*, 55, 602-608 (1977).

Tan, T.B. and J.P. Letkeman, "Application of D4 Ordering and Minimization in an Effective Partial Matrix Inverse Iterative Method", paper SPE 10493 presented at the 1982 SPE Symposium on Reservoir Siumulation, San Antonio, TX (1982).

Tan, T.B. and N. Kalogerakis, "A Three-dimensional Three-phase Automatic History Matching Model: Reliability of Parameter Estimates", *J. Can Petr. Technology*, 31(3), 34-41 (1992).

Tan, T.B. and N. Kalogerakis, "Improved Reservoir Characterization Using Automatic History Matching Procedures", *J. Can Petr. Technology*, 32(6), 26-33 (1993).

Tan, T.B., "Parameter Estimation in Reservoir Engineering", *Ph.D. Thesis,* Dept. of Chemical & Petroleum Engineering, University of Calgary, AB, Canada, 1991.

Tassios, D.P., *Applied Chemical Engineering Thermodynamics*, Springer-Verlag, Berlin, 1993.

Thaller, L.H., and G. Thodos, "The Dual Nature of a Catalytic-Reaction: The Dehydrogenation of sec-Butyl Alcohol to Methyl Ethyl Ketone at Elevated Pressures", *AIChE Journal*, 6(3), 369-373, 1960.

Thiessen, D.B., and Wilson, "An Isopiestic method for Measurement of Electrolyte Activity Coefficients", *AIChE J.,* 33(11), 1926-1929, 1987.

Trebble, M.A.; Bishnoi, P.R. Development of a New Four-Parameter Equation of State, *Fluid Phase Equilibria,* 35, 1-18 (1987).

Trebble, M.A.; Bishnoi, P.R. Extension of the Trebble-Bishnoi Equation of State to Fluid Mixtures. *Fluid Phase Equilibria,* 40, 1-21 (1988).

Valko, P. and S. Vajda, "An Extended Marquardt'-type Procedure for Fitting Error-In-Variables Models", *Computers Chem. Eng.,* 11, 37-43 (1987).

van Konynenburg, P.H. and R.L. Scott, "Critical Lines and Phase Equilibria in Binary van der Waals Mixtures", *Philos. Trans. R. Soc. London, 298*, 495-540 (1980).

Walters, F.H., Parker, J., Llyod, R., Morgan, S.L., and S.N. Deming, *Sequential Simplex Optimization: A Technique for Improving Quality and Productivity in Research, Development, and Manufacturing*, CRC Press Inc., Boca Raton, Florida, 1991.

Wang, B.C. and R. Luus, "Increasing the Size of the Region of Convergence for Parameter Estimation Through the Use of Shorter Data-Length", *Int. J. Control,* 31, 947-957 (1980)

Wang, B.C. and R. Luus, "Optimization of Non-Unimodal Systems", *Int. J. Numer. Meth. Eng.,* 11, 1235-1250 (1977).

Wang, B.C. and R. Luus, "Reliability of Optimization Procedures for Obtaining the Global Optimum", *AIChE J.*, 19, 619-626 (1978).

Watts, D.G., "Estimating Parameters in Nonlinear Rate Equations", *Can. J. Chem. Eng.,* 72, 701-710 (1994).

Wee, W., and N. Kalogerakis, "Modelling of Drilling Rate for Canadian Offshore Well Data", *J. Can. Petr. Technology*, 28(6), 33-48 (1989).

Wellstead, P.E. and M. B. Zarrop, *Self-Tuning Systems: Control and Signal Processing,* J. Wiley, New York, NY, 1991.

Yokoyama C., H. Masuoka, K. Aval, and S. Saito, "Vapor-Liquid Equilibria for the Methane-Acetone and Ethylene-Acetone Systems at 25 and 50°C", *J. Chem. Eng. Data*, 30, 177-179 (1985).

Zatsepina, O.Y., and B.A. Buffett, "Phase Equilibrium of Gas Hydrate: Implications for the Formation of Hydrate in the Deep Sea Floor", *Geophys. Res. Letters*, 24, 1567-1570 (1997).

Zeck S., and H. Knapp, "Vapor-Liquid and Vapor-LIquid-Liquid Phase Equilibria for Binary and Ternary Systems of Nitrogen, Ethane and Methanol: Experiments and Data Reduction", *Fluid Phase Equilibria*, 25, 303-322 (1986).

Zhu, X. D., G. Valerius, and H. Hofmann, "Intrinsic Kinetics of 3-Hydroxypropanal Hydrogenation over $Ni/SiO_2/Al_2O_3$ Catalyst", *Ind. Eng. Chem. Res*, 36, 3897-2902 (1997).

Zygourakis, K. Bizios, R. and P. Markenscoff, "Proliferation of Anchorage Dependent Contact Inhibited Cells: Development of Theoretical Models Based on Cellular Automata", *Biotechnol. Bioeng.*, 36, 459-470 (1991).

Appendix 1

A.1.1 THE TREBBLE-BISHNOI EQUATION OF STATE

We consider the Trebble-Bishnoi EoS for a fluid mixture that consist of N_c components. The equation was given in Chapter 14.

A.1.2 Derivation of the Fugacity Expression

The Trebble-Bishnoi EoS (Trebble and Bishnoi, 1988a;b) for a fluid mixture is given by equation 14.6. The fugacity, f_j, of component j in a mixture is given by the following expression for case 1 ($\tau \geq 0$)

$$\ln f_j = \ln\left(x_j P\right) + \frac{b_d}{b_m}(Z-1) - \ln\left|Z - B_m\right| + \Psi\lambda X - \Psi\frac{K}{\Lambda} \tag{A.1.1}$$

where

$$\Psi = \frac{A_m}{B_m\theta} \tag{A.1.2}$$

$$X = \frac{a_d / n}{a_m} - \frac{b_d}{b_m} - \frac{\eta\theta_d}{\theta} \tag{A.1.3}$$

$$K = ZB_m \eta\theta_d + 0.5B_m^2 \left(u\eta\theta_d - \theta\eta u_d\right) \tag{A.1.4}$$

$$\Lambda = Z^2 + \left(B_m + C_m\right)Z - \left(B_m C_m + D_m^2\right) \tag{A.1.5}$$

$$\theta = \tau^{0.5}, \quad \eta\theta_d = Q \text{ and } \lambda = \ln\left|\frac{2Z + B_m(u-\theta)}{2Z + B_m(u+\theta)}\right| \quad \text{if } \tau \geq 0 \tag{A.1.6a}$$

$$\theta = (-\tau)^{0.5}, \quad \eta\theta_d = -Q \text{ and } \lambda = 2\tan^{-1}\left(\frac{2Z + uB_m}{B_m\theta}\right) - \pi \quad \text{if } \tau < 0 \tag{A.1.6b}$$

$$u = 1 + c_m/b_m \tag{A.1.7}$$

$$\tau = 1 + \frac{6c_m b_m + c_m^2 + 4d_m^2}{b_m^2} \tag{A.1.8}$$

$$Q = \frac{1}{2\theta b_m^2}\left[6b_m c_d - 6b_d c_m + 2c_m c_d + 8d_m d_d - \frac{2b_d c_m^2}{b_m} - \frac{8b_d d_m^2}{b_m}\right] \tag{A.1.9}$$

$$A_m = \frac{a_m P}{R^2 T^2} \tag{A.1.10a}$$

$$B_m = \frac{b_m P}{RT} \tag{A.1.10b}$$

$$C_m = \frac{c_m P}{RT} \tag{A.1.10c}$$

$$D_m = \frac{d_m P}{RT} \tag{A.1.10d}$$

where a_m, b_m, c_m and d_m are given by Equations 14.7a, b, c and d respectively. The quantities a_d, b_d, c_d, d_d and u_d are given by the following equations

$$a_d \equiv \frac{\partial\left(a_m n^2\right)}{\partial n_j} = 2\sum_{i=1}^{N_c} n_i a_{ij} \qquad \text{(A.1.11a)}$$

$$b_d \equiv \frac{\partial\left(b_m n\right)}{\partial n_j} = 2\sum_{i=1}^{N_c} x_i b_{ij} - b_m \qquad \text{(A.1.11b)}$$

$$c_d \equiv \frac{\partial\left(c_m n\right)}{\partial n_j} = 2\sum_{i=1}^{N_c} x_i c_{ij} - c_m \qquad \text{(A.1.11c)}$$

$$d_d \equiv \frac{\partial\left(d_m n\right)}{\partial n_j} = 2\sum_{i=1}^{N_c} x_i d_{ij} - d_m \qquad \text{(A.1.11d)}$$

$$\theta_d \equiv \frac{\partial \theta}{\partial n_j} \qquad \text{(A.1.11e)}$$

$$u_d \equiv \frac{\partial(u)}{\partial n_j} \qquad \text{(A.1.11f)}$$

Also

$$nu_d = \frac{c_d}{b_m} - \frac{b_d c_m}{b_m^2} \qquad \text{(A.1.12)}$$

A.1.3 Derivation of the Expression for $\left(\frac{\partial \ln f_j}{\partial x_j}\right)_{T,P,x_{i\neq j}}$

Differentiation of Equation A1.1. with respect to x_j at constant T, P and x_i for $i=1,2,\ldots,N_C$ and $i\neq j$ gives

$$\frac{\partial \ln f_j}{\partial x_j} = \frac{1}{x_j} + \frac{b_d}{b_m}\left(\frac{\partial Z}{\partial x_j}\right) + \frac{(Z-1)}{b_m}\left(\frac{\partial b_d}{\partial x_j} - \frac{b_d}{b_m}\frac{\partial b_m}{\partial x_j}\right) - FX_1 + FX_2 \qquad \text{(A.1.13)}$$

where

$$FX_1 = \frac{\dfrac{\partial Z}{\partial x_j} - \dfrac{P}{RT}\dfrac{\partial b_m}{\partial x_j}}{Z - \dfrac{b_m P}{RT}} \tag{A.1.14}$$

$$FX_2 = \left(\lambda X - \frac{K}{\Lambda}\right)\frac{\partial \Psi}{\partial x_j} + \Psi\left(X\frac{\partial \lambda}{\partial x_j} + \lambda\frac{\partial X}{\partial x_j}\right) - \frac{\Psi}{\Lambda}\left(\frac{\partial K}{\partial x_j} - \frac{K}{\Lambda}\frac{\partial \Lambda}{\partial x_j}\right) \tag{A.1.15}$$

A number of partial derivatives are needed and they are given next,

$$\frac{\partial \Psi}{\partial x_j} = \frac{1}{B_m \theta}\left[\frac{\partial A_m}{\partial x_j} - \Psi\left(B_m\frac{\partial \theta}{\partial x_j} + \theta\frac{\partial B_m}{\partial x_j}\right)\right] \tag{A.1.16}$$

$$\begin{aligned}\frac{\partial X}{\partial x_j} = {} & \frac{1}{a_m}\left(\frac{\partial\, {a_d}/{n}}{\partial x_j} - \frac{{a_d}/{n}}{a_m}\frac{\partial a_m}{\partial x_j}\right) - \frac{1}{b_m}\left(\frac{\partial b_d}{\partial x_j} - \frac{b_d}{b_m}\frac{\partial b_m}{\partial x_j}\right) \\ & - \frac{1}{\theta}\left(\frac{\partial n\theta_d}{\partial x_j} - \frac{n\theta_d}{\theta}\frac{\partial \theta}{\partial x_j}\right)\end{aligned} \tag{A.1.17}$$

$$\begin{aligned}\frac{\partial K}{\partial x_j} = {} & \left(ZB_m + 0.5B_m^2 u\right)\frac{\partial n\theta_d}{\partial x_j} + \left(Zn\theta_d + un\theta_d B_m - \theta nu_d B_m\right)\frac{\partial B_m}{\partial x_j} \\ & + B_m n\theta_d\frac{\partial Z}{\partial x_j} + 0.5B_m^2\left(n\theta_d\frac{\partial u}{\partial x_j} - \theta\frac{\partial nu_d}{\partial x_j} - nu_d\frac{\partial \theta}{\partial x_j}\right)\end{aligned} \tag{A.1.18}$$

$$\begin{aligned}\frac{\partial \Lambda}{\partial x_j} = {} & 2Z\frac{\partial Z}{\partial x_j} + (B_m + C_m)\frac{\partial Z}{\partial x_j} + Z\left(\frac{\partial B_m}{\partial x_j} + \frac{\partial C_m}{\partial x_j}\right) - B_m\frac{\partial C_m}{\partial x_j} \\ & - C_m\frac{\partial B_m}{\partial x_j} - 2D_m\frac{\partial D_m}{\partial x_j}\end{aligned} \tag{A.1.19}$$

$$\frac{\partial A_m}{\partial x_j} = \frac{P}{R^2T^2}\frac{\partial a_m}{\partial x_j}, \quad \frac{\partial B_m}{\partial x_j} = \frac{P}{RT}\frac{\partial b_m}{\partial x_j} \tag{A.1.20a}$$

$$\frac{\partial C_m}{\partial x_j} = \frac{P}{RT}\frac{\partial c_m}{\partial x_j}, \quad \frac{\partial D_m}{\partial x_j} = \frac{P}{RT}\frac{\partial d_m}{\partial x_j} \tag{A.1.20b}$$

If $\tau \geq 0$ then λ is given by equation A.1.6a and the derivative is given by

$$\frac{\partial \lambda}{\partial x_j} = \frac{1}{\left(\dfrac{g-h}{g+h}\right)}\left[\frac{1}{g+h}\left(\frac{\partial g}{\partial x_j} - \frac{\partial h}{\partial x_j}\right) - \frac{(g-h)}{(g+h)^2}\left(\frac{\partial g}{\partial x_j} + \frac{\partial h}{\partial x_j}\right)\right] \tag{A.1.21a}$$

where $g = 2Z + B_m u$, $h = B_m \theta$ and

$$\frac{\partial g}{\partial x_j} = 2\frac{\partial Z}{\partial x_j} + B_m\frac{\partial u}{\partial x_j} + u\frac{\partial B_m}{\partial x_j} \tag{A.1.21b}$$

$$\frac{\partial h}{\partial x_j} = B_m\frac{\partial \theta}{\partial x_j} + \theta\frac{\partial B_m}{\partial x_j} \tag{A.1.21c}$$

If τ is negative then λ is given by equation A.1.6b and the derivative is given by

$$\frac{\partial \lambda}{\partial x_j} = \frac{2h}{\left(g^2 + h^2\right)}\left(\frac{\partial g}{\partial x_j} - \frac{g}{h}\frac{\partial h}{\partial x_j}\right) \tag{A.1.21d}$$

In addition to the above derivatives we need the following

$$\frac{\partial u}{\partial x_j} = \frac{\partial\left(c_m/b_m\right)}{\partial x_j} = \frac{1}{b_m}\left(\frac{\partial c_m}{\partial x_j} - \frac{c_m}{b_m}\frac{\partial b_m}{\partial x_j}\right) \tag{A.1.22}$$

$$\frac{\partial \tau}{\partial x_j} = 6\frac{\partial\left(c_m/b_m\right)}{\partial x_j} + 2\frac{c_m}{b_m}\frac{\partial\left(c_m/b_m\right)}{\partial x_j} + 8\frac{d_m}{b_m}\frac{\partial\left(d_m/b_m\right)}{\partial x_j} \tag{A.1.23}$$

$$\frac{\partial nu_d}{\partial x_j} = \frac{\partial\left(c_d / b_m\right)}{\partial x_j} - \frac{c_m}{b_m}\frac{\partial\left(b_d / b_m\right)}{\partial x_j} - \frac{b_d}{b_m}\frac{\partial\left(c_m / b_m\right)}{\partial x_j} \tag{A.1.24}$$

$$\frac{\partial Q}{\partial x_j} = \frac{1}{Q_C}\left[\frac{\partial Q_A}{\partial x_j} - \frac{\partial Q_B}{\partial x_j} - \frac{Q_A - Q_B}{Q_C}\frac{\partial Q_c}{\partial x_j}\right] \tag{A.1.25a}$$

where

$$Q_A = 6b_m c_d - 6b_d c_m + 2c_m c_d + 8d_m d_d \tag{A.1.25b}$$

$$Q_B = \frac{2b_d c_m^2}{b_m} + \frac{8b_d d_m^2}{b_m} \tag{A.1.25c}$$

$$Q_C = 2\theta b_m^2 \tag{A.1.25d}$$

$$\begin{aligned}\frac{\partial Q_A}{\partial x_j} &= 6\left(b_m \frac{\partial c_d}{\partial x_j} + c_d \frac{\partial b_m}{\partial x_j} - b_d \frac{\partial c_m}{\partial x_j} - c_m \frac{\partial b_d}{\partial x_j}\right) \\ &\quad + 2\left(c_m \frac{\partial c_d}{\partial x_j} + c_d \frac{\partial c_m}{\partial x_j}\right) + 8\left(d_m \frac{\partial d_d}{\partial x_j} + d_d \frac{\partial d_m}{\partial x_j}\right)\end{aligned} \tag{A.1.25e}$$

$$\begin{aligned}\frac{\partial Q_B}{\partial x_j} &= 2c_m^2 \frac{\partial\left(b_d / b_m\right)}{\partial x_j} + 4\frac{b_d}{b_m} c_m \frac{\partial c_m}{\partial x_j} + 8d_m^2 \frac{\partial\left(b_d / b_m\right)}{\partial x_j} \\ &\quad + 16\frac{b_d}{b_m} d_m \frac{\partial d_m}{\partial x_j}\end{aligned} \tag{A.1.25f}$$

$$\frac{\partial Q_C}{\partial x_j} = 4\theta b_m \frac{\partial b_m}{\partial x_j} + 2b_m^2 \frac{\partial \theta}{\partial x_j} \tag{A.1.25g}$$

$$\frac{\partial \theta}{\partial x_j} = \frac{1}{2\theta} \frac{\partial \tau}{\partial x_j} \quad \text{if } \tau \geq 0 \tag{A.1.25h}$$

or

$$\frac{\partial \theta}{\partial x_j} = -\frac{1}{2\theta} \frac{\partial \tau}{\partial x_j} \quad \text{if } \tau \text{ is negative} \tag{A.1.25i}$$

Appendix 2

A.2.1 LISTINGS OF COMPUTER PROGRAMS

In this chapter we provide listing of two computer programs with the corresponding input files. One is for a typical algebraic equation model (Example 16.1.2) and the other for an ODE model (Example 16.3.2). These programs can be used with a few minor changes to solve other parameter estimation problems posed by the user. As our intention was not to produce a full-proof commercial software package, a few things are required from the user.

In order to run the given examples, no modification changes need be made. If however, the user wants to use the Gauss-Newton method for an algebraic or ODE model, basic programming FORTRAN 77 skills are required. The user must change the subroutine MODEL and the subroutine JX (only for ODE models) using the existing ones as a template. In addition the user must select the value of a few program parameters which are detailed in the first set of comment cards in each program. Subsequently, the program needs to be compiled.

The input data file should also be constructed using any of the existing ones as a template.

All the programs have been written using FORTRAN 77. The programs have been tested using the MICROSOFT DEVELOPER STUDIO—FORTRAN POWER STATION. The programs call one or more IMSL MATH library routines which is available with above FORTRAN compiler. In general any FORTRAN compiler accompanied with the IMSL MATH library should be capable to run the enclosed programs without any difficulty.

These programs are provided as an aid to the user with the following disclaimer.

> *We make no warranties that the programs contained in this book are free of errors. The programs in this book are intended to be used to demonstrate parameter estimation procedures. If the user wishes to use the programs for solving problems related to the design of processes and products we make no warranties that the programs will meet the user's requirements for the applications. The user employs the programs to solve problems and use the results at his/her own risk. The authors and publisher disclaim all liability for direct or indirect injuries to persons, damages or loss of property resulting from the use of the programs.*

A.2.2 CONTENTS OF ACCOMPANYING CD

In the enclosed CD the computer programs used for the solution of selected problems presented throughout the book are provided as an aid to the user. Each example is provided in a different folder together with a typical input and output file. The *.exe file is also provided in case one wishes to run the particular examples and has no access to a FORTRAN compiler.

Programs are provided for the following examples dealing with algebraic equation models:

1. Example 4.4.1
2. Example 4.4.2
3. Example 14.2.7.1
4. Example 16.1.1
5. Example 16.1.2
6. Example 16.1.3
8. Example 17.1.1
9. Example 17.1.2

Programs are also provided for the following examples dealing with ODE models:

1. ODE_Example 16.3.1
2. ODE_Example 16.3.2
3. ODE_Example 16.3.3
4. ODE_Example 17.3.1

In addition, the program used for data smoothing with cubic splines for the short-cut methods are given for the example:

1. SC_Example 17.1.4

A.2.3 COMPUTER PROGRAM FOR EXAMPLE 16.1.2

```
c-------------------------------------------------------------------+
c
c     EXAMPLE 16.1.2 (Isomerization of Bicyclo Hexane)
c     in the book:
c     APPLIED PARAMETER ESTIMATION FOR CHEMICAL ENGINEERS
c     by Englezos & Kalogerakis / Marcel Dekker Inc. (2000)
c     COPYRIGHT: Marcel Dekker 2000
c
c-------------------------------------------------------------------+
c
c     PARAMETER ESTIMATION ROUTINE FOR ALGEBRAIC EQUATION MODELS
c     Based on Gauss-Newton method with PseudoInverse and Marquardt's
c     Modification. Hard BOUNDARIES on parameters can be imposed.
c
c     GENERALIZED LEAST SQUARES Formulation (when IGLS=1)
c        with the DIAGONAL WEIGHTING MATRIX, Q(i)/Ydata(i)**2, i=1,NY
c     or
c
c     WEIGHTED LEAST SQUARES Formulation (when IGLS=0)
c        with the DIAGONAL WEIGHTING MATRIX, Q(i), i=1,NY
c
c     Written in FORTRAN 77 by Dr. Nicolas Kalogerakis, Aug. 1999
c     Compiler: MS-FORTRAN using the IMSL library (routine:DEVCSF)
c
c-------------------------------------------------------------------+
c     The User must provide Subroutine MODEL [that describes the
c        mathematical model Y=f(X,P)]
c
c     The Following Variables MUST be Specified in Main Program:
c
c     IGLS = Flag to use Generalized LS (IGLS=1) or Weighted LS (IGLS=0)
c     NX = Number of independent variables in the model
c     NY = Number of depenent variables in the model
c     NPAR = Number of unknown parameters
c     NXX(i), i=1,NX = columns in the data corresponding to x(1), x(2),...
c     NYY(i), i=1,NY = columns in the data corresponding to y(1), y(2),...
c     FILEIN = Name of Input file
c     FILOUT = Name of Output file
c     Q(i), i=1,NY = Constant weights for each measured variable (Use 1 for LS)
c     IBOUND = 1 to enable hard constraints on the parameters (0 otherwise)
c     NSTEP = Maximum number of reductions allowed for Bisection rule (default is 10)
c     KIF, KOS, KOF = Input file unit, Screen output unit, Output file unit.
c     NSIG = Approx. Number of Significant digits (Tolerence EPS = 10**(-NSIG))
c     EPSPSI = Minimum Ratio of eigenvalues before the Pseudo-inverse approx. is used.
c     EPSMIN = A very small number (used to avoid division by zero).
c
c     For Space Allocation Purposes Use:
c
c     NVAR = Greater or equal to the maximum number of colums in Datafile
c     NROWS = Greater or equal to the maximum number of experiments (=rows) in Datafile
c     NPARMX = Greater or equal to the number of unknown parameters in the model.
c
c-------------------------------------------------------------------+
      USE MSIMSLMD
      PARAMETER (nvar=15, nrows=300, nparmx=15)
      DOUBLE PRECISION ydata(nvar,nrows),xdata(nvar,nrows)
      DOUBLE PRECISION alldata(nvar,nrows)
      DOUBLE PRECISION y(nvar),q(nvar),x(nvar),dydp(nvar,nparmx)
      DOUBLE PRECISION p0(nparmx),dp(nparmx),p(nparmx)
      DOUBLE PRECISION pmin(nparmx),pmax(nparmx), v(nparmx,nparmx)
      DOUBLE PRECISION a(nparmx,nparmx),b(nparmx),s(nparmx)
      DOUBLE PRECISION stdev(nparmx),bv(nparmx),bvs(nparmx)
      INTEGER nyy(nvar), nxx(nvar)
      CHARACTER*4 dummyline
      CHARACTER*24 filein, filout
c
```

```
      COMMON /gn/imode,ny,nx,npar
c
c
c-------------------------------Set convergence tolerence & other parameters
c
      DATA kos/6/,kif/2/,kof/7/,epsmin/1.e-80/,igls/0/
      DATA nsig/5/,epspsi/1.e-30/,nstep/10/,q/nvar*1.0/,ibound/1/
c
      eps = 10.**(-nsig)
      npar = 2
      ny = 1
      nx = 2
c-------------------------------Set database pointers:
c-------------------------------Select which columns of the data correspond
c-------------------------------to x(1), x(2),... y(1), y(2),... in the file
      nxx(1)=1
      nxx(2)=2
      nyy(1)=3
c
c-------------------------------Set Filenames and Open the files for I/O
      filein = 'DataIN_Ex16.1.2.txt'
      filout = 'DataOUT_Ex16.1.2.txt'
      open(kif,file=filein)
      open(kof,file=filout)
      write(kof, 71)
      write(kos, 71)
71    format(//20x,'PARAMETER ESTIMATION PROGRAM FOR ALGEBRAIC MODELS',
     & /20x,'Gauss-Newton Method with PseudoInverse/Marquardt/Bisection'
     & ,/20x,'Bounded parameter search/Weighted LS formulation...',/)
     if (igls .eq. 1) then
        write(kos,72)
        write(kof,72)
72      format(//20x,'GENERALIZED LEAST SQUARES Formulation...')
     else
        write(kos,73)
        write(kof,73)
73      format(//20x,'WEIGHTED LEAST SQUARES Formulation...')
     end if
      write(kos,74) filein
      write(kof,74) filein
74    format(/40x,'Data Input Filename: ',a24)
      write(kos,77) filout
      write(kof,77) filout
77    format(40x,'Program Output File: ',a24)
c
c-------------------------------Read Initial Guess for Parameters
     read(kif,*) dummyline
     read(kif,*) (p(j),j=1,npar)
c
c-------------------------------Read MIN & MAX parameter BOUNDS
     read(kif,*) dummyline
     read(kif,*) (pmin(j),j=1,npar)
     read(kif,*) dummyline
     read(kif,*) (pmax(j),j=1,npar)
     read(kif,*) dummyline
c
c-------------------------------Read NITER, IPRINT & EPS_Marquardt
     read(kif,*) niter, iprint, epsmrq
     write(kos,83) (p(j),j=1,npar)
     write(kof,83) (p(j),j=1,npar)
 83  format(//5x,'Initial-Guess P(j) =',6g12.4//)
     write(kof, 87) (pmin(j),j=1,npar)
 87  format(5x,'MIN Bounds Pmin(j) =',6g12.4//)
     write(kof, 88) (pmax(j),j=1,npar)
```

```
 88   format(5x,'MAX Bounds Pmax(j) =',6g12.4//)
c
c-------------------------------Read Input Data (Experimental Data Points)
      read(kif,*) dummyline
       read(kif,*) ncolr
       if (ncolr .lt. nx+ny) then
           ncolr = nx + ny
           write(kof,91) ncolr
91         format(/40x,'Number of Columns CANNOT be less than =',i3,
     &                    /40x,'Check you Model Equations again...')
          stop
       end if
       write(kof,92) ncolr
92     format(/40x,'Number of Columns =',i3)
       np = 0
      read(kif,*) dummyline
       do  100  i=1,nrows
       read(kif,*,err=105,end=108) (alldata(j,i),j=1,ncolr)
       np = np + 1
100   continue
105   continue
       write(kof,107)
107   format(/1x,' Input STOPPED when CHARACTERS were encountered...',/)
108   continue
       write(kof,109) np
       write(kos,109) np
109    format(40x,'Data Points Entered =',i3)
       close(kif)
c------------------------------Assign variables to Ydata and Xdata & print
them
       do 120 i=1,ny
       do 120 j=1,np
          ydata(i,j) = alldata(nyy(i),j)
 120  continue
       do 125 i=1,nx
       do 125 j=1,np
          xdata(i,j) = alldata(nxx(i),j)
 125  continue
      write(kof,131) ny, nx
 131  format(/1x,'Each Column corresponds to:  Y(1),...,Y(',i2,
     &                  '),  X(1),...,X(',i2,'):')
      do 130 j=1,np
      write(kof,132) (ydata(i,j),i=1,ny),(xdata(i,j),i=1,nx)
 130  continue
 132  format(1x,8g14.5)
      dfr = ny*np - npar
c
c-------------------------------Main iteration loop
       do  700  loop=1,niter
c------------------------------Initialize matrix A, b and Objective Function
       do  220  i=1,npar
          b(i)=0.0
       do  220  l=1,npar
          a(i,l)=0.0
 220  continue
       SSE=0.0
       if (iprint .eq. 1) write(kof,222)
 222  format(/1x,'Current MODEL output vs. DATA',
     &          /1x,'Data#   Y-data(i) & Y-model(i),  i=1,NY')
c
c-------------------------------Scan through all data points
      do  300  i=1,np
c
c-------------------------------Compute model output Y(i) & dydp(i,j)
```

```
      imode=1
      do  230  j=1,nx
          x(j)=xdata(j,i)
 230  continue
      call model(nvar,nparmx,y,x,p,dydp)
      if (iprint .ge. 1) then
          write(kof,238) i,(ydata(k,i),y(k),k=1,ny)
 238  format(1x,i3,2x,8g12.4)
      end if
c-------------------------------Compute LS Obj. function and matrix A, b
      do  240  j=1,ny
c-------------------------------Select weighting factor (GLS vs WLS)
          if (igls .eq. 1) then
                    qyj= q(j)/(ydata(j,i)**2 + epsmin)
          else
                    qyj= q(j)
          end if
          SSE=SSE + qyj*(ydata(j,i) - y(j))**2
      do  240  l=1,npar
          b(l)=b(l) + qyj*dydp(j,l)*(ydata(j,i) - y(j))
      do  240  k=1,npar
          a(l,k)=a(l,k) + qyj*dydp(j,l)*dydp(j,k)
 240  continue
 300  continue
c
c-------------------------------Keep current value of p(i)
      do 310 i=1,npar
          p0(i)=p(i)
 310  continue
c
c-------------------------------Decompose matrix A  (using DEVCSF from IMSL)
      call  devcsf(npar,a,nparmx,s,v,nparmx)
c
c-------------------------------Compute condition number & (V**T)*b
      conda=s(1)/s(npar)
      do 312 i=1,npar
          bv(i)=0.0
          do 312 j=1,npar
          bv(i)=bv(i) + v(j,i)*b(j)
 312  continue
 315  continue
c
c-------------------------------Use pseudoinverse (if cond(A) > 1/epspsi)
      ipsi=0
      do  320  k=1,npar
      if (s(k)/s(1) .lt. epspsi)  then
          bvs(k)=0.0
          ipsi=ipsi + 1
      else
c-------------------------------Include MARQUARDT'S modification
          bvs(k)=bv(k)/(s(k) + epsmrq)
      end if
 320  continue
      write(kos,325) loop,epsmrq,epspsi,ipsi,conda
      write(kof,325) loop,epsmrq,epspsi,ipsi,conda
 325  format(//1x,'ITERATION=',i3,/1x,'EPS_Marq.=',g10.4,4x,
     &  'EPS_PseudoInv.=',g10.4,4x,'No. PseudoInv. Apprx.=',i2,
     &  /1x,'Cond(A)=',e12.4)
      write(kof,326) (s(j),j=1,npar)
 326  format(1x,'Eigenvalues=',8g12.4)
c-------------------------------Compute  dp = (V**T)*b
      do  330  i=1,npar
          dp(i)=0.0
      do  330  j=1,npar
```

```
          dp(i)=dp(i) + v(i,j)*bvs(j)
 330  continue
c------------------------------Compute new vector p and ||dp||
      dpnorm=0.0
      do  340  i=1,npar
          dpnorm=dpnorm + dabs(dp(i))
          p(i)=p0(i)*(1. + dp(i))
 340  continue
      dpnorm = dpnorm/npar
      write(kos,345) SSE
      write(kof,345) SSE
 345  format(1x,'LS-SSE =',g15.6/)
      write(kos,346) (p(j),j=1,npar)
      write(kof,346) (p(j),j=1,npar)
 346  format(1x,'P(j) G-N      =',8g12.5)
c------------------------------Test for convergence
      if (dpnorm .lt. eps ) then
      if (ipsi .eq. 0) then
c------------------------------Check if Marquardt's EPS is nonzero
      if (epsmrq .gt. s(npar)*0.01) then
      write(kof,347) epsmrq
      write(kos,347) epsmrq
 347  format(//1x,'>>>>>>    Converged with EPS_Marquardt =',g14.5)
      epsmrq=epsmrq*0.01
      goto  700
      else
      if (epsmrq .gt. 0.0) then
      write(kos,347) epsmrq
      write(kos,349)
      write(kof,347) epsmrq
      write(kof,349)
 349  format(//1x,'>>>>>>     From now on EPS_Marquardt = 0.0')
      epsmrq=0.0
      goto  700
      end if
      end if
c------------------------------If CONVERGED go and compute standard deviation
      sigma=sqrt(SSE/dfr)
      write(kos,385) sigma,dfr
      write(kof,385) sigma,dfr
 385  format(///5x,'++++++   CONVERGED   ++++++',/5x,'Sigma=',
      &          g14.5,/5x,'Degrees of Freedom=',f4.0,/)
      write(kos,386) (p(j),j=1,npar)
      write(kof,386) (p(j),j=1,npar)
 386  format(5x,'Best P(i)  =',8g12.5)
      goto 800
      else
c------------------------------If converged with ipsi nonzero, include one
more eigenvalue
      epspsi=epspsi*1.e-1
      goto  315
      end if
      end if
c
c------------------------------Enforce HARD MIN & MAX parameter boundaries
      if (ibound .eq. 1) then
c----------------------------Use bisection to enforce:  Pmin(i)<P(i)<Pmax(i)
      do  460  j=1,npar
 448      continue
          if (p(j).le.pmin(j) .or. p(j).ge.pmax(j)) then
          do  450  i=1,npar
                   dp(i)=0.50*dp(i)
                   p(i)=p0(i) * (1. + dp(i))
 450      continue
```

```
        end if
        if (p(j).le.pmin(j) .or. p(j).ge.pmax(j)) goto 448
 460 continue
     write(kof,465) (p(j),j=1,npar)
 465    format(1x,'P(j) Bounded =',8g12.5)
     end if
c-------------------------------STEP-SIZE COMPUTATIONS:
c-------------------------------Use full step if not needed or nstep=0
     if (dpnorm .le. 0.010 .or. nstep .eq. 0) then
        goto  700
     end if
c-------------------------------Compute step-size by the BISECTION RULE
     do  600  kks=1,nstep
c-------------------------------Compute Model Output at new parameter values
     SSEnew=0.0
     imode=0
     do  580  ii=1,np
     do  530  j=1,nx
        x(j)=xdata(j,ii)
 530 continue
     call model(nvar,nparmx,y,x,p,dydp)
c-------------------------------Compute NEW value of LS Obj. function
     do  540  j=1,ny
c-------------------------------Select weighting factor (GLS vs WLS)
        if (igls .eq. 1) then
                qyj= q(j)/(ydata(j,ii)**2 + epsmin)
        else
                qyj= q(j)
        end if
        SSEnew=SSEnew + qyj*(ydata(j,ii) - y(j))**2
 540 continue
c--------------------------if LS Obj. function is not improved, half step-size
     if (SSEnew .ge. SSE)  then
     do  550  i=1,npar
        dp(i)=0.50*dp(i)
        p(i)=p0(i) * (1. + dp(i))
 550      continue
     goto  600
     end if
 580 continue
     write(kof,696) (p(jj),jj=1,npar)
     kkk=kks - 1
     write(kof,585) kkk,SSE,SSEnew
 585 format(/2x,'Stepsize= (0.5)**',i2,3x,'LS_SSE_old=',g14.6,4x,
     &      'LS_SSE_new=',g14.6)
     goto  700
 600 continue
     write(kof,696) (p(jj),jj=1,npar)
 696 format(1x,'P(j) Stepped =',8g12.5)
 700 continue
c-------------------------------Alert user that G-N did not converge...
     sigma=sqrt(SSE/dfr)
     write(kof,750) sigma,dfr
 750 format(///5x,'*****  Did NOT converged yet  *****',/5x,
     &          'LS-Sigma=',g11.5,/5x,'Degrees of freedom=',f4.0)
     write(kof,755) (p(j),j=1,npar)
 755 format(/5x,'Last P(j) =',8g12.5)
c---------------------------Compute sigma & stand. dev. for current parameters
 800 continue
     sigma=sqrt(SSE/dfr)
     do  870  i=1,npar
        stdev(i)=0.0
     do  870  j=1,npar
        stdev(i)=stdev(i) + v(i,j)**2/s(j)
```

```
 870 continue
     do  880  i=1,npar
        stdev(i)=sqrt(stdev(i))*sigma
 880 continue
     write(kos,886)  (stdev(j)*100,j=1,npar)
     write(kof,886)  (stdev(j)*100,j=1,npar)
 886 format(//5x,'St.Dev.(%) =',8g12.4)
c
c-----------------------------Alert User whether GLS or WLS was used...
     if (igls .eq. 1) then
      write(kof,887)
 887  format(//1x,50('-'),/5x,
     &   'GENERALIZED LEAST SQUARES Formulation Used',/1x,50('-'),/)
     else
      write(kof,888)
 888  format(//1x,50('-'),/5x,'WEIGHTED LEAST SQUARES Formulation Used'
     &           ,/1x,50('-'),/)
     end if
     write(kof,890)
 890 format(/1x,'------------------PROGRAM END----------------',/)
     write(kos,891) filout
 891 format(//1x,60('-'),/5x,'Program OUTPUT stored in file: ',
     &           a24,/1x,60('-'),/)
     close(kof,status='keep')
     pause
     stop 100
     end
c
c---------------------------------------------------------------------MODEL
c
c    Algebraic Equation Model of the form Y=f(X,P)
c
c    where   X(i), i=1,NX is the vector of independent variables
c            Y(i), i=1,NY is the vector of dependent variables
c                P(i), i=1,NPAR is the vector of unknown parameters
c
c-------------------------------------------------------------------------+
c
c    The User must specify the MODEL EQUATIONS
c                Y(1) = f1(X(1),...X(NX); P(1),...P(NPAR))
c                Y(2) = f2(X(1),...X(NX); P(1),...P(NPAR))
c                Y(3) = f3(X(1),...X(NX); P(1),...P(NPAR)),... etc
c
c    and the Jacobean matrix dYDP(i,j)=[dY(i)/dP(j)]
c
c
c---------------------------------------------------------------------MODEL
     SUBROUTINE model(nvar,nparmx,y,x,p,dydp)
     DOUBLE PRECISION y(nvar),dydp(nvar,nparmx),x(nvar),p(nparmx)
c
     common /gn/imode,ny,nx,npar
c
c--------------------Compute model output
      y(1) = dexp(-p(1)*x(1)*dexp(-p(2)*(1./x(2)-1./620.)))
c
c--------------------Compute Jacobean dydp(i,j)
     if (imode .eq. 0)  return
c
     dydp(1,1)=-x(1)*dexp(-p(2)*(1./x(2)-1./620.))*y(1)
     dydp(1,2)=-(1./x(2)-1./620.)*dydp(1,1)*p(1)
c
c--------------------Normalise sensitivity coefficients
     do 200 i=1,ny
     do 200 j=1,npar
```

```
        dydp(i,j)=dydp(i,j)*p(j)
 200 continue
c
     return
     end
```

INPUT FILE

```
Enter Intial Guess for the parameters...
1.      100000.
Enter Min values for the parameters...(Pmin(i))
0.      0.
Enter MAX values for the parameters...(Pmax(i))
1000000.        1000000.
Enter NITER (Max No. Iter.), IPRINT (=1 for complete Output), EPSMRQ
20   0    0.0
Enter number of columns of DATA
3
Enter DATA points (one line per experiment)
120.    600     .900
60.0    600     .949
60.0    612     .886
120.    612     .785
120.    612     .791
60.0    612     .890
60.0    620     .787
30.0    620     .877
15.0    620     .938
60.0    620     .782
45.1    620     .827
90.0    620     .696
150.    620     .582
60.0    620     .795
60.0    620     .800
60.0    620     .790
30.0    620     .883
90.0    620     .712
150.    620     .576
90.4    620     .715
120.    620     .673
60.0    620     .802
60.0    620     .802
60.0    620     .804
60.0    620     .794
60.0    620     .804
60.0    620     .799
30.0    631     .764
45.1    631     .688
30.0    631     .717
30.0    631     .802
45.0    631     .695
15.0    639     .808
30.0    639     .655
90.0    639     .309
25.0    639     .689
60.1    639     .437
```

```
60.0    639    .425
30.0    639    .638
30.0    639    .659
60.0    639    .449
```

A.2.4 COMPUTER PROGRAM FOR EXAMPLE 16.3.2

```
c---------------------------------------------------------------------+

c
c       EXAMPLE 16.3.2 (Pyrolytic Dehydrogenation of Benzene to...)
c       in the book:
c       APPLIED PARAMETER ESTIMATION FOR CHEMICAL ENGINEERS
c       by Englezos & Kalogerakis / Marcel Dekker Inc. (2000)
c       COPYRIGHT: Marcel Dekker 2000
c
c---------------------------------------------------------------------+
c
c       PARAMETER ESTIMATION ROUTINE FOR ODE MODELS
c       Based on Gauss-Newton method with PseudoInverse and Marquardt's
c       Modification. Hard BOUNDARIES on parameters can be imposed.
c       Bayesian parameter priors can be used as an option.
c
c       GENERALIZED LEAST SQUARES Formulation (when IGLS=1)
c             with the DIAGONAL WEIGHTING MATRIX, Q(i)/Ydata(i)**2, i=1,NY
c       or
c
c       WEIGHTED LEAST SQUARES Formulation (when IGLS=0)
c             with the DIAGONAL WEIGHTING MATRIX, Q(i), i=1,NY
c
c       Written in FORTRAN 77 by Dr. Nicolas Kalogerakis, Aug. 1999
c       Compiler: MS-FORTRAN using the IMSL library (routines: DEVCSF,
c                                    DIVPAG,SETPAR)
c
c---------------------------------------------------------------------+
c       The User must provide Subroutine MODEL [that describes the
c            mathematical model dx/dt=f(x,p)]
c            and subroutine JX [computes the Jacobeans (df/dx) & (df/dp)]
c            where x(i), i=1,NX is the State vector.
c
c            The Output vector y(i), i=1,NY is assumed to correspond to
c                 the first NY elements of the State vector
c
c
c       The Following Variables MUST be Specified in PARAMETER statement:
c
c       NXZ (=NX) = Number of state variables in the model
c       NYZ (=NY) = Number of measured variables in the model
c       NPARZ (=NPAR) = Number of unknown parameters
c       NRUNZ = Greater or equal to the maximum number of runs (experiments)
c                                                                  to be
regressed simultaneously
c       NPZ = Greater or equal to the maximum number of measurements per run
c
c
c       The Following Variables MUST be Specified in Main Program:
c
c       IGLS = Flag to use Generalized Least Squares (IGLS=1) or Weighted LS (IGLS=0)
c       FILEIN = Name of Input file
c       FILOUT = Name of Output file
c       Q(i), i=1,NY = Constant weights for each measured variable (Use 1.0 for Least
Squares).
```

```
c         IBOUND = 1 to enable hard constraints on the parameters (0 otherwise)
c         NSTEP = Maximum number of reduction allowed for Bisection rule (default is 10)
c         KIF, KOS, KOF = Input file unit, Screen output unit, Output file unit.
c         NSIG = Approx. Number of Significant digits (Tolerence EPS = 10**(-NSIG))
c         EPSPSI = Minimum Ratio of eigenvalues before the Pseudo-inverse approx. is used
c         EPSMIN = A very small number (used to avoid division by zero).
c         TOL = Tolerence for local error required by ODE solver (default is 1.0e-6)
c
c-----------------------------------------------------------------------------+
        USE MSIMSL
        EXTERNAL jx,model
        PARAMETER (nxz=2,nyz=2,nparz=2,ntot=nxz*(nparz+1),npz=100,nrunz=5)
        DOUBLE PRECISION ydata(nrunz,nyz,npz),tp(nrunz,npz),p(nparz)
        DOUBLE PRECISION x0(nrunz,nxz),q(nyz),t0(nrunz),nobs(nrunz)
        DOUBLE PRECISION p0(nparz),dp(nparz),x(ntot),v(nparz,nparz)
        DOUBLE PRECISION a(nparz,nparz),b(nparz),s(nparz),ag(1,1)
        DOUBLE PRECISION stdev(nparz),bv(nparz),bvs(nparz),pivpag(50)
        DOUBLE PRECISION pmin(nparz),pmax(nparz)
        DOUBLE PRECISION pprior(nparz),vprior(nparz)
        DOUBLE PRECISION t, tend, tol
        COMMON /gn/imode,nx,npar,p
        CHARACTER*4 dummyline
        CHARACTER*24 filein, filout
c
c------------------------------Set convergence tolerence & other parameters
c
      DATA kos/6/,kif/2/,kof/7/,epsmin/1.e-80/,tol/1.d-6/,igls/0/
      DATA nsig/5/,epspsi/1.e-30/,nstep/10/,q/nyz*1.0/,ibound/1/
c
      eps = 10.**(-nsig)
        nx=nxz
        ny=nyz
        npar=nparz
        nxx=nx + 1
        n=nx*(npar + 1)
c
c------------------------------Set Filenames and Open the files for I/O
c
      filein = 'DataIN_Ex16.3.2.txt'
      filout = 'DataOUT_Ex16.3.2.txt'
      open(kif,file=filein)
      open(kof,file=filout)
c------------------------------Print header information
        write(kos,60)
        write(kof,60)
  60    format(/20x,'PARAMETER ESTIMATION PROGRAM FOR ODE MODELS using',
     & /20x,'the GAUSS-NEWTON Method with Marquardt`s Modification,',
     & /20x,'Pseudo-Inverse Approximation, Bisection Rule,',
     & /20x,'Bounded Parameter Search, Bayesian Parameter Priors')
        if (igls .eq. 1) then
              write(kos,65)
              write(kof,65)
  65          format(//20x,'GENERALIZED LEAST SQUARES Formulation...')
        else
              write(kos,67)
              write(kof,67)
  67          format(//20x,'WEIGHTED LEAST SQUARES Formulation...')
        end if
        write(kos,70) nx, ny, npar
        write(kof,70) nx, ny, npar
  70    format(//20x,'Number of State Variables (NX)   = ',i3,
     &  /20x,'Number of Output Variables (NY) = ',i3,
     &  /20x,'Number of Parameters (NPAR)     = ',i3)
      write(kos,71) filein
      write(kof,71) filein
```

```
  71  format(//40x,'Data Input Filename: ',a24)
      write(kos,72) filout
      write(kof,72) filout
  72  format(40x,'Program Output File: ',a24)
c
c------------------------------Read PRIOR,  MIN  & MAX parameter values
c
         read(kif,*) dummyline
         read(kif,*) (pprior(j),j=1,npar)
         read(kif,*) dummyline
         read(kif,*) (vprior(j),j=1,npar)
         read(kif,*) dummyline
         read(kif,*) (pmin(j),j=1,npar)
         read(kif,*) dummyline
         read(kif,*) (pmax(j),j=1,npar)
c
c------------------------------Read NITER, IPRINT & EPS_Marquardt
c
         read(kif,*) dummyline
         read(kif,*) niter, iprint, epsmrq
c
c------------------------------Read Number of Runs to be Regressed
c
         read(kif,*) dummyline
         read(kif,*) nrun
         if (nrun .gt. nrunz) then
                 write(kos,74) nrun
  74         format(///20x,'NRUNZ in PARAMETER must be at least',i3//)
                 stop 111
         end if
         write(kos,75) nrun
         write(kof,75) nrun
  75     format(//20x,'Input Data for',i3,'  Runs',//)
         dfr= - npar
c
c------------------------------Read Measurements for each Run
c
         do  90  irun=1,nrun
         read(kif,*) dummyline
         read(kif,*) nobs(irun),t0(irun),(x0(irun,j),j=1,nx)
         np=nobs(irun)
         dfr=dfr + ny*np
c
         write(kos,85) irun,np,t0(irun),(x0(irun,j),j=1,nx)
         write(kof,85) irun,np,t0(irun),(x0(irun,j),j=1,nx)
  85     format(//5x,'Run:',i2,4x,'NP=',i3,4x,
     &       'T0 =',g12.4,4x,'X(0) =',10g11.4/)
c
         read(kif,*) dummyline
         do  90  j=1,np
         read(kif,*)     tp(irun,j),(ydata(irun,i,j),i=1,ny)
         write(kof,91)   tp(irun,j),(ydata(irun,i,j),i=1,ny)
  91     format(5x,'Time=',g12.4,10x,'Y-data(i)=',10g12.4)
  90     continue
c
         do 92 j=1,npar
                p(j)=pprior(j)
  92     continue
         write(kof,93) (p(j),j=1,npar)
  93     format(///2x,'Pprior(j)=',8g12.5)
         write(kos,94) (p(j),j=1,npar)
  94     format(//2x,'Pprior(j)=',8g12.5)
         write(kos,96) (vprior(j),j=1,npar)
         write(kof,96) (vprior(j),j=1,npar)
```

```
  96     format(1x,'1/VARprior=',8g12.5//)
         write(kof,97) (pmin(j),j=1,npar)
         write(kos,97) (pmin(j),j=1,npar)
  97     format(1x,'  Pmin(j) =',8g12.5//)
         write(kof,98) (pmax(j),j=1,npar)
         write(kos,98) (pmax(j),j=1,npar)
  98     format(1x,'  Pmax(j) =',8g12.5//)
         close(kif)
c
c------------------------------Main iteration loop
c
         do  500  nloop=1,niter
c------------------------------Initialize matrix A, b and Objective function
         SSEprior=0.0
         do  104  i=1,npar
                  SSEprior=SSEprior + vprior(i)*(pprior(i)-p(i))**2
                  b(i)=vprior(i)*(pprior(i)-p(i))
         do  104  l=1,npar
                  a(i,l)=0.0
 104     continue
         do  106  i=1,npar
                  a(i,i)=vprior(i)
 106     continue
         SSEtotal=SSEprior
c
c------------------------------Go through each Run
         do  210  irun=1,nrun
      if (iprint .eq. 1) write(kof,108)  irun
 108     format(/4x,'Current MODEL output vs. DATA for RUN#',i2,
     &        /5x,' Time        Y-data(i) & Y-model(i),  i=1,NY')
c
c------------------------------Initialize ODEs for current Run
         do  112  j=1,nx
                  x(j)=x0(irun,j)
 112     continue
         do  114  j=nxx,n
                  x(j)=0.0
 114     continue
c
c------------------------------Initialize ODE solver: DIVPAG
         np=nobs(irun)
         call SETPAR(50,pivpag)
         index=1
         t=t0(irun)
         imode=1
c
c------------------------------Integrate ODEs       for this Run
         do  200  i=1,np
                  tend=tp(irun,i)
         call DIVPAG(index,n,model,jx,ag,t,tend,tol,pivpag,x)
         if (iprint .eq. 1) then
           write(kof,135) t,(ydata(irun,k,i),x(k),k=1,ny)
 135       format(1x,g11.4,3x,8g12.5)
         end if
c
c------------------------------Update Objective function and matrix A, b
        do  140  j=1,ny
c------------------------------Select weighting factor (GLS vs WLS)
                  if (igls .eq. 1) then
                          qyj= q(j)/(ydata(irun,j,i)**2 + epsmin)
                  else
                          qyj= q(j)
                  end if
                  SSEtotal=SSEtotal + qyj*(ydata(irun,j,i) - x(j))**2
```

```
      do  140  l=1,npar
        ll=l*nx
        b(l)=b(l) + qyj*x(j+ll)*(ydata(irun,j,i) - x(j))
        do  140  k=1,npar
          kk=k*nx
          a(l,k)=a(l,k) + qyj*x(j+ll)*x(j+kk)
 140  continue
c------------------Call DIVPAG again with INDEX=3 to release MEMORY from IMSL
        if (i.eq.np) then
              index=3
              call DIVPAG(index,n,model,jx,ag,t,tend,tol,pivpag,x)
        end if
 200  continue
 210  continue
c
c--------------------------------Keep current value of p(i)
      do 211 i=1,npar
              p0(i)=p(i)
 211  continue
c--------------------------------Add Bayes influence of A & b
      do 212 i=1,npar
              b(i) = b(i) + vprior(i)*(pprior(i)-p0(i))
              a(i,i) = a(i,i) + vprior(i)
 212  continue
c
c--------------------------------Print matrix A & b
      if (iprint .eq. 1) then
         write (kof,214)
 214     format(/1x,'   Matrix A and vector b....')
         do 218 j1=1,npar
           write(kof,216) (a(j1,j2), j2=1,npar),b(j1)
 216       format(11g12.4)
 218     continue
      end if
c
c--------------------------------Decompose matrix A  (using DEVCSF from IMSL)
c
      call  devcsf(npar,a,nparz,s,v,nparz)
c
c--------------------------------Compute condition number & (V**T)*b
      conda=s(1)/s(npar)
      do 220 i=1,npar
              bv(i)=0.0
      do 220 j=1,npar
              bv(i)=bv(i) + v(j,i)*b(j)
 220  continue
 225  continue
c
c--------------------------------Use Pseudo-inverse (if Cond(A) > 1/epspsi)
      ipsi=0
      do  230  k=1,npar
      if (s(k)/s(1) .lt. epspsi)  then
      bvs(k)=0.0
      ipsi=ipsi + 1
      else
c--------------------------------Include MARQUARDT'S modification
      bvs(k)=bv(k)/(s(k) + epsmrq)
      end if
 230  continue
c
      write(kos,235) nloop,epsmrq,epspsi,ipsi,conda
      write(kof,235) nloop,epsmrq,epspsi,ipsi,conda
 235  format(//1x,'ITERATION=',i3,/1x,'EPS_Marq.=',g10.4,4x,
     &  'EPS_PseudoInv.=',g10.4,4x,'No. PseudoInv. Apprx.=',i2,
```

```
     &     /1x,'Cond(A)=',e12.4)
           write(kof,237) (s(j),j=1,npar)
 237       format(1x,'Eigenvalues=',10g11.4)
c
c-----------------------------Compute  dp = (V**T)*b
           do  240  i=1,npar
                    dp(i)=0.0
           do  240  j=1,npar
                    dp(i)=dp(i) + v(i,j)*bvs(j)
 240       continue
c
c-----------------------------Compute new vector p and ||dp||
           dpnorm=0.0
           do  250  i=1,npar
                    dpnorm=dpnorm + abs(dp(i))
                    p(i)=p0(i) * (1. + dp(i))
 250       continue
           dpnorm = dpnorm/npar
                    if (dpnorm .le. 10*eps*npar) then
              istep=0
            else
              istep=1
            end if
           write(kos,263) SSEtotal,SSEprior/SSEtotal
 263       format(1x,'SSE_total =',g11.5,5x,'SSEprior/SSEtot =',f14.4,/)
           write(kof,264) SSEtotal,SSEprior
 264       format(1x,'SSE_total =',g11.5,3x,'Prior_SSE =',g11.5,/)
           write(kos,265) (p(j),j=1,npar)
           write(kof,265) (p(j),j=1,npar)
 265       format(1x,'P(j) by G-N =',8g12.5)
c
c-----------------------------Enforce HARD MIN & MAX parameter boundaries
           if (ibound .eq. 1) then
c---------------------------Use bisection to enforce:  Pmin(i)<P(i)<Pmax(i)
           do  460  j=1,npar
 440       continue
                    if (p(j).le.pmin(j) .or. p(j).ge.pmax(j)) then
                    do  450  i=1,npar
                             dp(i)=0.50*dp(i)
                             p(i)=p0(i) * (1. + dp(i))
 450                continue
                    end if
                    if (p(j).le.pmin(j) .or. p(j).ge.pmax(j)) goto 440
 460                continue
                    write(kos,465) (p(j),j=1,npar)
                    write(kof,465) (p(j),j=1,npar)
 465                format(1x,'P(j) Bounded=',8g12.5)
                    end if
c-----------------------------Test for convergence
           if (dpnorm .lt. eps ) then
           if (ipsi .eq. 0) then
c-----------------------------Check if Marquardt's EPS is nonzero
           if (epsmrq .gt. s(npar)*0.01) then
           write(kof,347) epsmrq
           write(kos,347) epsmrq
 347       format(//1x,'>>>>>>    Converged with EPS_Marquardt =',g14.5)
           epsmrq=epsmrq*0.01
           goto  500
           else
           if (epsmrq .gt. 0.0) then
           write(kos,347) epsmrq
           write(kos,349)
           write(kof,347) epsmrq
           write(kof,349)
```

```
 349     format(//1x,'>>>>>>     From now on EPS_Marquardt = 0.0')
         epsmrq=0.0
         goto  500
         end if
         end if
c
                  sigma=sqrt((SSEtotal-SSEprior)/dfr)
                  write(kos,285) sigma,dfr
                  write(kof,285) sigma,dfr
  285    format(///5x,'++++++ CONVERGED ++++++',/5x,'LS-Sigma=',g11.5,
     &    /5x,'Degrees of Freedom=',f4.0,//)
                  write(kos,286) (p(j),j=1,npar)
                  write(kof,286) (p(j),j=1,npar)
  286             format(5x,'Best P(j) =',8g12.5)
c
c------------------------------Go and get Standard Deviations & Sigma
                  goto 600
      else
c
c------------------------------Include one more singular values
                  epspsi=epspsi*1.e-1
                  goto  225
                  end if
                  end if
c------------------------------STEP-SIZE COMPUTATIONS:
c------------------------------Use full step if it is not needed or NSTEP=0
          if (istep .eq. 0 .or. nstep .eq. 0)  then
             goto  500
          end if
c------------------------------Compute step-size by the bisection rule
          do  400  kks=1,nstep
c------------------------------Initialize Objective function, ODEs and DIVPAG
         SSEprior=0.0
         do 301 i=1,npar
                  SSEprior=SSEprior + vprior(i)*(pprior(i)-p(i))**2
 301     continue
         SSEnew=SSEprior
         do  390  irun=1,nrun
         do  300  i=1,nx
                  x(i)=x0(irun,i)
 300     continue
         np=nobs(irun)
         call SETPAR(50,pivpag)
         index=1
         t=t0(irun)
         imode=0
         do  340  ii=1,np
            tend=tp(irun,ii)
         call DIVPAG(index,nx,model,jx,ag,t,tend,tol,pivpag,x)
c
c------------------------------Compute NEW value of the Objective function
      do  320  j=1,ny
c
c------------------------------Select weighting factor (GLS vs WLS)
                  if (igls .eq. 1) then
                           qyj= q(j)/(ydata(irun,j,ii)**2 + epsmin)
                  else
                           qyj= q(j)
                  end if
                  SSEnew=SSEnew + qyj*(ydata(irun,j,ii) - x(j))**2
 320  continue
c
c------------------------If LS Obj. function is not improved, half step-size
          if (SSEnew .ge. SSEtotal)  then
```

```
          do  330  i=1,npar
            dp(i)=0.50*dp(i)
            p(i)=p0(i) * (1. + dp(i))
 330      continue
c------------------------------Release MEMORY from ISML (call with index=3)
        index=3
        call DIVPAG(index,nx,model,jx,ag,t,tend,tol,pivpag,x)
          goto  400
         end if
        if (ii.eq.np) then
          index=3
          call DIVPAG(index,nx,model,jx,ag,t,tend,tol,pivpag,x)
        end if
 340    continue
 390    continue
      write(kof,394)  (p(j),j=1,npar)
      write(kos,394)  (p(j),j=1,npar)
 394  format(1x,'P(j) stepped=',8g12.5/)
                kkk=kks - 1
                write(kof,395) kkk,SSEtotal,SSEnew
                write(kos,395) kkk,SSEtotal,SSEnew
 395    format(/2x,'Stepsize= (0.5)**',i2,4x,'SSE_total_old=',g11.5,3x,
     &     'SSE_total_new=',g11.5/)
          goto  500
 400  continue
      write(kof,496)  (p(jj),jj=1,npar)
 496  format(/1x,'P(j)laststep=',8g12.5)
 500  continue
c
c------------------------------Alert user that It did not converge...
        sigma=sqrt((SSEtotal-SSEprior)/dfr)
        write(kof,585) sigma,dfr
 585    format(///5x,'*****  Did NOT converged yet  *****',/5x,
     &          'LS-Sigma=',g11.5,/5x,'Degrees of freedom=',f4.0)
        write(kof,586)  (p(j),j=1,npar)
 586    format(/5x,'Last P(j) =',8g12.5)
c
c---------------Compute Sigma & Stand. Deviations of the Parameter
c---------------CALCULATIONS ARE BASED ON THE DATA ONLY (Prior info ingnored)
600     continue
        do  670  i=1,npar
                stdev(i)=0.0
                do  670  j=1,npar
                stdev(i)=stdev(i) + v(i,j)**2/s(j)
 670    continue
        do  680  i=1,npar
                stdev(i)=sqrt(stdev(i))*sigma
 680    continue
                write(kos,686)  (stdev(j)*100,j=1,npar)
                write(kof,686)  (stdev(j)*100,j=1,npar)
 686            format(/5x,'St.Dev.(%)=',8g12.5)
c
c------------------------------Alert user whether GLS or WLS was used...
        if (igls .eq. 1) then
         write(kof,688)
 688     format(//1x,50('-'),/5x,
        &        'GENERALIZED LEAST SQUARES Formulation Used',/1x,50('-'),/)
        else
         write(kof,689)
 689     format(//1x,50('-'),/5x,'WEIGHTED LEAST SQUARES Formulation Used'
        &                  ,/1x,50('-'),/)
        end if
        write(kof,890)
 890    format(/1x,'------------------PROGRAM END----------------',/)
```

```
      write(kos,891) filout
 891  format(//1x,60('-'),/5x,'Program OUTPUT stored in file: ',
     &                a24,/1x,60('-'),/)
      close(kof,status='keep')
      stop
      end
c
c---------------------------------------------------------------------MODEL
c
c     Ordinary Differential Equation (ODE) Model of the form dX/dt=f(X,P)
c
c     where   X(i), i=1,NX is the vector of state variables
c                   P(i), i=1,NPAR is the vector of unknown parameters
c
c-------------------------------------------------------------------------+
c
c     The User must specify the ODEs to be solved
c              dX(1)/dt = f1(X(1),...X(NX); P(1),...P(NPAR))
c              dX(2)/dt = f2(X(1),...X(NX); P(1),...P(NPAR))
c              dX(3)/dt = f3(X(1),...X(NX); P(1),...P(NPAR)),... etc
c
c     and the Jacobean matrices: dfdx(i,j)=[df(i)/dX(j)] i=1,NX & j=1,NX
c                 dfdp(i,j)=[df(i)/dP(j)] i=1,NX & j=1,NPAR
c
c---------------------------------------------------------------------MODEL
      SUBROUTINE model(n,t,x,dx)
      DOUBLE PRECISION x(n),dx(n),fx(2,2),fp(2,2),p(2)
      DOUBLE PRECISION t,r1,r2,dr1dx1,dr1dx2,dr2dx1,dr2dx2
      COMMON/gn/imode,nx,npar,p
c
c-----------------------------------------Model  Equations (dx/dt)
        r1=(x(1)**2 - x(2)*(2-2*x(1)-x(2))/0.726)
        r2=(x(1)*x(2) - (1-x(1)-2*x(2))*(2-2*x(1)-x(2))/3.852)
        dx(1)=-p(1)*r1 - p(2)*r2
        dx(2)=0.5*p(1)*r1 - p(2)*r2
c
c-----------------------------------------Jacobean (df/dx)
      if (imode.eq.0) return
        dr1dx1=2*x(1) + 2*x(2)/0.726
        dr1dx2=-(2-2*x(1)-x(2))/0.726 + x(2)/0.726
        dr2dx1=x(2) + (2-2*x(1)-x(2))/3.852 + 2*(1-x(1)-2*x(2))/3.852
        dr2dx2=x(1) + 2*(2-2*x(1)-x(2))/3.852 + (1-x(1)-2*x(2))/3.852
        fx(1,1)=-p(1)*dr1dx1 - p(2)*dr2dx1
        fx(1,2)=-p(1)*dr1dx2 - p(2)*dr2dx2
        fx(2,1)=0.5*p(1)*dr1dx1 - p(2)*dr2dx1
        fx(2,2)=0.5*p(1)*dr1dx2 - p(2)*dr2dx2
c
c-----------------------------------------Jacobean (df/dp)
        fp(1,1)=-r1
        fp(1,2)=-r2
        fp(2,1)=0.5*r1
        fp(2,2)=-r2
c
c-----------------------------------------Set up the Sensitivity Equations
        do 10 k=1,npar
              kk=k*nx
              do 10 j=1,nx
              dx(j+kk)=fp(j,k)*p(k)
              do 10 l=1,nx
              dx(j+kk)=dx(j+kk)+fx(j,l)*x(l+kk)
 10     continue
        return
        end
c
```

```
c-----------------------------------------------------------------------JX
c        JACOBEAN of the
c        Ordinary Differential Equation (ODE) Model
c        required by the ODE solver (version for stiff systems)
c
c-----------------------------------------------------------------------+
c
c        The User must specify the Jacobean matrix:
c                          dfdx(i,j)=[df(i)/dX(j)] i=1,NX & j=1,NX
c
c-----------------------------------------------------------------------JX
         SUBROUTINE jx(n,t,x,fx)
         DOUBLE PRECISION x(n),fx(n,n),p(2)
         DOUBLE PRECISION t, r1,r2,dr1dx1,dr1dx2,dr2dx1,dr2dx2
         common/gn/imode,nx,npar,p
c
c------------------------------Initialize Jacobian matrix Fx
         do  20  i=1,n
         do  20  j=1,n
                 fx(i,j)=0.0
  20     continue
c
c------------------------------Jacobian matrix Fx
         r1=(x(1)**2 - x(2)*(2-2*x(1)-x(2))/0.726)
         r2=(x(1)*x(2) - (1-x(1)-2*x(2))*(2-2*x(1)-x(2))/3.852)
         dr1dx1=2*x(1) + 2*x(2)/0.726
         dr1dx2=-(2-2*x(1)-x(2))/0.726 + x(2)/0.726
         dr2dx1=x(2) + (2-2*x(1)-x(2))/3.852 + 2*(1-x(1)-2*x(2))/3.852
         dr2dx2=x(1) + 2*(2-2*x(1)-x(2))/3.852 + (1-x(1)-2*x(2))/3.852
         fx(1,1)=-p(1)*dr1dx1 - p(2)*dr2dx1
         fx(1,2)=-p(1)*dr1dx2 - p(2)*dr2dx2
         fx(2,1)=0.5*p(1)*dr1dx1 - p(2)*dr2dx1
         fx(2,2)=0.5*p(1)*dr1dx2 - p(2)*dr2dx2
c
c------------------------------Set up the expanded Jacobian
         if (imode .eq. 0) return
         ll=nx
         do  60  kk=1,npar
                 do  50  j=1,nx
                 jj=ll+j
                 do  50  i=1,nx
                         ii=ll+i
                         fx(ii,jj)=fx(i,j)
  50             continue
                 ll=ll+nx
  60     continue
         return
         end
c
c---------------------------------------------------------------SETPAR
c
         subroutine SETPAR(nn,param)
c
c------------------------------Initialize vector PARAM for DIVPAG
c
         double precision param(nn)
         do 10 i=1,nn
                 param(i)=0
  10     continue
         param(1)=1.0d-8
         param(3)=1000.
         param(4)=500
         param(5)=50000
         param(6)=5
```

```
      param(9)=1
      param(11)=1
      param(12)=2
      param(13)=1
      param(18)=1.0d-15
      param(20)=1
      return
      end
```

INPUT FILE

```
Enter Prior Values for the Parameters (=Initial Guess) (Pprior(i))...
10000.          10000.   355.55     402.91
Enter Prior Values for the INVERSE of each Parameter VARIANCE (Vprior(i))...
0.1d-18 0.1d-18
Enter MIN values for the Parameters (Pmin(i))...
0.0             0.0
Enter MAX values for the Parameters (Pmax(i))...
1.d+8   1.d+8
Enter NITER (Max No. Iterations), IPRINT (=1 to print Output vector), EPSMRQ
40   0     0.0
Enter Number of Runs in this dataset (NRUN)...
1
Enter for this Run: No. of Datapoints (NP), Init. Time (t0), Init. state (XO(i))
8               0.0              1.0              0.0
Enter NP rows of Measurements for this Run: Time (TP) and OUTPUT (ym(i))...
0.000563        0.828    0.0737
0.001132        0.704    0.113
0.001697        0.622    0.1322
0.002262        0.565    0.1400
0.00340         0.499    0.1468
0.00397         0.482    0.1477
0.00452         0.470    0.1477
0.01697         0.443    0.1476


COMMENTS regarding Datafile (not read by the program):

If one wishes to use NO PRIOR INFORMATION on a particular parameter
simply use Vprior = 0 (i.e., variance is infinity - no prior info).
```

Index